T0138419

Terra Incognita

TERRA INCOGNITA
MAPPING THE ANTIPODES BEFORE 1600

Alfred Hiatt

THE UNIVERSITY OF CHICAGO PRESS
CHICAGO AND LONDON

For LRH

Alfred Hiatt is a senior lecturer in Old and Middle English Literature at the School of English at the University of Leeds. He is also the author of *The Making of Medieval Forgeries*, published by the British Library.

The University of Chicago Press, Chicago 60637
The University of Chicago Press, Ltd., London

Printed in Great Britain

17 16 15 14 13 12 11 10 09 08 1 2 3 4 5

ISBN-13: 978-0-226-33303-8 (cloth)
ISBN-10: 0-226-33303-5 (cloth)

Library of Congress Cataloging-in-Publication Data

Hiatt, Alfred.
Terra incognita : mapping the Antipodes before 1600 / by Alfred Hiatt.
p. cm.
Includes bibliographical references.
ISBN-13: 978-0-226-33303-8 (cloth : alk. paper)
ISBN-10: 0-226-33303-5 (cloth : alk. paper) 1. Early maps. 2. Cosmography—Early works to 1800. 3. World maps, Manuscript—Early works to 1800. 4. Geography, Medieval—Maps.
I. Title.
GA221.H53 2008
912—dc22

2007047091

∞ The paper used in this publication meets the minimum requirements of the American National Standard for Information Sciences—Permanence of Paper for Printed Library Materials, ANSI Z39.48-1992.

Contents

Acknowledgements

I first became seriously interested in medieval and early modern maps when I was finishing my PhD on the reception of forged documents in the later Middle Ages. Volume one of J. B. Harley and David Woodward's *The History of Cartography* provided an immensely stimulating introduction to the subject, but my particular interest in *terrae incognitae* and the antipodes developed from browsing Rodney Shirley's compendious *The Mapping of the World.* At an embryonic stage of my research, I sat in the Map Room of the University Library, Cambridge, staring at map after map in Shirley's collection, bewildered by the array of material that sixteenth- and seventeenth-century cartographers chose to deposit in their Great Southern Lands. What interested me was the use of a space that was a part of the represented world yet not a part, on the map but not mapped. How had this *Terra Australis Incognita* arisen – this conjectured land, with its capacity to be filled with animals, peoples, allegories, satires, portraits of great men, pompous declarations of cartographic method? Where had it come from, and should it be seen as an expression of the plenitude of sixteenth-century European culture, or as a sign of some profound restlessness?

The contrast with the spare quality of medieval representations of unknown land in the southern hemisphere – and part of the early excitement of this project came from the realization that medieval visual and written culture *did* contain such representations – was also striking and stimulating. How and why had representation of *terrae incognitae* moved from restraint – an inscription of a few words, a blank space – to something that looked like abundance? Just as there was something thrilling about the richness of display in the sixteenth century's *Terra Australis,* so there was something poignant and oddly dignified about a tradition that required the acknowledgement of land beyond the known world – and its demarcation – but that refused to go further. By the seventeenth century the southern land was a staple of fantasy literature, utopic, dystopic and all stops in between. But fantasies of only the most muted kind can be found in medieval invocations of the other side of the earth, of antipodal spaces where men might – or might not – stand with their feet against ours. Indeed it was precisely as a warning against fantasy, against both intellectual and political overreach, that the antipodes frequently made their appearance in classical and medieval literature.

So I was encouraged to write this book not simply by the intrinsic idea of *terra incognita* as a space that could be represented, but also by its history as a space that persisted across the period divide between the medieval and the early modern. The history of a non-place.

It would not have been possible to begin work on *terrae incognitae* and to get as far as I did in a relatively short space of time without the very generous support provided by Trinity College, Cambridge, in the form of a Junior Research Fellowship from 1998 to 2002. During this time of extensive, uninterrupted research time I was also fortunate to receive a Visiting Fellowship in the History of Cartography from the Newberry Library, Chicago, in the spring of 2001. I am particularly grateful for the assistance and hospitality shown to me at the Newberry by Jim Akerman, Arthur Holzheimer, and Robert Karrow. More recently, as a lecturer in the School of English at the University of Leeds, I have been lucky to find an environment in which research is well supported through the generous award of study leave. In combination with the School's provision, an Arts and Humanities Research Council grant of matching research leave enabled me to spend a full year on this project in 2005. This was augmented by a Small Research Grant from the British Academy, which allowed me to examine manuscripts in Italy, Germany, the Netherlands and Belgium.

I have used many libraries in the course of researching and writing this book, but I owe a particular debt to three: the University Library, Cambridge, the British Library, and the Warburg Institute Library, London, without which a very different – and much poorer – monograph would have emerged. In addition to these places, I must thank the Institute of Classical Studies Library in the School of Advanced Study, University of London, for providing an invaluable environment for a non-classicist to venture into new territory. I would also like to thank staff at the following institutions: in Germany, the Bayerische Staatsbibliothek, Munich, the Staatsbibliothek, Bamberg, and the Dombibliothek, Cologne; in Italy, the British School at Rome, the Biblioteca Apostolica Vaticana, the Biblioteca Laurenziana and the Biblioteca Nazionale Centrale, Florence, the Biblioteca Nazionale, Naples, and the Biblioteca Comunale, Trent; in France, the Bibliothèque nationale de France, and the Bibliothèque municipale, Orléans; in the Netherlands, the Universiteitsbibliotheek, Leiden; in Belgium, the Bibliothèque royale de Belgique; in England, the Bodleian Library, Oxford, and the library of Worcester Cathedral; and in Sweden, the Kungliga Biblioteket, Stockholm.

Many people have helped me to write this book. In particular I would like to thank William Flynn for help with Latin translations, and Ananya Jahanara Kabir, Richard Serjeantson, and Chet Van Duzer for advice, encouragement, and invaluable references to some of the more arcane reaches of antipodia. Two readers were especially helpful. Andy Merrills provided astute and amusing comments on drafts of the first three chapters, for which I am very grateful. Above all calls of duty, Catherine Delano Smith read and commented extensively on a full draft of this book, and encouraged me to revise and rethink many aspects of it. The final version is undoubtedly better for the immense care with which she engaged with, and responded to, my work.

An outline of some of the themes and arguments of this book appeared as 'Blank Spaces on the Earth', *The Yale Journal of Criticism* 15 (2002), 223–50. Expanded versions of my discussions of the medieval transmission of the world map of Macrobius, and Petrarch's use of the idea of the antipodes, have been published, respectively, as 'The Map of Macrobius before 1100', *Imago Mundi* 59 (2007), 149–76, and 'Petrarch's Antipodes', *Parergon* 22 (2005), 1–30. I am grateful to these journals for permission to reproduce material.

At British Library Publications David Way has been a source of guidance from the very inception of this project; he, Lara Speicher, Belinda Wilkinson and Kate Hampson have made the process of completing this book smoother than I thought possible.

List of Plates and Figures

Plates

Figures

Chapter 1

Chapter 2

Chapter 3

Chapter 4

Chapter 5

Chapter 6

Chapter 7

Chapter 8

Chapter 9

Frequently used Abbreviations

BNF Bibliothèque nationale de France
BAV Biblioteca Apostolica Vaticana
BL British Library
PL Patrologia Latina
Geographia 1513 Claudius Ptolemy, *Geographię opus nouissima traductione e Gręcorum archetypis castigatissime pressum* (Strasbourg, 1513; facsimile edition Amsterdam: Theatrum Orbis Terrarum, 1966)

Chapter One

Beyond the Known World

Few sixteenth-century maps are better known than Abraham Ortelius' 'Typus Orbis Terrarum' of 1570 (fig. 1). Ortelius' world image appeared in his *Theatrum Orbis Terrarum* (Theatre of the World), the first systematically-compiled printed atlas, and as a result it was swiftly disseminated throughout Europe. At first glance the image is a celebration of the sixteenth-century expansion of European knowledge. The map nestles snugly in a bed of clouds, affording the viewer a quasi-divine perspective. On the right of the image are the Old World continents of Europe, Africa, and Asia, but on its left appears the New World of 'America or New India, first discovered by Christopher Columbus in the year 1492 in the name of the King of Castile'. In the far east, meanwhile, the Spice Islands, objects of Portuguese and Spanish commercial and political competition, are dotted around the equator. A dramatic departure from its classical and medieval predecessors, Ortelius' map is recognizable to a twenty-first century audience in its extent and in much of its geographical detail. And yet obvious differences strike the eye. North America is largely unmapped, and South America clearly misshapen; the interiors of Old and New World continents are sketchy at best; and four large, unmarked islands occupy the Arctic region. Most striking of all is the presence of a vast southern continent, blank apart from a few inscriptions on its coast and a temporary title: 'Terra Australis Nondum Cognita'.

The 'southern land not yet known' was a commonplace of the sixteenth-century world image. Its jagged coastline and nascent topography were as standard and recognizable as the Strait of Magellan, the boot of Italy, and the Cape of Good Hope. Unlike those places, of course, *Terra Australis* was a cartographic fiction, the product of cosmological theory and the confusing welter of travel narratives that flooded into Europe during the sixteenth century, from which the land's faltering toponyms were taken by Ortelius and others. 'Terra del Fuego' (Land of Fire) was the coastline sighted by Magellan as he passed beneath South America; 'Prom. Terrae Australis' (Promontory of the Southern Land) and 'Psitacorum regio' (Region of Parrots) were similarly derived from Portuguese voyages of exploration; 'Beach', 'Lucach' and 'Maletur' (marked at the far right of the map beneath Java Maior) were places allegedly visited by Marco Polo. Stitched together, the traces of disparate explorations added verisimilitude to the thesis, in existence since classical times, of a vast Antarctic continent. Indeed, as Ortelius explained in a caption beneath New Guinea, the emergent continent had even been tentatively christened: 'some call this

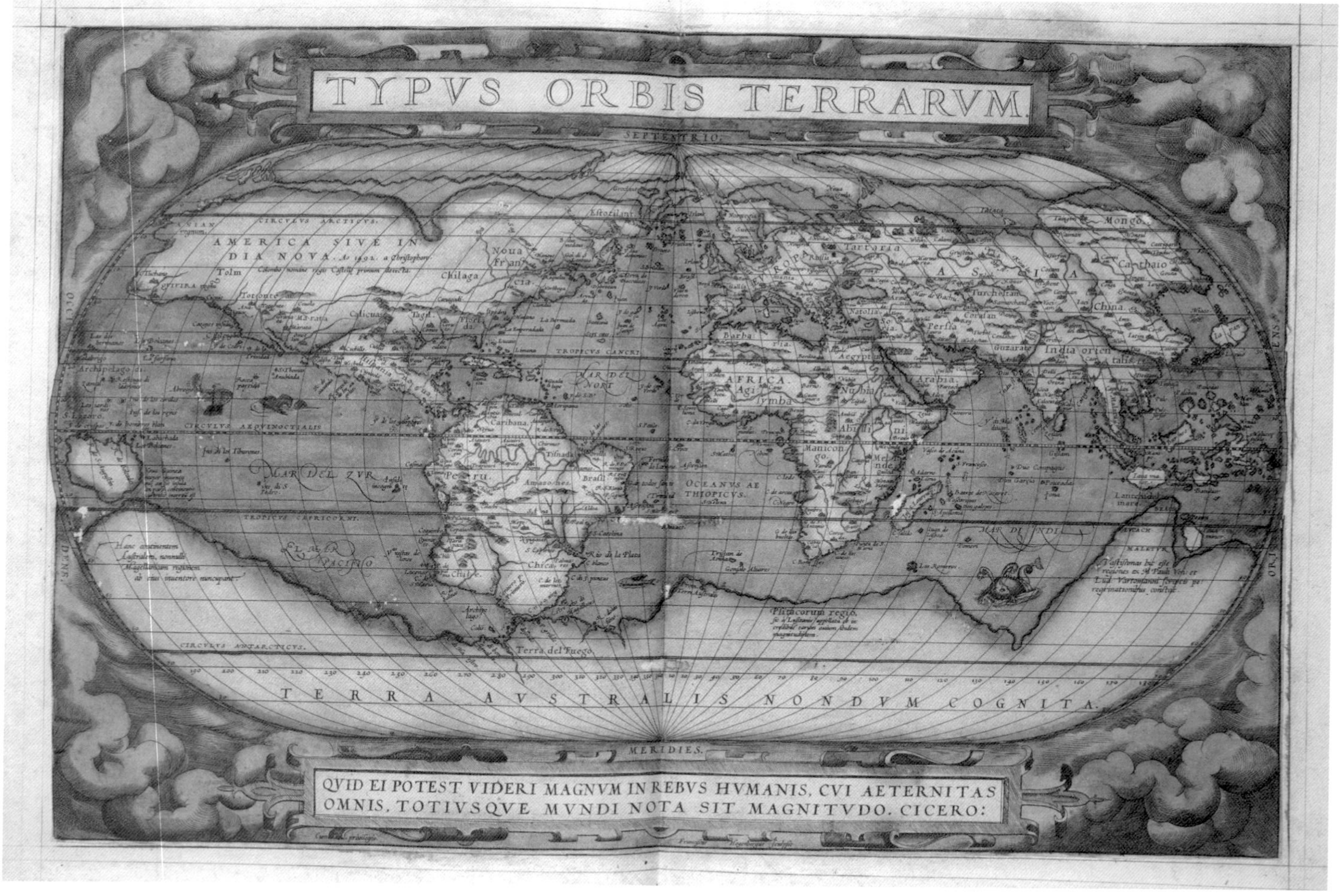

Fig. 1. Abraham Ortelius, 'Typus Orbis Terrarum', in *Theatrum Orbis Terrarum* (Antwerp, 1570). London, BL, Maps C.2.c.3. An emblem of early modern cartography, Ortelius' world map shows the increased knowledge of both Old and New Worlds available to mapmakers in the sixteenth century. At the same time, its vast southern continent, marked 'Terra Australis Nondum Cognita', draws attention to the incomplete nature of the map. *Terra Australis* was the product of the theory that a vast landmass must exist in the southern hemisphere to balance the land of the northern hemisphere; its jagged coastline was constructed from a mixture of medieval and early modern travel narrative.

southern continent "Magellanica" after its finder'. *Terra Australis* or Magellanica was, then, a work in progress, but one that gave every appearance of occupying the same level of reality as the Americas. The southern land's incompleteness, its state of partial discovery, shows that the sixteenth-century world image combined frank admissions of ignorance with a sense of cartographic and political expansion. The unknown was not only present, but fundamental to the map's statement of geographical knowledge on the move: *Terra Australis* was land not simply 'unknown', but '*not yet* known'.

Terra incognita is now a metaphor. The twenty-first century map of the world contains no unknown land, and in the age of Google Earth it is inconceivable that a world map of the future will bear the words 'land not yet known'. Yet it takes only the slightest acquaintance with the history of European cartography to realize that the absence of unknown land from the world map is a very recent phenomenon. A glance at nineteenth- and even early twentieth-century atlases reveals considerable tracts of land marked 'unknown', or at least 'unexplored'. Moreover, as Ortelius' 'Typus Orbis Terrarum' suggests, the history of cartography is not simply a narrative of the gradual documentation of the earth's surface; it is also the story of non-places, of lands that are not and never were, but that – often for considerable periods – existed on maps. The disappearance of *terrae incognitae* from the world image in the first half of the twentieth century marked the conclusion of a process of predominantly European-sponsored exploration and colonization, underway since the thirteenth century, which resulted in the mapping of large portions of the globe. But the cartographic record makes it possible to consider the nature and significance of *terrae incognitae* in their historical contexts – as spaces integral to world maps, located outside of geographical experience, yet not beyond the bounds of geographical reasoning and imagination.

The purpose of this book is to trace the representation of a particular version of *terra incognita* – the antipodes – from classical antiquity until the end of the sixteenth century. Before outlining my approach to this topic, it is necessary to give a brief overview of the range of responses to the antipodes expressed by classical, medieval, and post-medieval authors. The term 'antipodes' initially referred to people dwelling opposite to – literally with feet against – the known world. The concept was the product of classical Greek geometry, which calculated the size and shape of the earth with a remarkable degree of accuracy, and argued that unknown lands and peoples were likely to exist in parts of the world beyond the land mass constituted by Europe, Asia, and Africa. Antipodal lands and peoples were posited not only in the southern hemisphere, but also in the northern hemisphere, on the 'underside' of a spherical earth. Several different names were used for these putative spaces and peoples, all of which described them in terms of their relationship to the Greek *oikoumenē*, the known, or inhabited, world. As well as 'antipodes', dwellers of parts of the world beyond the ecumene could, for example, be termed 'antoikoi' (opposite the *oikoumenē*), 'antikthones' (on the opposite side of the earth), and 'perioikoi' (around from the *oikoumenē*).[1] The world known in detail to classical Greek and Roman geography was centred around the Mediterranean basin, but general knowledge of places beyond the Greco-Roman domain was extensive. By the first century BC the geographer Strabo was able to describe the world from the Atlantic coast of Europe and northern Africa in the west as far east as India, and from Ireland, the Caucasus and Scythia in the north to Ethiopia and 'Taprobana' (modern day Sri Lanka) in the south.[2] The outermost limits of the *oikoumenē* were usually defined by the presence of a vast Ocean, encircling all three continents as a river around an island.[3]

Medieval authors inherited classical descriptions of the earth and adapted them to new religious and political frameworks. *Mappaemundi*, medieval European world maps that represented the ecumene, consisted of the three *partes* of Europe, Asia, and Africa. This schematic tripartite division was marked on many maps by the course of the Don river (Tanais), which separated Europe from Asia; the Nile, dividing Africa from Asia; and the Mediterranean, which acted as the boundary between Europe and Africa. The tripartite scheme could be aligned with the population of the world by the three sons of Noah: Sem (Asia); Japheth (Europe); and Cham (Africa). In general, the area of the known world depicted on medieval European maps extended little further than the classical world image: west to east from the British Isles to India, and north to south from Scythia or the Rhipaean Mountains to Ethiopia. An encircling ocean, often filled with islands ranging from Thule and the Orkneys in the far north-west to Taprobana in the far east, was usually depicted on world maps; in its most schematic form, this ocean, in combination with the tripartite division, constituted the 'T-O map', usually oriented to the east. Amongst the many functions of medieval *mappaemundi*, recent research has emphasized their historiographical significance, with the representation of biblical history a specific concern, although not to the exclusion of a considerable amount of secular material. Of particular importance for medieval European geographical representation in the Middle Ages was the theory of the six ages of world history prior to Revelation; *mappaemundi* such as the monumental thirteenth-century Hereford and Ebstorf maps correspondingly represented events and landmarks from the Garden of Eden to the Resurrection of Christ, and looked forward towards the end of history.[4]

Antipodal spaces stood outside this context of ecumenical mapping and geographical description. They were purely theoretical, and usually regarded as unreachable from the known world due to the barriers posed by the intense heat of equatorial regions and the size of the encircling Ocean. For this reason the antipodes and their peoples formed no part of the historical record, and did not feature on maps with a primarily historiographical function. Instead, they were represented as part of a tradition of zonal maps that showed known alongside unknown parts of the world, ecumene and *antoecumene*. Zonal maps represented a theory of classical geometry that posited the division of the earth into five latitudinal zones: two zones of extreme cold at the far south and north; an equatorial ocean within a central band of extreme heat; and two temperate zones, one in the northern hemisphere, and one in the southern (fig. 2). As a consequence, zonal maps showed a band of temperate land in the southern hemisphere, cut off from the temperate zone of the northern hemisphere by a combination of the equatorial region of extreme heat and the equatorial ocean. Although some authors believed that antipodal lands contained similar, even identical, features to those of the known world (a southern-hemisphere Mediterranean, for example), none was prepared to authorize the representation of such features in the absence of sure proof. And so one of the key features of the medieval representation of the antipodes took shape: the contrast between, on the one hand, a known

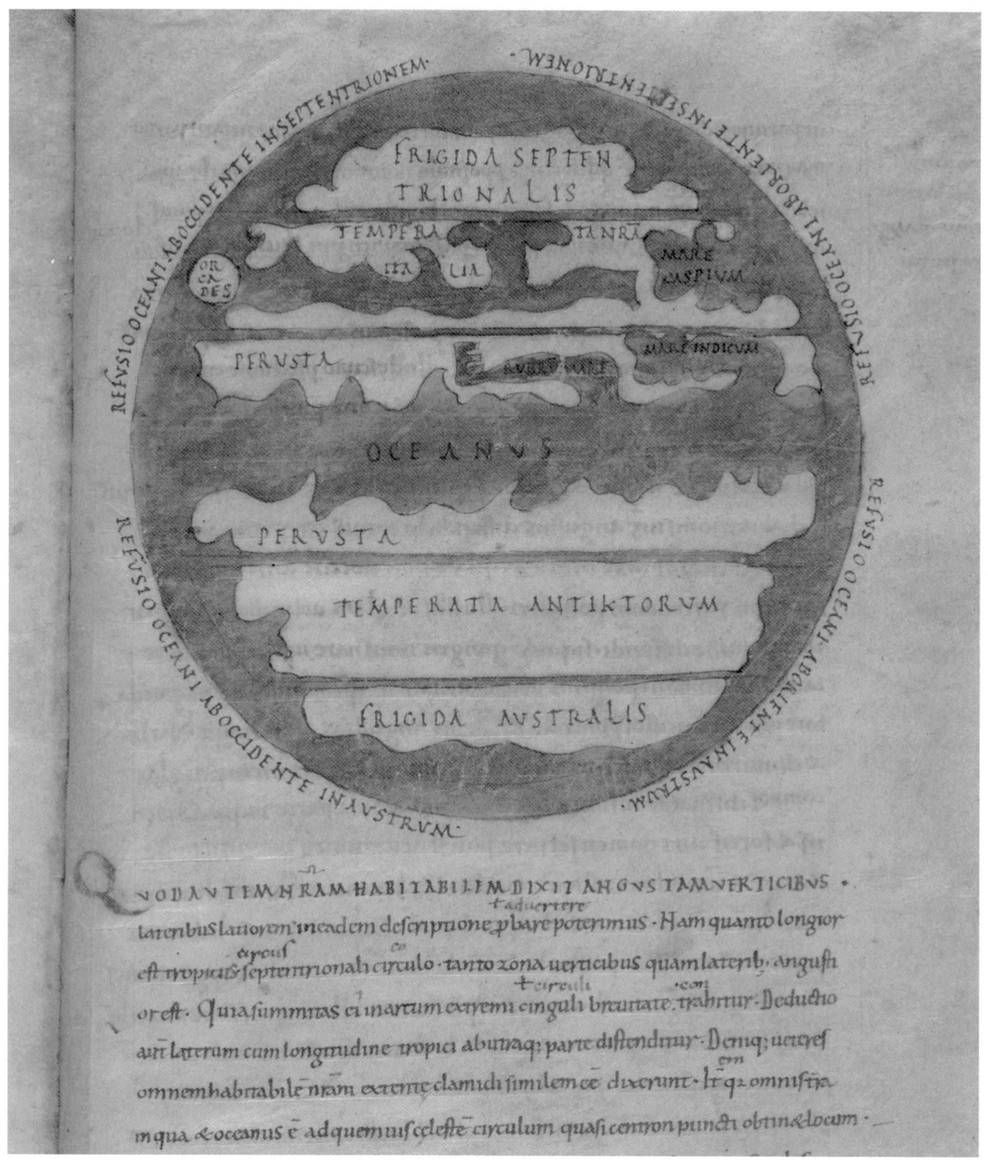

Fig. 2. Oxford, Bodleian Library, MS D'Orville 77, f. 100r. Zonal world map illustrating a c. 1000 copy of Macrobius' *Commentarii in Somnium Scipionis* (Commentary on the Dream of Scipio). North at the top. The zones at the far north and south are noted for their extreme cold ('Frigida septentrionalis', 'Frigida australis'); the central zone is marked 'Perusta' (burned up) above and beneath an equatorial ocean ('Oceanus'). The known world of the northern hemisphere is marked 'Temperata nostra', while the temperate zone in the southern hemisphere is marked 'Temperata Antiktorum', 'temperate zone of the antikthones'. Ecumenical features include the 'Orcades' (Orkneys), represented by a circle, 'Italia', and an unmarked but prominent indentation representing the Mediterranean, with a northern extension leading into the (similarly unmarked) Palus Maeotis (i.e. the Sea of Azov). In accordance with Macrobius' original design, the map shows the Caspian, Indian, and the Red seas, the latter appropriately coloured. Ocean currents from the equator to the poles are represented by four inscriptions around the diagram: 'Refusio Oceani ab occidente in septentrionem' (flow of Ocean from the west to the north) etc.

world that is mapped, replete with cities, rivers, seas and mountain ranges, and on the other the unknown world, present on the map but devoid of any topography.

Antipodal places and peoples quickly acquired significance beyond that of a dry scientific theory. From Plato onwards they held a broader philosophical meaning and, transferred into the corpus of Roman learning, they were deployed as a political sign. For Cicero the idea of people standing beyond the reach of Roman rule served to show the limited nature of empire, and the transience of earthly glory. Poets such as Virgil, Ovid, and Lucan invoked the antipodes to signify imperial ambition, but also, especially in the figure of Alexander the Great, political overreach to the point of madness. The position of the antipodes opposite to the known world made them ideal for satire, as a trope of reversal, the world turned upside down, and as a means of mirroring the absurdities and pretensions of the satirist's own society. By the same token, the concept of the antipodes and antipodeans existing beneath the feet of inhabitants of the known world led from time to time to their confusion with infernal regions: such confusion, often creative, found its ultimate resolution in Dante's *Commedia*, where purgatory and the earthly paradise are located in the southern hemisphere diametrically opposite to Jerusalem.

The question of habitation constituted, as many commentators have noted, a critical breach between classical and Christian responses to the antipodes. Classical authors disagreed over the plausibility of the thesis that lands existed beyond the known world, and they argued about the possibility that such lands could be inhabited. At base, however, their argument was about the shape and nature of the universe, and consequently their writings on the topic were scientific and philosophical and only by extension theological. But for Christian writers such as Augustine of Hippo the proposition that antipodal spaces might be inhabited by humans was untenable. The notion that the antipodes were unreachable meant that any putative inhabitants could not have received the word of God, and that Christ's injunction to the apostles to 'teach all nations' could not be fulfilled.[5] Augustine's comments on the antipodes in *De civitate dei* (On the City of God) had a profound impact on Christian attitudes towards the concept, but they did not disable discussion or representation. Not long after Augustine denounced the antipodes, another author – the non-Christian Macrobius – went to great lengths to explicate the idea and defend it against doubters in his commentary upon Cicero's 'Somnium Scipionis' (Dream of Scipio). The serious discussion of antipodal theory by Macrobius – and by other late antique authors such as the commentator on Virgil, Servius, and the encyclopedist Martianus Capella – was important because it ensured the preservation of the idea and its transmission as part of a body of classical cosmological learning. Interest in the antipodes flourished wherever Neoplatonic thought was revived, first during the Carolingian Renaissance, later in the French schools of the twelfth century, and subsequently as part of the medieval university syllabus. The existence of the antipodes became a *quaestio* for scholastic debate in the thirteenth century, and it was in this guise that it appears to have first attracted the interest of Dante.

Two profound changes occurred to antipodal theory, and to the representation of unknown land more generally, at the beginning and end of the fifteenth century. These were caused by the translation of Ptolemy's *Geographia* into Latin in 1406–7, and the news of European discovery of the New World following Columbus' landfall in 1492. The popularity of Ptolemy's world and regional images encouraged mapmakers to depict unknown land as contiguous with the known world, rather than separated by impenetrable barriers of heat and ocean. Columbus' voyages achieved, as several of his contemporaries were quick to observe, the discovery of the western antipodes. Discovery meant a change from lands unknown to lands recently found, and the emergence of a new category: land *not yet* discovered. In the sixteenth century, the antipodes were reformulated as *Terra Australis*, a land mass that mimicked and mirrored New World discoveries by means of its position on the brink of exploration. It was not until the eighteenth century that the idea of the southern land was conclusively demolished, and with it the last vestiges of a virtually unbroken tradition of geographical hypothesis.

Scholarship on the question of the antipodes has traditionally been the preserve of two interlinked sub-disciplines: the history of cartography and the history of discoveries. Over one hundred years ago the magisterial survey of Armand Rainaud traced the two strands of hypothesis and discovery of the antipodes, 'one of the greatest problems of the history of geography', from Greek and Roman antiquity to the voyages of James Cook.[6] Not unlike Joseph Conrad, who divided European geography into three phases – 'geography fantastic' from the Middle Ages to Columbus; 'geography militant', from Columbus to the mid nineteenth century; and the 'geography triumphant' of his own day – Rainaud identified three distinct periods in the history of the antipodes.[7] The first period, spanning antiquity and the Middle Ages, was based on the *a priori* conjecture that a great continent existed in the southern hemisphere. This 'confused, uncertain' conjecture derived from the presumption that the globe was 'constructed after the laws of a rigorous symmetry'; it had no foundation whatsoever in experience.[8] The second period identified by Rainaud was that of the great age of Portuguese and Spanish exploration, in which new discoveries of land in the southern hemisphere were assimilated into classical theory: each new promontory or island was fitted onto the imagined coastline of a monumental austral continent. The final and decisive period in this story encompassed subsequent exploration, as a result of which the hypothesis of a great southern land was 'condemned without appeal': the last two voyages of Cook constituted its death blow.[9] All the same, Rainaud concluded, the flawed hypothesis of the southern continent drove exploration and the discovery of half of the globe: the chimera had generated a large measure of scientific truth.[10]

In its scope and detail Rainaud's survey has not been bettered, but recent years have seen the development of approaches to the antipodes from outside the domains of map and exploration history. The deployment of the antipodes as a theme or motif in classical literature has been revealingly analysed by James Romm and Gabriella Moretti; their works have emphasized the multiple function of the antipodes as emblem of imperial ambition, scientific curiosity, theological problem, or vehicle for biting satire or polemic.[11] Anna-Dorothee von den Brincken has provided a useful survey of the treatment of the 'ends of the earth' in medieval encyclopedic and historiographical, as well as cartographic, texts.[12] And, while medieval literary adaptations of the antipodal theme remain under-explored, recent discussions of post-medieval literary traditions have drawn attention to the particular allure of *Terra Australis* as a site for utopic literature.[13]

However, until now little attention has been given to creative uses of the southern land on world maps themselves. In particular, few scholars have considered the relationship between word and image in the depiction of antipodal space, and, for sixteenth-century material, the range of extraneous matter that was often positioned on *Terra Australis* (including cartouches, coats of arms, ethnographic drawings, portraits, and scientific disquisition). One obvious problem of scholarship on the antipodes post-Rainaud has been the tendency to remain within period boundaries. Discussion of sixteenth-century representations of unknown land has tended to pass quickly over the medieval legacy, in the assumption that a combination of the impact of exploration and a revitalized classicism overturned medieval geographical models.[14] Conversely, discussions of the medieval 'ends of the earth', such as von den Brincken's, have given only fleeting consideration to post-medieval revisions and adaptations of *terrae incognitae*. Yet the degree of continuity within discussions of the antipodes makes it very limiting to restrict analysis to a particular period. As a result, I have extended the chronological scope of my discussion of the antipodes from antiquity to the turn of the seventeenth century – thereby crossing more than one border. Equally, in the course of writing this book it became clear that to have any depth such an investigation had to cross disciplinary boundaries. The question of the theorization and representation of antipodal lands is one that should be set not simply within the frameworks of map, travel, or exploration history. Instead it should be addressed in terms of the range of functions that the antipodes played as a space outside of history, faith, and politics that nevertheless interacted with these forces in curious and unpredictable ways.

Four argumentative themes develop these points and run through the analysis contained in the following chapters. The first concerns the political significance of spaces and peoples beyond the known world. The idea of the antipodes in its classical formulation worked to consolidate power by positing a space beyond the geographical extent of European polities, most obviously the Roman empire. It did so in at least two ways. One was to emphasize celestial vision; texts in which a chosen individual, and by extension the reader, was afforded a privileged vision of the entire earth used antipodal spaces as a means of emphasizing the position of the individual as representative and manifestation of the state *within* the known world. Cicero's narrative of the 'Dream of Scipio' in *De re publica* is the *locus classicus*: Scipio is able to see non-Roman parts of the world from his celestial vantage point, but only in order to return his attention to Rome, Carthage – and the music of the spheres. The other mode of consolidation was to imagine passage to unknown spaces, and to show by means of that breach of established geographical boundaries the overextension and dissipation of power. Here the Alexander tradition contributes the classic statement of imperialist insanity: the Macedonian, inflamed by conquest, unsatisfied by his mastery of the known world, plans an expedition to the antipodes – and promptly dies, leaving a legacy of violence and the state in chaos. The function of the antipodes, whether within celestial vision or as sign of imperial overreach, was the same: antipodal spaces mark the boundaries of political power and in so doing consolidate it. Early Christian thinkers reacted to the idea of the antipodes in a variety of ways, but one response was to replicate the function of classical formulations: denial of the antipodes consolidated the space and the history of Christianity by showing where the faith had not gone, and could not go.

The expansion of the known world in the fifteenth and sixteenth centuries critically changed the function of consolidation. The antipodes were increasingly described in terms of *terra incognita*, and that shift, prompted by the translation of Ptolemy's *Geographia*, was important. *Terra incognita* marks the frontier, but also the interior. It is, literally, inchoate – less fixed in geometrical thinking than the antipodes, and more subject to revision. So did the representation of *terra incognita* act also to consolidate power? It seems rather to have marked its incompletion.

The second argumentative theme of this book addresses the nature of tradition. It seems legitimate to speak of a 'tradition' of unknown land, constituted by a corpus of classical and late antique texts, which provided theories about land beyond the known world, as well as modes of understanding and representing such land. As part of the broader reception of classical tradition, these theories were transmitted throughout the period from the fifth century AD to the sixteenth. Reception was dynamic, not static: the tradition – or preferably, traditions – of the antipodes were constantly adjusted and adapted, at times dormant, at others reanimated. They were also subject to change as the result of interaction with what might be termed 'new' knowledge: information derived from travel and exploration. The interaction of new and old knowledge can be conceived, I will argue, in terms of supplementation, a process that involved not simply the addition of information, but also the active construction of an opposition between 'ancient' and 'modern' knowledge. In this regard, however, the antipodes confuse clear oppositions: in the conflation of the tradition of the antipodes to the tradition of *terra incognita* – the transition from spaces beyond the known world to spaces at its edge – newly discovered and unknown lands appeared as both old and new knowledge.[15] New discoveries made in the fifteenth and sixteenth centuries were ones foretold, since antipodal land had been for so long a part of the geographical imagination. Was the representation of *terrae incognitae* in the New World – and in particular the representation of a large and fictitious southern land – an old tradition lingering, or part of a new one? The antipodes, existing outside structures of faith and verifiable knowledge, reveal tradition to be dependent upon acts of commentary and reinvention.

The issue of periodization is related to these questions about the constitution of traditions. The idea of the antipodes shows that the medieval period should be seen not as a conduit but as a period of intervention in, and reconstruction of, antique traditions. The relationship of medieval visual and literary culture to unknown space was not one of closure, apprehension, or uniform rejection: it was fundamentally dialogic. And that relationship was transformed but not erased by processes of political, cultural, and intellectual change that took place from the fourteenth to sixteenth centuries. It is legitimate, even necessary, to define periods, because change did occur, undeniably in the case of the world image. But an explanation of how and why that change occurred will not be enlightening if it falls back on banalities about inherently 'antique', 'medieval', or 'modern' ways of viewing the world. Medieval theories of the antipodes can, for instance, easily be assimilated to a narrative that insists that the conception of space in the Middle Ages was structured around the binarisms inside/outside and here/there.[16] The overarching binarism is that of open/closed: according to this view, it was not until the thirteenth century that the closed medieval European world started to embrace the outside and finally became able to reach *there* from *here*. Even scholars, such as Edward Casey, who show a refreshing willingness to acknowledge the vigour of medieval debate about abstract space from the thirteenth century, ultimately recur to this distinction between the Old World of Place (ends c. 1400) and 'the vista of a New World of Space [which] began to captivate the ablest minds of the succeeding period'.[17] The antipodes can be made to fit rather nicely into this structure, since inhabitable spaces beyond the known world were characterized by their unknowability, by their essential alterity, their *there*-ness. And indeed it was in the thirteenth century that some strain began to be placed on the notion of the unreachability of distant parts of the earth, strain that by the end of the fifteenth century had grown so to constitute its collapse. But binaries need not always be reductive. Even to represent *there* was significant, because to represent antipodal space was to make decisions about it. To represent requires first conceptualization, then acts of description, and these acts were fundamentally creative. Crucially, representation of unknown parts of the world de-centres, especially when the unknown is represented as an unreachable other part of the world. The dynamic of the known world and the antipodes was not that of centre and periphery, nor was it that of inside and outside. It was, on the contrary, the product of a geometrical notion of symmetry and balance, in which the known was matched, even exceeded by equivalent land masses in other parts of the world. The known world was a part of the whole, but it was not the largest, nor the most central part, and it could be conceived with full acknowledgement of its partial nature.

All three lines of argument converge in the question of representation. How was it possible to represent the unknown, either alongside or in opposition/apposition to the known? Alessandro Scafi's *Mapping Paradise*, a study of the representation of the earthly paradise on maps from the early Middle Ages to the present, asks the same question of a sacred, rather than secular, space. He describes mapping paradise on earth as 'one of the most powerful expressions of the fundamental tension between the locative and utopian tendencies in Christianity', an act that 'pointed to both the reality and the loss of a perfect human nature in paradise'.[18] Precisely the same answer cannot be given for the antipodes, however, because this was a space without the biblical and patristic authorization possessed by paradise. The terrestrial paradise may have been located in the antipodes, as some thirteenth- and fourteenth-century theologians dared to propose, but this was a solution to the question of the representation of paradise, not to that of the antipodes. Instead, the representation of the antipodes should be explained not in terms of tendencies in Christianity, but in terms of the political, historiographical, and literary appeal and *necessity* of the idea. Consolidation of political or religious identities depends on acts of representation: in their Ciceronian deployment, lands beyond the known world must be seen – and seen to be seen – in order to focus attention on *patria*; even to deny unknown lands and peoples, in the manner of Augustine, was nevertheless to advert to their theoretical presence.

Traditions of representation of the antipodes cut across boundaries of religion and time. To take perhaps the most significant example, Cicero's late antique commentator Macrobius illustrated his exposition of the 'Dream of Scipio' with a world map that showed unknown, antipodal, land in the southern hemisphere. The map of Macrobius was copied from the ninth to the fifteenth century. Yet the scribes who copied the map altered its form, expanding ecumenical areas, occasionally juxtaposing the comments of Augustine, and from time to time relocating and reconfiguring the image. Is it possible to consider such a mode of representation in terms of period? The map of Macrobius illustrates Cicero's theories; it was produced for a text written in the fifth century; there is reason to think that it was wholly or partially reconstructed in the tenth. It underwent significant adaptations in the twelfth century, and a revival of interest in the fifteenth as a result of humanist interest in Cicero. Is the map classical, late antique, medieval, or Renaissance? Does it not rather belong to any period in which it was reproduced? Contemplation of the broader questions of *terra incognita* and the antipodes suggests that we might do better to think in terms of dialogic relations in more than one temporal direction. To show unknown, antipodal, land on a map – whether of the third, ninth, thirteenth, or sixteenth centuries – was to show an ancient land, the product of long-held theories; it was simultaneously a land of the present, for those theories were still considered valid; and it was a land of the future, since it always carried the possibility of the contact and conquest to come. To mark unknown land on the map was to use a different order of representation. It is not simply that such representation was fictive, or imaginative, since elements of fiction and imagination were to be found also within *terra cognita*. Rather, *terra incognita* constituted an a-cartographic mode of representation within the map, uncharted land that nevertheless appeared on the chart. Such land was stripped to its raw essentials, to its fundamental idea: *terra incognita* was land unknown but not unthought.

Notes

1 Throughout this book I use the term 'antipodes' to refer to the general concept of land beyond the world known to ancient and medieval Europeans, whether in the southern hemisphere, or in the northern hemisphere 'beneath' the ecumene. I use 'antipodeans' to refer to the (hypothetical) inhabitants of such lands. The Latin 'antipodes' could be used to refer either to land or inhabitants, or both. As I discuss in greater detail in chapters 2–4, a more refined vocabulary was developed by classical authors for describing different antipodal places and peoples, in which, along with other terms, 'antipodes' was used to designate only one quarter of the inhabited world. That level of specificity was often blurred by medieval authors and commentators – and by certain antique and late antique authors themselves.

2 Strabo disputed the view of earlier geographers that human habitation extended as far north as the island of Thule (66°N), preferring to limit its northernmost extent to Ireland (54°N): Germaine Aujac, 'Greek Cartography in the Early Roman World', in *The History of Cartography*, 6 vols, vol. 1: *Cartography in Prehistoric, Ancient, and Medieval Europe and the Mediterranean*, ed. J. B. Harley and David Woodward (Chicago: University of Chicago Press, 1987), pp. 161–76, p. 174.

3 See James S. Romm, *The Edges of the Earth in Ancient Thought* (Princeton: Princeton University Press, 1992), pp. 9–44, who also discusses the periodic objections to this theory by commentators such as Herodotus, Aristotle, and Ptolemy of Alexandria.

4 On *mappaemundi* generally see David Woodward, 'Medieval *Mappaemundi*', in *The History of Cartography*, vol. 1, pp. 286–370; on the historiographical function of *mappaemundi* valuable discussions are provided by Evelyn Edson, *Mapping Time and Space: How Medieval Mapmakers viewed their World* (London: British Library, 1997), pp. 97–144; Anna-Dorothee von den Brincken, 'Mappa mundi und chronographia: Studien zur imago mundi des abendländischen Mittelalters', *Deutsches Archiv für Erforschung des Mittelalters* 24 (1968), 118–86; and most recently Alessandro Scafi, *Mapping Paradise: A History of Heaven on Earth* (London: British Library, 2006), esp. pp. 84–159. On the Ebstorf map, see Jürgen Wilke, *Die Ebstorfer Weltkarte*, 2 vols (Bielefeld: Verlag für Regionalgeschichte, 2001), who revises the dating of the map, traditionally assigned to the first half of the thirteenth century, and argues that it was produced around 1300.

5 Augustine of Hippo, *De civitate dei*, ed. B. Dombart and A. Kalb, Corpus Christianorum, Series Latina, 2 vols (Turnhout: Brepols, 1955), 16.9. Matthew 28:19; Mark 16:15; Luke 24:47.

6 Armand Rainaud, *Le Continent Austral: Hypothèses et Découvertes* (Paris: Colin, 1893). For the history of discoveries, the classic work is Richard Hennig, *Terrae incognitae*, 2nd edn, 4 vols (Leiden: Brill, 1944–56). Part anthology, part commentary, Hennig attempts to document exploration before Columbus in Asia and the Arab world, as well as in Europe.

7 See Conrad's 'Geography and Some Explorers', *National Geographic Magazine* 45 (1924), 239–74.

8 Rainaud, *Le continent austral*, pp. 475–8.

9 Rainaud, *Le continent austral*, p. 475. Rainaud nevertheless pointed out that for the nineteenth century the hypothesis had simply been transferred to the question of the existence of an Antarctic continent (p. 476). And, writing in 1893, he noted that the problem of the poles, especially that of the south pole, awaited a definitive solution (p. 6).

10 Rainaud, *Le continent austral*, p. 479. Not long after Rainaud, Giuseppe Boffito, a historian of ideas rather than geography, divided literature concerning the antipodes into four periods: the first and most ancient up to the early Christian era, in which antipodeans were held to be a truth by most and a fable by some; a second in which they risked becoming a heresy; a third (roughly the high Middle Ages) when they become a fable; and a fourth in which they passed into the dominion of art and poetry, being adapted most notably in Dante's *Inferno* and *Purgatorio*: Giuseppe Boffito, 'La leggenda degli antipodi', in *Miscellanea di studi critici edita in onore di Arturo Graf* (Bergamo: Istituto Italiano d'Arti Grafici, 1903), pp. 583–601. The only really sustainable element of this periodization is the distinction between classical non-Christian and post-classical Christian attitudes to the antipodes.

11 Romm, *Edges of the Earth*, pp. 124–40; Gabriella Moretti, *Gli antipodi: avventure letterarie di un mito scientifico* (Parma: Pratiche Editrice, 1994). Moretti's study includes medieval and early modern texts, but provides only a cursory treatment of maps. A shorter version of this work was published as *Agli antipodi del mondo* (Trent: Dipartimento di Scienze Filologiche e Storiche, 1990) and appeared in English as 'The Other World and the "Antipodes". The Myth

of the Unknown Countries between Antiquity and the Renaissance', in *The Classical Tradition and the Americas*, 6 vols, vol. 1: *European Images of the Americas and the Classical Tradition*, ed. W. Haase and M. Reinhold (Berlin: de Gruyter, 1994), pp. 241–84.

12 *Fines Terrae: Die Enden der Erde und der vierte Kontinent auf Mittelalterlichen Weltkarten* (Hanover: Hahnsche Buchhandlung, 1992). For consideration specifically of the meaning and representation of *terra incognita* in the Middle Ages, see von den Brincken's '*Terrae Incognitae.* Zur Umschreibung empirisch noch unerschlossener Räume in lateinischen Quellen des Mittelalters bis in die Entdeckungszeit', in *Raum und Raumvorstellungen im Mittelalter*, ed. Jan A. Aertsen and Andreas Speer (Berlin: de Gruyter, 1998), pp. 557–72. Two articles by Danielle Lecoq survey the classical and medieval tradition of the antipodes and the 'ends of the earth': 'Au delà des limites de la terre habitée. Des îles extraordinaires aux terres antipodes (XIe–XIIIe siècles)', in *Terre à découvrir, terres à parcourir: Exploration et connaissance du monde XIIe–XIXe siècles*, ed. Danielle Lecoq and Antoine Chambard (Paris: L'Harmattan, 1998), pp. 14–41, and 'Des antipodes au Nouveau Monde ou de la difficulté de l'Autre', in *La France-Amérique (XVIe–XVIIIe siècles)*, ed. Frank Lestringant (Paris: Champion, 1988), pp. 65–102.

13 On allusions to unknown and antipodal lands in medieval French texts see Jill Tattersall, '"Terra incognita": allusions aux extrêmes limites du monde dans les anciens textes français jusqu'en 1300', *Cahiers de civilisation médiévale* 24 (1981), 247–55. On post-medieval literary responses to the unknown southern land (with little discussion of maps) see David Fausett, *Writing the New World: Imaginary Voyages and Utopias of the Great Southern Land* (Syracuse: Syracuse University Press, 1993); Neil Rennie, *Far-Fetched Facts: The Literature of Travel and the Idea of the South Seas* (Oxford: Clarendon Press, 1995); and Glyndwr Williams, *The Great South Sea: English Voyages and Encounters 1570–1750* (New Haven: Yale University Press, 1997), pp. 48–75. On the related phenomenon of the 'lost' land of Lemuria, subject to various nineteenth- and twentieth-century fantasies, see Sumathi Ramaswamy, *The Lost Land of Lemuria: Fabulous Geographies, Catastrophic Histories* (Berkeley: University of California Press, 2004).

14 See Günter Schilder, *Australia Unveiled: the share of the Dutch navigators in the discovery of Australia*, trans. Olaf Richter (Amsterdam: Theatrum Orbis Terrarum, 1976), esp. pp. 7–31; W. A. R. Richardson, 'Enigmatic Indian Ocean Coastlines on Early Maps and Charts', *The Globe* 46 (1998), 21–41, and Richardson, 'Mercator's Southern Continent: Its Origins, Influence and Gradual Demise', *Terrae Incognitae* 25 (1993), 67–98; Helen Wallis, 'Visions of Terra Australis in the Middle Ages and Renaissance', in *Terra Australis: The Furthest Shore* (Sydney: International Cultural Corporation of Australia, 1988), pp. 35–8; William Eisler, *The Furthest Shore: Images of Terra Australis from the Middle Ages to Captain Cook* (Cambridge: Cambridge University Press, 1995), pp. 8–43, Robert Clancy, *The Mapping of Terra Australis* (Sydney: Universal Press, 1995); Peter Whitfield, *New Found Lands: Maps in the History of Exploration* (London: British Library, 1998), pp. 90–126.

15 On the capacity of the map to 'get[] ahead of geographical discoveries, provoke[] them in imagining the world such as it could be' see the richly suggestive essay by Christian Jacob, 'Il faut qu'une carte soit ouverte ou fermée', *Revue de la Bibliothèque Nationale* 45 (1992), 34–41. The significance of blank spaces in post-medieval, particularly eighteenth- and nineteenth-century, cartography, is the subject of the essays collected in *Combler les blancs de la carte: Modalités et enjeux de la construction des savoirs géographiques (XVIe–XXe siècle)*, ed. Isabelle Laboulais-Lesage (Strasbourg: Presses Universitaires de Strasbourg, 2004). Amongst the sparse literature on this topic J. K. Wright's '*Terrae Incognitae*: The Place of the Imagination in Geography', in *Human Nature in Geography: Fourteen Papers, 1925–1965* (Cambridge: Harvard University Press, 1966), pp. 68–88, is a sharp if frankly idiosyncratic and self-deprecating essay. J. B. Harley's seminal 'Silences and Secrecy: the Hidden Agenda of Cartography in Early Modern Europe', *Imago Mundi* 40 (1988), 57–76, deals with the suppression and/or erasure of information on maps, rather than the open display of ignorance on them.

16 Summarized by Paul Zumthor, *La Mesure du monde: Représentation de l'espace au Moyen Âge* (Paris: Éditions du Seuil, 1993), pp. 58–62.

17 *The fate of place: a philosophical history* (Berkeley: University of California Press, 1997), p. 15.

18 Scafi, *Mapping Paradise*, p. 153.

Chapter Two

The Antipodes in Antiquity

Classical Greek and Roman discussion of spaces beyond the known world extended over a very large range of texts and authors, but at its foundation was a principle of analogy. Geometrical calculations insisted that the classical ecumene did not occupy the entire surface of the earth, so the conjecture arose that, beyond the ends of the earth, there must exist places, spaces, and peoples analogous to those of the known world. Within analogy there existed two different modes of conceptualizing the relationship of known to unknown land: one was to duplicate the known world, and to assume an inverse identity with the unknown; the other was to envisage multiple versions of the known world, and to posit variations of identity and difference. These two modes were not, I will suggest, incompatible, but tensions nevertheless existed between the conception of a binary opposition between known and unknown worlds, ecumene and *antoecumene*, and the conception of the known world as one among several. In the case of binary opposition, the unknown world could act as a mirror, reflecting, often unflatteringly, the image of the known world; where unknown spaces were conceived as multiple, identity was not reflected so much as refracted, broken up rather than repeated. Whether by duplication or multiplication, however, both modes of analogy de-centred the known world. For that reason, lands beyond the ecumene could operate not simply as geometrical conjecture; they could also function as political sign.

To conceive of the state meant conceiving of its limits, literal as well as metaphorical. At its most crudely propagandistic level, the antipodes could be invoked as part of a fantasy of world domination, in which the limits of the ecumene were extended across the entire surface of the earth. More ambivalently, as in Virgil's *Georgics* and *Aeneid*, the antipodes could hover at the extreme limit of imperial ambition. Against these formulations, however, there existed a powerful tradition of using the antipodes as a reproof to power. In its most striking and influential formulation, Cicero's 'Somnium Scipionis' (Dream of Scipio), the integrity of the state was seen to depend upon its separation from antipodal parts of the earth. Furthermore, a corollary of the function of unknown land as a means of concentrating political energy within the ecumene was its deployment as a sign of unhinged political ambition, and it is with this signification that antipodal spaces appear in Lucan's *De bello civili*. The threat of unknown lands was that they could tempt rulers towards expanding and thereby disintegrating the state. Those insane, boundary-breaking, attractions of unknown lands eventually became a target of satirists and moralists such as Seneca and Lucian, a means by which pretensions and credulous assumptions could be revealed, and vice reproved.

In the following analysis I will outline the ways in which these three modes of writing – the scientific, the political, and the satiric – deployed *terrae incognitae* from the earliest known references to this topic in the sixth century BC until the second century AD. At no point did these literary strands constitute stable and separate traditions. Indeed they were entwined around two dominant motifs: that of inversion (the world turned upside-down); and relativization (the known world viewed from a different, external perspective).[1]

A geometrical conjecture

The motifs of inversion and relativization can both be found in the works of Plato, who appears to have adapted pre-existing scientific theory for philosophical purposes.[2] The late classical biographer Diogenes Laertius ascribed to Pythagoras the doctrine that the spherical earth is inhabited 'all around', and that 'there are also antipodes and our "down" is their "up"'.[3] But Plato, according to Diogenes, was the first to use the term 'antipodes' in philosophical discussion.[4] Although Plato's use of the term antipodes as a noun is not attested by extant texts,[5] he clearly refers to the possibility of lands beyond the known world in four surviving works: the *Timaeus* and *Critias*, the *Republic*, and *Phaedo*. Each reference can be characterized by its defamiliarizing purpose. In the *Timaeus*, a discussion of the terms 'above' (κάτω) and 'below' (ἄνω) leads the speaker to claim that, given the spherical nature of the cosmos:

> even were a man to travel round it in a circle he would often call the same part of it both 'above' and 'below', according as he stood now at one pole, now at the opposite ... the assertion that [the universe] has one region 'above' and one 'below' does not become a man of sense.[6]

Timaeus' insistence on the relativity of the concepts 'above' and 'below' anticipates and challenges a perpetual convention of descriptions of the antipodes, in which the antipodeans are dwellers below, or beneath, those in the known world. For, as Timaeus sees here, this convention carries with it the potential for its own reversal. Elsewhere in the *Timaeus*, as in *Critias*, the notion of land to the west of the *oikoumenē* was invoked to construct the Atlantis myth. Atlantis in the *Timaeus* and *Critias* is described as an island beyond the Pillars of Hercules (the westernmost extent of the known world), once bigger than Africa and Asia, but now sunk into Ocean, and cut off from the known world by impenetrable mud.[7] Nine-thousand years ago the inhabitants of Atlantis fought a war against those within the Pillars of Hercules; the memory of the war is the memory of a lost geography as well as a lost people. The Atlantis myth operates within the mode of duplication, although it does not rest on the opposition between above and below: Atlantis is equivalent to the known world, its rival and also its double.

In the *Phaedo*, by contrast, Socrates is made to refer not to one 'other world' but to many. His famous statement, in which he compares the inhabitants of the area between the Phasis river and the Pillars of Hercules to 'ants or frogs around a marsh', and states that 'there are many others living elsewhere in many such places', again defamiliarizes through its insistence on the position of the Mediterranean world relative to 'many other' such regions.[8] Here topographical, rather than global, imagery is deployed to offer not a vision, nor a theory, but a 'pseudo-geography',[9] a conjecture all the more gripping for its fleeting and undeveloped nature. In the *Phaedo*, relativization of perspective expresses contempt for the world, yet it necessarily occurs by reference to known geography, and the apposition of conjectured other worlds to the known. The perspective of Socrates is lateral, looking beyond the Phasis to the east, and the Pillars of Hercules to the west. In the *Republic*'s Myth of Er, on the other hand, Plato established the trope of a cosmic vision accorded to a privileged mortal. The vision of Er, a dead soldier permitted to return to life in order to narrate the passage of his soul to heavenly regions, where it witnesses a system of divine penalties and rewards and the allotment of souls to bodies, contains a lengthy description of the celestial spheres. The vision offers, then, the means by which the earth's position, and the preoccupations of its inhabitants, may be put into the broader perspective of the workings of cosmic order.

Plato's works contained the seeds for the classical discussion of the antipodes precisely because they offered more than one model for conceiving of unknown spaces: as opposite to the known world, as multiple other worlds, or seen from above, defining the known world and life on earth. All of these instances depend on the operation of analogy. Unknown land is assumed to exist parallel to the known world, and by extension to share at least some of its characteristics. These assumptions of proportion, symmetry, and parallel identity were made explicit in Aristotle's *Meteorologica*, where he posited two 'drum-shaped' habitable sectors of the earth's surface, one towards the north pole, the other towards the south: 'a region which bears to the other pole the same relation as that which we inhabit bears to our pole, ... *analogous to* (ὡς ἀνάλογον) ours in the disposition of winds as well as in other respects'.[10]

The notion of a bi-partite division between known and unknown worlds was consolidated by the development of a theory of habitation that divided the world into zones. The theory of zonal division, which enjoyed enormous popularity in antiquity and in the Middle Ages, was apparently invented by Parmenides (fifth century BC), although modifications were attempted by numerous later writers, such as Aristotle and Posidonius.[11] In essence zonal theory posited five zones: two frigid, uninhabitable, zones at either end of the earth, extending from the poles (though the extent of the frigid zones was a matter of frequent debate); a central zone of intolerable heat (though some commentators, including Posidonius, thought habitation of this region possible);[12] and two temperate zones, one in each hemisphere, both inhabitable by humans.[13]

The equivalence of the two temperate zones meant that it was possible, even attractive, to envisage inhabitants of the southern, unknown temperate zone, and in the case of the cosmic vision, to see antipodeans in relation to dwellers in the known world. In his poem 'Hermes', Eratosthenes (c. 285–194 BC) used colour to characterize the zones seen by the eponymous god when he ascends to the heavenly spheres: the frozen polar zones are darker than dark blue-green, and the central torrid zone red and burned. Colour serves to emphasize inaccessible cold and scorching heat, but the poem then turns from the extremes of fire and ice to the temperate zones, described in identical terms, 'both ... growing corn, the fruit of Eleusinian Demeter; in which men live as antipodeans' (ἐν δέ μιν ἄνδρες/ ἀντίποδες ναίουσι).[14] The use of the term 'antipodes' is interesting here: an adjective, accompanying 'men' (ἄνδρες), it serves to describe habitation on either side of the equator.[15] The symmetry is total: north and south are not mentioned in the poem, nor up and down. Instead the temperate zones are located between hot and cold zones, fertile, identical, indistinguishable, and foot-to-foot. Men in the temperate zones exist in a state of mutual relation, they are each other's antipodeans – and more than that: to be a man is to be an antipodean.

The state of opposition and mutual relation between known and unknown worlds and their inhabitants exploited in poems such as 'Hermes' was complicated by a theory, usually credited to Crates of Mallos (fl. c. 150 BC), that posited the division of the earth's surface into four rather than two regions.[16] Crates, a Stoic commentator on Homer, took the logic of the sphere a step further, by concluding that there must be a second habitable landmass in the northern hemisphere, on the 'underside' of the *oikoumenē*, and two corresponding landmasses in the southern hemisphere. His image of the world was that of a sphere divided into four separate inhabitable segments, irrevocably isolated from each other due to two encircling bands of ocean, one vertical, the other horizontal (fig. 3).[17] Crates' theory, which seems to have been a development of the theory of zonal division, meant that the distinction between the known and unknown worlds could be made not only in terms of temperature (temperate, frigid, and torrid zones), but also in terms of the relations between inhabitants of the various parts of the earth. Following the Cratesian divisions, commentators such as the first-century AD geographer Geminus described dwellers in unknown parts of the earth as *perioikoi* (around from the known world, i.e. the underside of the northern hemisphere), *antoikoi* (in the southern hemisphere opposite, i.e. due south of, the *oikoumenē*); and *antipodes* (on the underside of the southern hemisphere).[18] All three other worlds are represented from the perspective of the known, and in all at least the possibility of habitation is assumed. A fourth term, *antikthones*, was derived from Pythagorean theories of 'another earth, lying opposite our own',[19] and tended to be used to refer to those furthest away from, and having least in common with, the inhabitants of the known world.[20]

In addition to the characterization of inhabitants of other parts of the world in terms of stance, peoples and places beyond the ecumene could also be understood in terms of shadow. Posidonius was credited with the division of zones on this basis, predicated on the notion that the passage of the sun ran between the two tropics: hence 'periskioi' described those places under the poles where the sun did not set for six months of the year, and which were thus 'encircled' by shadow; 'heteroskioi' were found in zones beyond the tropics where shadows only fell in one direction – to the north or (in the southern hemisphere beneath the tropic of Capricorn) to the south; and 'amphiskioi' were places in between the tropics in which shadows fell either to the north or south, depending on the position of the sun.[21] Despite such conceptualization of people living in places beyond the known world, no attempt appears to have been made to assimilate antikthonal, antipodal, peri- or antoecumenical peoples with those in the *oikoumenē*. They were not, for example, included in the discussions of ethnic groups in works

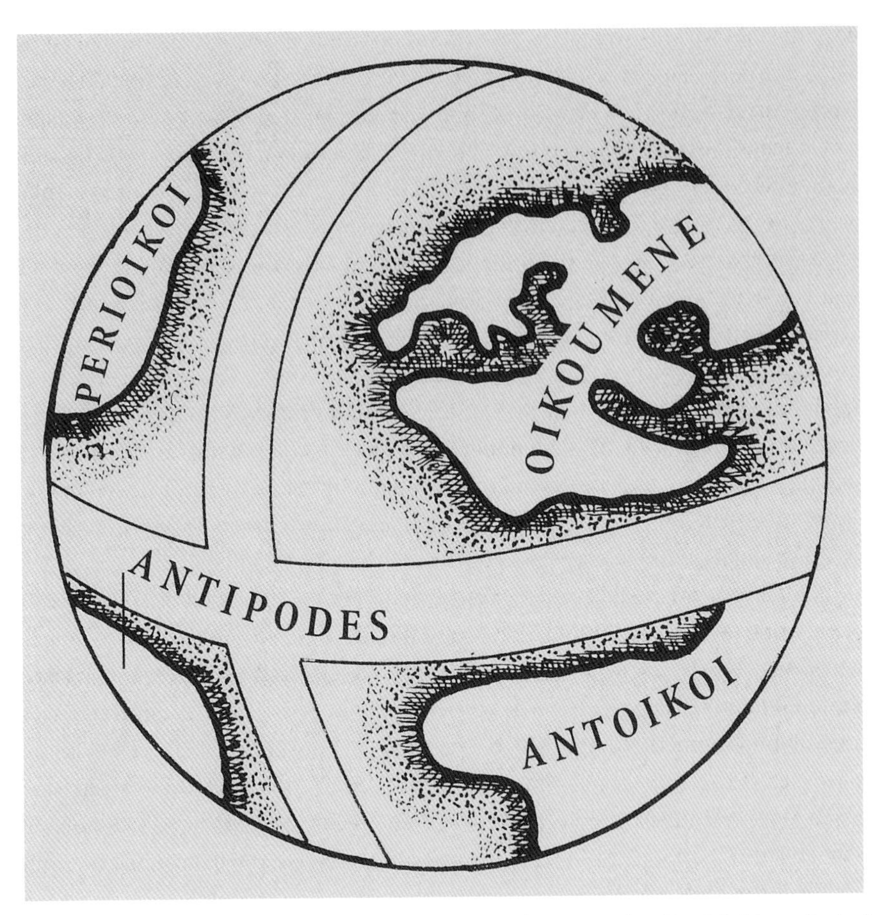

Fig. 3. Reconstruction of the division of the earth into four parts by Crates of Mallos. Inhabitants in the known world of Europe, Africa, and Asia (the *oikoumenē*) are divided from the 'antoikoi' to their south by an equatorial ocean. The 'perioikoi' dwell within the same parallels of latitude as those in the *oikoumenē* but are located 180°E or W of them, separated by an encircling ocean, flowing from pole to pole. The 'antipodes' dwell in the southern hemisphere beneath the 'perioikoi'.

such as Strabo's *Geography*, precisely because the antipodes did not belong to geography as defined by Strabo – that is, a description of the known world. Instead, discussion of the antipodes belonged, primarily, within the field of a mathematically-based astronomy in which the concept of a spherical earth was fundamental, and, secondarily, to philosophers and writers of fiction for whom the position of antipodal regions beyond the reach of empirical knowledge gave them an emblematic appeal. Even in these fields, though, authors emphasized the absence of information about antipodal inhabitants, and the uncertainty surrounding their very existence.[22]

The chief opposition to theories of other parts of the world propagated by the works of Plato, Aristotle, and the Stoics seems to have come from Epicurean natural philosophy, at least as disseminated to the Roman world through *De rerum natura* of Lucretius (c. 94–55 BC). In Lucretius' work, rejection of the notion of antipodal spaces is prefaced by a brief summary of the theory that 'all weights which are beneath the earth press upwards and repose on the earth upside-down, like the images of things we see when we look into water' (quae pondera sunt sub terris omnia sursum/ nitier in terraque retro requiescere posta,/ ut per aquas quae nunc rerum simulacra videmus), that animals can therefore move about under the earth without falling into the sky, and that they experience day and night and the seasons at different times to inhabitants of the ecumene.[23] These ideas are immediately denounced as 'empty' (vanus) and 'for the stupid' (stolidis) because there is no middle (the Epicureans conceived of the universe as infinite), and because it is not acceptable to believe that bodies could be held in a state of attraction towards the middle.[24] Nevertheless it is important to note the element of dialogue internal to the passage: the presentation of argument and counter-argument allows for the expression of certain motifs of antipodal description, such as the notion of the antipodeans as 'upside-down', the putative equivalence of 'upper' and 'lower' regions and the striking use of the simile of reflection in water. The same idea of aquatic reflection was later exploited by the satirist Lucian: behind it lurks a mockery of the self-reflexive nature of antipodal theory, the construction of the *antoecumene* as the image, the simulacrum, of the ecumene.

There is little evidence to suggest that the views expressed by Lucretius on regions 'beneath' the known world were accepted by later writers, but there are substantial grounds for thinking that they offered a useful summation of opposition to antipodal theory. They found an echo, for example, in Plutarch's dialogue 'On the Face on the Moon' where, as an example of the pointless paradoxes of philosophers, Lamprias cites the proposition 'that people live on the opposite hemisphere (ἀντίποδας οἰκεῖν) clinging to the earth like wood-worms or geckos turned bottom side up – and that we ourselves in standing remain not at right angles to the earth but at an oblique angle, leaning from the perpendicular like drunken men'.[25] Indeed the conjectural and contested nature of the *antoecumene* made necessary a more complex theorization of the relationship between known and unknown peoples. One late antique commentator, Cleomedes, distinguished between the part of the world inhabited by 'people known to history' and those inhabited by others beyond the historical record: the 'perioikoi', 'antoikoi' or 'antomoi', and the 'antipodes'.[26] Relationships between known and the three unknown peoples were described in terms of convertibility or *reciprocality*. The Greek verb ἀντιστρέφω, used by Cleomedes to characterize the positions of the four parts of the world, had applications in a variety of contexts (including grammar, logic, rhetoric, and poetry) and carried connotations of opposite, inverse, converse, and interdependent relationships. Crucially, its appearance in Cleomedes' treatise was explicitly designed to describe a principle that excluded hierarchy. The disposition of the four groups of peoples to each other resembled, Cleomedes explained, 'those of friends and brothers, rather than those of fathers and children, or of slaves and masters'.[27] Cleomedes meant that 'they are reciprocal (ἀντιστρέφουσι/reciprocae), for we are indeed the perioikoi of our perioikoi, and the antipodes of our antipodes, and similarly the antoikoi of our antoikoi'.[28] As it had done for Plato, spherical logic continued to demand the possibility of reversals of perspective, or more precisely a persistent relativization of perspective. This did not mean that differences were not identified between the various non-ecumenical peoples. On the contrary, as Cleomedes went on to explain, inhabitants of the ecumene have seasons in common with the 'perioikoi' (since they inhabit the other side of the northern temperate zone), but experience nights and days at different times and of different lengths. Similarly, those in the ecumene have days and nights at the same time as the 'antoikoi' (since they occupy the upper hemisphere in the southern temperate zone), but the seasons are reversed; with the antipodeans, however, nothing is in common, everything – the seasons, the times of day – reciprocal (ἀντέστραπται).[29] The verb signifies a state of permanent alternation, of inverse identity. Within the motif of inversion, Cleomedes exploits the possibility of a lateral instead of a vertical and hierarchical understanding of spatial identity (brother, friend rather than father/child, master/slave), with the result that inversion is not restricted to a binary but is the sign under which a sequence of relationships are conceived between 'us', perioikoi, antoikoi, and antipodeans. Evidently this mode of theorization could amount to a quasi-ethnography, the construction of named groups of broadly-delineated hypothetical peoples. Such an ethnography, based on the location of three other, unknown parts of the world, carried with it a powerful political charge precisely because it could invert assumed hierarchies (down is up), and at the same time relativize perspectives. Perhaps most strikingly of all, the hypothetical ethnography of *terrae incognitae* encouraged the idea of mutual ignorance, the notion that on the other side(s) of the earth there existed peoples who had never heard of the glory and dominion of European powers – and who never would.

The antipodes as political sign

In Cicero's 'Somnium Scipionis', the final section of *De re publica*, the Roman military tribune and future consul Scipio Aemilianus narrates a dream in which, from 'a certain high place' (de excelso ... quodam loco),[30] his adoptive grandfather, Scipio Africanus, shows him the stars, the nine spheres that constitute the universe, the celestial harmonies, the sweet music of the spheres (too swift to be heard by human ears) – and the earth. The elder Scipio uses the shape of the world, its smallness, and the brevity of a human life-span to show the dreamer the transience of worldly renown and glory, and the relative unimportance of pursuing them. The earth, he points out, is sparsely inhabited, and where inhabitants are found they are separated by large areas of wasteland ('vastae solitudines').[31] In a passage that caused considerable difficulty to later commentators, the elder Scipio then describes the relationship of inhabitants of three other quarters of the earth to the Romans in terms of their stance:

> eosque qui incolunt terram non modo interruptos ita esse ut nihil inter ipsos ab aliis ad alios manare possit, sed partim obliquos, partim transversos, partim etiam adversos stare vobis. a quibus expectare gloriam certe nullam potestis.[32]
>
> [you see] those who inhabit the earth are not only divided, so that nothing can pass between them from one group to another, but some stand obliquely, some transversely, and some even directly opposite to you; from these you can certainly expect no glory.

The act of looking down has revealed a divided earth, in which no region is more central than any other, and in which peoples are inevitably and unalterably remote from each other. The earth, Scipio goes on to explain, is encircled by five belts ('cinguli'):

> e quibus duos maxime inter se diversos et caeli verticibus ipsis ex utraque parte subnixos obriguisse pruina vides, medium autem illum et maximum solis ardore torreri. duo sunt habitabiles, quorum australis ille, in quo qui insistunt adversa vobis urgent vestigia, nihil ad vestrum genus; hic autem alter subiectus aquiloni quem incolitis cerne quam tenui vos parte contingat.[33]

> of those the two furthest apart from each other, supported at each end by the very poles of heaven, you see stiffened with frost; that middle one however, the biggest, you see burn with the heat of the sun. Two belts are habitable: those who dwell in the southern one press their feet against you, and have nothing to do with your people; and this other one lying to the north, which you inhabit – look what a meagre portion has anything to do with you!

The principle at work in the 'Somnium Scipionis' is that of the augmentation of imperial knowledge of the world by means of a celestial perspective, one that simultaneously reveals the puny size of dominion, and which, crucially, seeks to demonstrate the impossibility and futility of expansion beyond its quarter.[34] Scipio's appreciation of the delights of the celestial spheres is predicated upon the recognition of the absurdity of earthly ambition that an ethereal perspective brings: 'the earth itself now seemed to me so small that I was ashamed of our empire, through which we possess just a single point of it' (iam vero ipsa terra ita mihi parva visa est, ut me imperii nostri quo quasi punctum eius attingimus paeniteret).[35] All the same, the 'Somnium Scipionis' by no means repudiates empire, so long as imperial rule is based on justice and the consent of the ruled. Elsewhere in *De re publica* Scipio's interlocutor Laelius warns of the danger that Roman *imperium* might become based on force rather than law ('ad vim a iure'), and cause subject peoples to be held by terror rather than free will: the consequent threat will be to the 'immortality' of the commonwealth, 'which could last for all time' (quae poterat esse perpetua).[36] The security of imperial place in Africa, a security which stands for the integrity of the Roman state as a whole, is what Scipio is destined to defend, as his grandfather makes clear.[37] The first words of Scipio's account of the dream, 'Cum in Africam venissem', foreground the entry of the Roman into Africa, and it is to Carthage – a city whose defeat was emblematic of Roman dominance – that the elder Scipio draws the attention of Aemilianus at the beginning of his tour of the universe, 'that city, forced by me to obey the Roman people ... this you will overthrow as consul two years from now ...' (illam urbem, quae parere populo Romano coacta per me ... hanc hoc biennio consul evertes).[38] Yet the imperial perspective must be informed by the celestial; and the check on imperial ambition, on excessive pride in worldly achievement, is constituted by the presence of other, unnamed spaces, the lands of others, whom the fame and glory of Rome will never reach.

Cicero's topos of worldly insignificance in the light of geographical expanse has clear echoes of Er's vision in Plato's *Republic* and Socrates' comments on spaces beyond the Mediterranean in *Phaedo*. The 'Somnium Scipionis' also testifies to Cicero's awareness of the post-Platonic elaboration of antipodal theory, in particular the division of the earth into five zones, and Crates' quadri-partite division of human habitation – a theory, as Claude Nicolet has noted, more popular with philosophers than geographers.[39] Cicero in fact adapts a number of different images of the world to emphasize the relatively small extent of Roman rule. Scipio is first directed towards the vast 'solitudines' lying between inhabitants of the earth;[40] the Cratesian model is then invoked, expressed in geometrical terms (oblique, transverse, adverse), in order to expand further the sense of distance from the Roman world. At this point Cicero switches models, deploying the zonal theory, although he uses the term 'cingulus', rather than the more technical 'zona'. The emphasis remains on the distance between human habitation: those who inhabit the southern 'cingulus' (still described in terms of their stance, adverse to 'your footprints') have 'nothing to do with your people', and an examination of the northern belt reveals the narrowness and insularity of 'the land which is inhabited by you'. From this point, Cicero adapts descriptions of the ecumene, citing the traditional boundaries of geographical knowledge: the elder Scipio describes the known world as a 'small island' (parva insula) surrounded by the Atlantic, itself a 'small' sea in spite of its grand name, and asks whether the name of the Romans can ever pass beyond the Ganges, or the Caucasus mountains.[41]

In adapting, and indeed mingling, cosmological and geographical description Cicero deliberately avoids deploying quasi-ethnographic terminology to describe the inhabitants of the three unknown quarters of the world. Elsewhere in his writing, Cicero showed a familiarity with precisely the kinds of terms used to describe inhabitants beyond the known world. In his *Academica* he uses the term 'antipodes' to describe 'those who stand with adverse feet against ours' (qui adversis vestigiis stent contra nostra vestigia), a locution that shows the term's currency within Latin, and also strongly resembles the description of the inhabitants of the southern belt in the 'Somnium Scipionis'.[42] In the *Tusculanae Disputationes* (Tusculan Disputations), Cicero refers to 'the other southern [part of the world], unknown to us, which the Greeks call 'antikthona' (altera australis, ignota nobis, quam vocant Graeci ἀντίχθονα).[43] A close reading of the 'Somnium Scipionis' indicates, then, not only that Cicero blurred the divisions between cosmological theories, moving between the model of binary division and the model of, to use Cleomedes' term, reciprocal relations between four sets of peoples, but also that he employed a deliberately opaque mode of reference to those peoples beyond the ecumene. In spite of this opacity, Cicero's adaptation of Cratesian theory illustrates one of its most interesting features. In the course of the movement of Scipio's vision from the known world to human habitation beyond, and then back again to the ecumene, attention leaves the Roman world. For just a moment habitation is not described in terms of the relationship of the 'we' of the known world to the 'they' of the unknown, but in terms of *their* relation to *them*: 'nothing can pass between them from one group to another' (nihil inter ipsos ab aliis ad alios manare possit). That vision of the relations of others *to* others is directly connected to the theory of quadri-partite division: positing more than one other world means that there is more than one 'they'.

Roman political power, one can conclude from the 'Somnium Scipionis', was conceived by Cicero in relation to conjectural *terrae incognitae* and their peoples, even when those spaces and peoples were invoked to show where *imperium* could not go. Subsequent expressions of Roman glory and destiny returned to the image of the world in its entirety, but they increasingly played with the idea of the expansion of fame and power beyond natural boundaries. Visions of imperial expansion to the other side of the earth were not necessarily laudatory, however: they could also offer scope for the critique of the disintegration caused by spreading power too far.

Or, as in the case of Virgil, the antipodes could function as a fundamentally ambiguous sign, poised between glory and folly.

In Virgil's *Georgics* and *Aeneid* brief, but telling, allusions to the antipodes certainly hint at a revision of Cicero's limits of renown. In the *Georgics*, global vision is subtly connected to imperial ambition. This work, which begins by imagining Augustus as a god, ends with a record of his triumph at the battle of Actium. It also contains a description of the heavenly regions that gained wide currency in the Middle Ages. In Book I, in the course of a discussion of the correct times for sowing and harvesting, Virgil explains the circuit of the sun, and the five celestial *zonae*:

> quinque tenent caelum zonae: quarum una corusco
> semper sole rubens et torrida semper ab igni;
> quam circum extremae dextra laeuaque trahuntur
> caeruleae, glacie concretae atque imbribus atris;
> has inter mediamque duae mortalibus aegris
> munere concessae diuum, et uia secta per ambas,
> obliquus qua se signorum uerteret ordo[44]

> five zones occupy the sky: one of them is always red with the gleaming sun and always hot with fire; around it gloomy extremities are stretched to the right and left, hard with ice and black with storms; between these and the central zone two have been granted by gift of the gods to feeble mortals, and a path cut between the two, along which the slanting procession of signs might revolve.

This passage provides a good example of the ease with which correlation could be made between the celestial and the terrestrial zones. The description of the torrid and frigid zones in terms of natural elements (fire, ice) and light and colour (bright, red; dark, black), contrasts with the description of the temperate zones in terms of human habitation and experience. Virgil's chief source for these lines seems to have been Eratosthenes' 'Hermes', but it is a sophisticated adaptation, in which many emphases of the Alexandrian's work are altered, and in which the influence of no less than six other authors has been discerned.[45] Most strikingly, Virgil makes only a relatively passing and unflattering reference to the temperate zones – an act that has surprised critics, given the focus of the *Georgics* on the cultivation of land. Where Eratosthenes described the growth of 'the fruit of Eleusinian Demeter' in the temperate regions, Virgil notes only the feeble nature of their inhabitants and does not use the adjective 'temperate'.[46] The description of humans as 'feeble mortals' seems to be a deliberate reference to Lucretius, who in book six of *De rerum natura* had Athens first bring crops 'mortalibus aegris'.[47] Nevertheless, as Monica Gale has noted, these lines mark Virgil's deliberate refashioning of Lucretius' poem as well as his adaptation of 'Hermes'.[48] Crucially, even as he diminishes the sense of symmetrical bounty conveyed by Eratosthenes, Virgil insists upon the divine basis of human habitation, an emphasis absent from Lucretius. That point is drawn out by the clear connection in lines 237–8 between divine providence and the 'order of signs' (i.e. the zodiac) that cuts between the zones.

The notion of the turning of zodiacal *ordo* prompts the poet to a further consideration of known and unknown worlds. After two lines that sketch the northern and southern extremities of the known world (Scythia and the Rhipaean mountains in the north, and Africa in the south), Virgil contrasts the view of the stars enjoyed by 'us' with the parts of the earth 'beneath our feet':

hic uertex nobis semper sublimis; at illum
sub pedibus Styx atra uidet Manesque profundi.
maximus hic flexu sinuoso elabitur Anguis
circum perque duas in morem fluminis Arctos,
Arctos Oceani metuentes aequore tingi.
illic, ut perhibent, aut intempesta silet nox
semper et obtenta densentur nocte tenebrae;
aut redit a nobis Aurora diemque reducit,
nosque ubi primus equis Oriens adflauit anhelis
illic sera rubens accendit lumina Vesper.[49]

for us this pole is always elevated; but beneath our feet black Styx and the infernal shades see the other pole. Here the great Snake glides along with sinuous coil around and between the two Bears in the manner of a river, the Bears refusing to be moistened by the smooth surface of Ocean. There, it is said, either dismal night is silent, and always the darkness is thick with enveloping night; or Dawn returns from us and brings back the day, and when the rising sun with panting steeds first breathes upon us, there bright Vesper kindles the evening lights.

This passage is significant for its conflation of celestial imagery with the iconography of the underworld. The underside of the earth is described in antipodal terms ('sub pedibus'), but it is associated with Styx and the shades, rather than with peoples and places analogous to the known world. Celestial imagery (the constellations of the Bears and the Snake) remains,[50] but the possibility of eternal night is given equal status to the theory that day and night alternate between the upper and lower hemispheres. What prompted Virgil to combine two contradictory representations of the antipodes? According to his late antique commentator Servius, Virgil aimed to contrast the views of philosophers. Against the Stoic view of a southern temperate zone inhabited by mortals, he set the Epicurean notion that the southern hemisphere remained in darkness.[51] Recent commentators have accepted Servius' further point that, in his location of 'Styx atra' in the southern hemisphere, Virgil was 'using poetic licence'. Although his remarks contradict 'the traditional view that Tartarus lies deep inside the earth' (a view expressed elsewhere in his own poetry),[52] he was, as Franz Cumont put it, pursuing a literary goal which was served by 'an impression that he judged poetic'.[53] This passage is essentially a digression from the compendium of practical advice contained in the *Georgics*, and its justification is that a study of celestial phenomena enables the prediction of changes in the weather. Just as the 'Somnium Scipionis' returns the attention of the dreamer to the affairs of state in the known and knowable world, here the poem returns to experience of the natural world: farming, foresting, and sailing. The reference to the underworld does contain a political echo, nonetheless, since an aside at the beginning of Book 1 mentions that Tartarus does not hope for Caesar as king, and prays that he be not overtaken by such a dire lust for power: 'nam te nec sperant Tartara regem/ nec tibi regnandi ueniat tam dira cupido'.[54] The equation of the regions 'beneath the feet' with the classical underworld, and the oblique reference to Augustus' potential desire to rule there, indicate Virgil's flirtation with the notion that the other side of the earth may be capable of receiving the renown of empire, and may excite an excess of imperial ambition.

Two asides, in books 6 and 7 of the *Aeneid*, elaborate on this vision of power and renown spilling over the boundaries of the known world. In the first of these, while in the underworld, Aeneas is informed by his father, Anchises, of the glories to come during the reign of Augustus Caesar. Not only will Augustus establish a new Golden Age in Latium, he will accompany it with a period of dramatic territorial expansion:

> [...] super et Garamantas et Indos
> proferet imperium; iacet extra sidera tellus,
> extra anni solisque uias, ubi caelifer Atlas
> axem umero torquet stellis ardentibus aptum.[55]

> ... he will extend empire beyond the Garamants [i.e. Africans] and the Indians; there is a land that lies beyond the stars, beyond the paths of the year and the sun, where Atlas the heaven bearer turns on his shoulder the firmament studded with blazing stars.

Again the reference to the antipodes is oblique: the movement of empire does not stop at the extremes of the known world, the Garamants and Indians, but heads beyond the path of the zodiac. The model here is still that of zonal division, however, and the 'tellus' Virgil has in mind is surely that of the southern temperate zone. The lines may be read as fulfilment of Jupiter's promise to Venus in book 1, that 'on [the Romans] I impose boundaries neither of space nor of time: I have given power without end' (his ego nec metas rerum nec tempora pono:/ imperium sine fine dedi).[56] In the same speech Jupiter foresees the birth of Julius Caesar, and states that he will take Roman power as far as the encircling Ocean (1.287); by implication Augustus may surpass even this boundary.

If Augustus' antipodal expansion is the potential fulfilment of Roman *imperium*, the *Aeneid* also attributes the same capacity to exceed territorial boundaries to the empire's origins. In Book 7, following the arrival of the Trojans in Italy, Ilioneus, ambassador of Aeneas to King Latinus, describes the Trojan war. It was, he says, a savage storm, indeed a battle between the worlds of Europe (Greeks) and Asia (Trojans), the news of which has travelled quite literally to the ends of the earth:

> quanta per Idaeos saeuis effusa Mycenis
> tempestas ierit campos, quibus actus uterque
> Europae atque Asiae fatis concurrerit orbis,
> audiit et si quem tellus extrema refuso
> summouet Oceano et si quem extenta plagarum
> quattuor in medio dirimit plaga solis iniqui.[57]

> How much of a storm sent forth from savage Mycenae across the fields of Ida, and by which fates the worlds of both Europe and Asia were driven to clash together – these things were heard, even by those banished to the furthest land flung back by Ocean, and even by those cut off by the region of the oppressive sun, stretched in the middle of the four.

Once more, Virgil seems to envisage passage of news in the first instance to the eastern and western extremities of the known world (those in the 'extrema tellus' reachable by Ocean), and then beyond, suggesting that news of the war may have reached as far as the southern hemisphere, across the central torrid zone, here described by the word 'plaga', instead of 'zona', or the Ciceronian 'cingulus'. This piece of hyperbole is given particular significance by its context: Latinus has heard prophesied the arrival of strangers who will 'bear our name to the stars with blood', and whose descendants will rule 'all beneath their feet [sub pedibus] that the sun views hastening from Ocean to Ocean'; Aeneas has subsequently recognized in Italy the 'unknown shores' promised to the Trojans ('hic domus, haec patria est'); and in his embassy Ilioneus emphasizes the kinship of the Trojans and the Latins, based on their shared descent from Jupiter.[58] Virgil stages the legitimization of the Trojan invasion of Italy by expressing the relationship of the Trojans to the land of Italy as that, apparently paradoxically, of both promised strangers and native race. This foundational moment of empire has its roots in a war that the entire world heard about: antipodal inhabitants serve here as unnamed aural witnesses to Rome's origins in epic conflict.[59] The subtle implication, following on from the paean to Augustus in book 6, may be that they could yet serve as witnesses to its imperial expanse.

Beyond Virgilian ambiguity, more explicit examples of the use of the antipodes to invoke military and imperial glory can be found in works such as the *Panegyricus Messallae* (after 27 BC), an anonymous eulogy of the military leader, consul, literary patron, and supporter of Octavian, Messalla Corvinus, and in Manilius' didactic poem, the *Astronomica*.[60] In the former Messalla is informed that 'Britain, unconqured by Roman power, awaits you, and so too the other part of the world beyond the path of the sun' (te manet invictus Romano Marte Britannus/ teque interiecto mundi pars altera sole).[61] The *Astronomica* emphasizes the position of the antipodes beyond knowledge: 'barred to us', they contain 'unknown races of men and unreachable kingdoms that draw light from one common sun'. However, Manilius figures Augustus as a star that conquers the stars of the southern hemisphere: 'they are defeated by one star alone, Augustus, the star that took hold of our world, now the greatest maker of laws on earth, hereafter in the heavens' (uno vincuntur in astro,/ Augusto, sidus nostro qui contigit orbi,/ legum nunc terris post caelo maximus auctor).[62]

Yet throughout the period from Augustus to Nero there was a persistent and countervailing tendency to represent imperial passage to the antipodes as a regressive *dis*ordering of natural boundaries, and therefore as the disordering of political structure. Ovid's *Metamorphoses* express very neatly the contrast between natural and political order, and the profoundly disruptive effects of human intervention therein. In book 1 of the *Metamorphoses* zonal division is an expression of divine control, the imposition of order on the 'congeries', the unshaped pile or mass of the elements, synonymous with primal chaos. The zones are designed by a deity (quisquis … deorum: 'some one of the gods'), a 'fabricator mundi' who marks the earth with the same pattern of zones as already exists in the heavens.

> utque duae dextra caelum totidemque sinistra
> parte secant zonae (quinta est ardentior illis),
> sic onus inclusum numero distinxit eodem
> cura dei, totidemque plagae tellure premuntur.
> quarum quae media est non est habitabilis aestu;
> nix tegit alta duas; totidem inter utrumque locauit
> temperiemque dedit mixta cum frigore flamma.[63]

> And just as two zones cut the heavens on the right, and two on the left (a fifth burns more than they do), so divine oversight divided the confined freight of the earth by the same number, and five tracts were impressed on the earth. The middle one of them is uninhabitable because of its heat; deep snow covers two; two more he placed inbetween and, mingling flame with cold, gave them a moderate temperature.

The language leading up to and including this description is of division and command: he [the god] cut (secuit) ... he assembled (coegit) ... he formed into a ball (glomeravit) ... he ordered (iussit) ... he added (addidit) ... he encircled (cinxit) ... he ordered (iussit) ... they cut (secant) ... divine oversight divided (distinxit) ... they were impressed (premuntur).[64] As the passage progresses, the subject of the verbs begins to shift from the god to the zones, and then to the earth itself, the two uses of 'est' in line 49 marking the transition from past to present tense, and from divine action to enduring terrestrial order. The repetition of 'totidem' (just as many, so many) three times in six lines emphasizes the symmetry between celestial and terrestrial zones, and the broader themes of regulation and proportion. Thus stamped, the earth is made habitable, two temperate zones imposed between the extremes of heat and snow. Imposed? Actually the final verbs in the passage are more delicate: 'he placed ... and he gave' (locavit ... dedit), as is the concluding image of the mutual moderation of flame and cold ('mixta cum frigore flamma'), a mixture given to the human race.

However, in a work preoccupied with the transgression of boundaries it could not be expected that such clear-cut order would remain intact and unchallenged: book 2 of the *Metamorphoses* sees the imagined dismantling of zonal organization in the ride of Phaethon, progeny of the sun, who desires to drive his father's chariot for a day. 'You will be safest on the middle way' (medio tutissimus ibis), says his anxious parent, urging the boy to stick to the accustomed route around the earth (i.e. the ecliptic), slanting but 'confined by the limits of the three [central] zones' (zonarumque trium contentus fine), and neither too high nor too low.[65] When Phaethon loses control of the chariot, his horses career 'through the airs of an unknown region' (per auras/ ignotae regionis),[66] and the work of creation begins to unravel: meadows are scorched, seas dried, rivers burned and put to flight, cities razed (yet even here a sub-creation myth is enacted – Ethiopians become black-skinned as a result of the excessive heat); all leading, as an aggrieved Earth points out to the almighty father, 'back to primeval chaos' (in Chaos antiquum).[67] The restoration of order entails the fall of Phaethon and his burial on foreign shores, far from *patria*, and his epigraph serves as the marker of transgression, ambiguously recording (celebrating or mocking?) the magnitude of his failure and the greatness of his daring: 'Here lies Phaethon, driver of the paternal chariot; although he did not control it, he perished with great daring' (Hic situs est Phaethon currus auriga paterni/ quem si non tenuit magnis tamen excidit ausis).[68]

Ovid envisaged a catastrophic dismantling of the zones, and above all the irruption of fierce, uncontrolled solar heat on the northern temperate zone: his account of Phaethon's ride is also a mini-geography of the known world in melt-down. That breach of the natural order is not desired, even by Phaethon, and its disastrous consequences could be read as a warning against human assumption of a divine status, and hence aimed squarely at Augustus. They also constitute an oblique commentary on the significance of the zones – and the antipodes – in the poetry of Virgil. Virgil established the zones as a sign of celestial and terrestrial order, and simultaneously a sign of poetic mastery. The description of the zones in the *Georgics*, described by one critic as 'the most densely allusive passage in ancient literature',[69] shows Virgil's control over an array of poetic material, at the same time as it represents distinct natural boundaries. Ovid's response in the *Metamorphoses* was to represent not only the making but also the unmaking of zonal order, with the result that the sign of poetic virtuosity was not the magisterial marshalling of learning and art, but rather the crafted – and always controlled – depiction of disorder. Further, Ovid amplified to the point of comedy Virgil's hints, in the *Georgics* and the *Aeneid*, that imperial expanse, however much the product of destiny, might breach order. Phaethon's ride transgresses order with consequences that are as much comic as dire, so that the charioteer ends not only a broken, but also a laughable figure.

Ovid's imagination of zonal disintegration was taken up by a literature that, developing Cicero's seminal use of the topic in the 'Somnium Scipionis', invoked global vision to criticize domestic discord, and in so doing used the antipodes as a sign of ambition gone mad. Two works from the third quarter of the first century AD, both of immense significance for medieval literary and scientific traditions, illustrate this connection: Lucan's *De bello civili*, and Pliny the Elder's *Naturalis historia*. The final book of Lucan's *De bello civili* contains a highly unflattering digression on Alexander the Great, in which the king appears as the embodiment of violent expansionism and the enemy of liberty. In a few deft strokes Lucan sketches Alexander's overreach in terms of geography: content with neither Macedonian birthright nor his Athenian inheritance, Alexander assaults the peoples of Asia, leaving a trail of carnage in his wake, mingling the waters of the Euphrates and the Ganges with, respectively, Persian and Indian blood. But the known world is not enough for this 'fortunate robber' (felix praedo):

> [...] Oceano classes inferre parabat
> exteriore mari. non illi flamma nec undae
> nec sterilis Libye nec Syrticus obstitit Hammon.
> isset in occasus mundi devexa secutus
> ambissetque polos Nilumque a fonte bibisset:
> occurrit suprema dies, naturaque solum
> hunc potuit finem vaesano ponere regi;
> qui secum invidia, qua totum ceperat orbem,
> abstulit imperium, nulloque herede relicto
> totius fati lacerandas praebuit urbes.[70]

> He was preparing to lead a fleet into the outermost Ocean. No heat or waves, neither barren Africa nor Syrtian Ammon impeded. Heading west he would have followed the slope of the world, made a circuit of the poles, and drunk from the Nile at its source. The final day came, and nature was able to fix on just one end for the mad king: he who had seized the entire world for envy, out of envy took empire with him, and having left no heir to inherit all his power, he left cities to be ripped to shreds.

Alexander seems to plan not merely latitudinal but even longitudinal circumnavigation, in the process unravelling the hidden source of the Nile. The antipodes are not mentioned by Lucan, although one twelfth-century commentator was left in little doubt: he glossed 'heading west' (in occasus mundi) with the words 'that is, as far as the antipodes' (id est usque ad antipodes),[71] and the implication seems clear that Alexander's fleet will search for new lands and peoples to conquer. In previous books of *De bello civili* the antipodes appear as a potential place of refuge for the defeated Pompey, and as a place of desolate – and bitterly ironic – wandering for his followers, unmoored in the African desert, led by Cato of Utica: 'even now perhaps Rome itself is beneath our feet' (nunc forsitan ipsa est/ sub pedibus iam Roma meis').[72] In book 10 the legacy of antipodal ambition is death and destruction: the attempt to reach the other side of the world is explicitly described as 'insane' (vaesanus), a word that recurs three times in the space of fifty lines.[73] The evocation of Alexander's expansionism is prompted by Julius Caesar's desire to visit the Macedonian's tomb in Egypt, and the parallel between the two is sustained to the end of the book: for Caesar the Roman world is not enough and, like Alexander, he seeks to conquer Egypt and learn the source of the Nile. Indeed a description of the Nile given to Caesar by an old retainer of Cleopatra midway through the book repeats much of the vocabulary of the earlier passage: Alexander, we are told, 'envied' (invidit) the Nile; he sent chosen men 'through the furthest lands of the Ethiopians' (per ultima terrae/ Aethiopum) in search of it; nature, in the form of the torrid zone, restrained them (illos rubicunda perusti/ zona poli tenuit). Other kings were similarly afflicted: Sesostris headed 'to the westernmost limits of the known world' (ad occasus mundique extrema), Pharios sought to drink from the Nile 'at its source' (de fonte), 'vaesanus' Cambysus sought the secret of the great river in the east and ended up eating the corpses of his followers – all in vain.[74] Power strains against nature and loses, at a terrible cost. Yet if expansionary violence seems to link *imperium* with *invidia* and a ferocious derangement, it is violence dircted inwards, towards the state itself, that is the ultimate target of Lucan's poem about the war between Caesar and Pompey. And it is this connection between imperial and civil bloodshed that finds amplification in a striking passage from Pliny's *Naturalis historia*.

In book two of his encyclopedia, Pliny first outlines the extent of geographical knowledge, and then demonstrates the capacity for global contemplation to be turned against imperial ambition. He notes that the seas that surround the globe 'take from us part of the world, neither is there passage to that region from here to there, nor from there to here. This reflection is apt for the exposure of the vanity of mortals ...' (partem orbis auferunt nobis, nec inde huc nec hinc illo peruio tractu. Quae contemplatio apta detegendae mortalium vanitati).[75] The invocation of the unknown part of the world as a check against vanity echoes, of course, Cicero's use of antipodal spaces to demonstrate the limited extent of Roman dominion, as well as a section from Seneca's *Naturales quaestiones* in which the world is viewed from above, and humans compared to ants. There Seneca envisages the rational soul ('animus') looking down on the world, and exclaiming in amazement 'this is that pin-prick [punctum] that is divided between so many peoples by sword and fire?', before denouncing the borders of mortals as 'ridiculous'.[76] Pliny also describes the earth as a 'punctus', and, like Cicero, amplifies the theme of the earth's comparative insignificance through an account of the zones. Cruel frost, eternal ice, and perpetual mist prevail at the polar zones, while flames burn up the central zone, across which passage between the two temperate zones is not possible.[77] The size of Ocean and other waters reduces the extent of land still further; indeed, such is the tiny size of land worthy of cultivation that human struggles for its possession are contemptible:

> conputetur etiamnum mensura tot fluminum, tantarum paludium, addantur et lacus, stagna, iam elata in caelum et ardua adspectu quoque iuga, iam silvae vallesque praeruptae et solitudines ac mille causis deserta; detrahantur hae tot portiones terrae, immo vero, ut plures tradidere, mundi puncto ... haec est materia gloriae nostrae, haec sedes. hic honores gerimus, hic exercemus imperia, hic opes cupimus, hic tumultuamur humanum genus, hic instauramus bella etiam civilia mutuisque caedibus laxiorem facimus terram.[78]

> Let the measure be taken of so many rivers, of the great marshes, and to that add the lakes, pools, and also the mountains that rise to the sky steep even to the sight, then the woods and precipitous valleys and the wastelands and those places deserted for a thousand causes. Subtract these many portions from the earth, or rather, as a great number teach, from this pin-prick of the world ... this is the matter of our glory, this the abode: here we achieve honours, here we administer empires, here we desire riches, here we stir up the human race, here we renew wars – even civil wars – and we make the earth more spacious by mutual slaughter.

Pliny's critique of worldly ambition is more profound than Cicero's, since all territorial ambition comes in for censure, from *imperium* to *domus* (he gives the example of the man who increases the boundaries of his land by stealing from his neighbour). The rhetorical force of this passage, along with the popularity of Pliny's text, ensured its longevity: its concluding statement became a popular epitaph on world maps in the sixteenth century. The connection with maps is not surprising, since description is a vital component of the *contemptus mundi* tradition; without the capacity to describe the world, there can be no contempt. Pliny's denunciation (hae ... haec ... haec ... hic etc.) requires a referent, requires that the world be shown.[79] The theory of zonal division, emphasizing uninhabitable hot and cold regions, allied with an emphasis on the size of oceans, lent itself to this simultaneous display of the world and diminution of the size and significance of territory within the world.

Satirical reversals

In function and technique antipodal satire was clearly related to political uses of the topic, serving to direct vision towards the imperfections of the known world by invoking that which lay beyond it. Satirists were able to criticize contemporary society by using the antipodes as the sign of a lack of self-reflection and self-knowledge: they aimed to mock rather than denounce the possibility of passage to unknown parts of the earth. Just as Scipio was made to see beyond the Roman world in order to understand its place in the world, so satirists aimed to expose the folly of looking beyond the known world and thereby failing to see one's immediate surroundings – including oneself.

The principal deployment of the antipodes within classical satire was the invocation of a world 'turned upside down' in order to mock folly and immorality in the known world.[80] In one of Seneca's letters to Lucilius, decadent people who sleep during the day and wake at night are described as 'antipodeans in the same city [as us], who, as Cato says, do not see the sun rise and do not see it fall' (in eadem urbe antipodes qui, ut M. Cato ait, nec orientem umquam solem viderunt nec occidentem).[81] The joke – that nocturnal Romans, like those on the other side of the earth, experience night when upright people are hard at work in the day – indicates the currency of antipodal theory during the first century AD, and the potency of the notion of reversal, here extended to those who live contrary not merely to the norms of society, but to nature itself.[82] More often the targets of satirical adaptations of the antipodes were fools, charlatans, and overreachers. In Lucian of Samosata's 'The Ship or Wishes' (second-century AD), the character Timolaeus wishes for a set of magical rings, one of which will enable him to fly. This ring will enable him to see the Phoenix bird in India, and learn the source of the Nile, as well as 'how much of the earth is uninhabited and if people live head-downwards (ἀντίποδες) in the southern half of the world'.[83] Timolaeus goes on to boast that he 'would be thought by all others a god', but the fantasy is primarily of universal knowledge rather than power. Here, as in other contexts, vision of the antipodes is associated with flight, and, as in Manilius' *Astronomica*, quasi-divine presence.[84] But Lucian's objective is to show that the result of antipodal speculation should be the 'good laugh' that is had at the end of the story at the expense of Timolaeus and others. Similarly, in Lucian's *Demonax*, a series of ripostes to fools and charlatans of various kinds includes the case of a natural philosopher (φυσικόυ) who talks of antipodeans. Demonax takes him to a well, shows him his reflection in the water, and asks whether this is the sort of antipodean he had in mind. As Gabriella Moretti has noted, this deflation of a charlatan enacts one of the most important features of the antipodal motif: reversal of perspective.[85] The natural philosopher looks at himself, momentarily, as an antipodean, and at the same time is brought sharply back to the *oikoumenē* from the world beyond it.

Satirical adaptations of the antipodes played on the fantasy of contact across the allegedly insurmountable barriers posed by the ocean, and the frigid and torrid zones – only to dissolve such notions in laughter. Perhaps the most tantalizing use of the antipodes was in Lucian's 'A True Story II', in which a nautical voyage ends in 'the world opposite the one which we inhabit', and with the promise to tell the reader all that happened there in (non-existent) succeeding books.[86] The late antique poet Tiberianus appears to have gone one better by composing a satirical letter from the antipodes. The letter, supposedly borne to the north on the breeze, begins: 'We above send greetings to you below' ('superi inferis salutem').[87] Aristotle had denied that the hot southern wind felt in the *oikoumenē* had its origins in the southern hemisphere (it came, he thought, from the torrid zone),[88] but many commentators argued the contrary, that winds travelled the length of the globe, heating and cooling as they went. It is this theory that Tiberianus appears to have seized upon and directed toward satirical ends. This text not only reverses perspectives,[89] it mocks the notion of equivalence: there are no linguistic barriers, documentary forms are identical, the antipodeans write the same language, think the same thoughts. However playful, the fundamentally conservative nature of such satirical manoeuvres is manifest: the unknown is the foil that returns the satire to its target, and allows it to express an insular response to the possibility of travel and contact.

Lucian's 'True Story II' and Tiberianus' letter touch on a matter fundamental to all representation of the antipodes, and of *terrae incognitae* generally. Conjectured spaces cannot speak: they can only be located outside, opposite, beyond. In the 'Somnium Scipionis', as the attention of both Scipio and the reader is quickly returned to Roman power and its frontiers, antipodal inhabitants remain exterior to dominion: 'from those you can certainly expect no glory'. So the antipodes constitute a representational problem, since, fictitious travellers aside, they cannot be described by first-hand experience. In such circumstances ecphrasis (literally 'speaking out', and in its literary usage a self-contained description of an object) is possible only by analogous inversion (there is here), or by the imposition on the antipodes of other spaces beyond the known world – in classical literature hell, and later purgatory and paradise. At the same time any global vision had either to acknowledge antipodal spaces and peoples or to deny their very existence; ignoring the question was not possible. If they could not speak, antipodal regions could nevertheless signify, however mutely, and in terms that could cut to the heart of European political and cultural identity. As I have argued, the complication of the binary opposition between known and unknown worlds, here and there, above and below, by the supposition of more than one unknown, *antoecumenical* region, carried important implications for the imagination of political power. A multiplicity of unknown spaces could be used to emphasize the small yet integral nature of dominion. Yet, once thought, the formulation of political power in spatial terms – in terms, that is, of the ends of the known world, and worlds beyond – provided a potent vision of possibility as well as limitation.

During the late antique and medieval periods the question of the place of the antipodes – their place in relation to learned discourse, and to knowledge *tout court* – was increasingly expressed in visual terms. In attempting to understand the theory of lands beyond the known world, to preserve and to elaborate upon a classical legacy, the commentators of late antiquity turned to diagrams. The representation of unknown lands as an image crystallized the motifs of opposition and inversion, the dichotomies between known and unknown, claimed and unclaimed, that I have outlined so far. More than that, the presence of the antipodes as a part of the world image subtly changed the terms in which unknown parts of the earth could be discussed and incorporated into languages of science, statecraft, and satire: mapping *terrae incognitae* gave them a practical as well as theoretical significance.

Notes

1 These motifs have been well explored by classicists in discussions of Greek and Roman constructions of barbarian others. For example, François Hartog, *The Mirror of Herodotus: The Representation of the Other in the Writing of History*, trans. Janet Lloyd (Berkeley: University of California Press, 1988), pp. 212–30, identifies inversion, comparison and analogy as figures in a classical 'rhetoric of otherness'. On the Greek-barbarian dichotomy see also Edith Hall, *Inventing the Barbarian: Greek Self-Definition through Tragedy* (Oxford: Clarendon Press, 1989). Yves Albert Dauge, *Le Barbare: Recherches sur la conception romaine de la barbarie et de la civilisation* (Brussels: Latomus, 1981) provides a compendious treatment of the Roman-barbarian polarity. Less consideration has been given to the use of such motifs in constructing antipodal spaces, aside from the analysis in Moretti's *Gli antipodi.*

2 Moretti, *Gli antipodi*, pp. 18–20; Romm, *Edges of the Earth*, pp. 124–8.

3 Diogenes Laertius, *Lives of Eminent Philosophers*, trans. R. D. Hicks, 2 vols (London: Heinemann, 1925), 8.26.

4 *Lives of Eminent Philosophers*, 3.24.

5 See Romm, *Edges of the Earth*, p. 129 *n*17.

6 Plato, *Timaeus, Critias, Cleitophon, Menexenus, Epistles*, trans. R. G. Bury (Cambridge: Harvard University Press, 1929), 63A. Diogenes may have had in mind Plato's use of 'antipous' here when he accorded him with the introduction of the term into philosophy. As Romm notes, this term designates a particular place on the earth's surface rather than a people: *Edges of the Earth*, p. 129 *n*17.

7 Plato, *Timaeus*, 24E–25D; *Critias*, 108E–109A. On mud see Romm, *Edges of the Earth*, pp. 125-6.

8 Plato, *Phaedo*, ed. C. J. Rowe (Cambridge: Cambridge University Press, 1993), 109B.

9 The phrase is Romm's, *Edges of the Earth*, p. 126.

10 Aristotle, *Meteorologica*, trans. H. D. P. Lee (London: Heinemann, 1952), 2.5 (362b), my emphasis. In *De Caelo*, in what he claims is a reversal of a Pythagorean concept, Aristotle described the southern pole as the 'upper' one, and the inhabitants of the northern hemisphere as those 'in the lower [hemisphere] and to the left': *De Caelo* (*On the Heavens*), trans. W. K. C. Guthrie (London: Heinemann, 1939), 2.2 (285b).

11 Strabo, *Geography*, trans. H. L. Jones (Cambridge: Harvard University Press, 1917), 2.2.2. According to Strabo, Parmenides' torrid zone was too large, since it extended beyond the two tropics; Aristotle regarded the tropics as the outer boundaries of the torrid zone; Posidonius reduced it still further, allowing that there was habitation beneath the northern (and, by implication, above the southern) tropic. Polybius' innovation was to add a sixth zone by dividing the torrid zone into two (2.3.1). On the significance of Posidonius' theory of the zones, see Claude Nicolet, *L'inventaire du monde: Géographie et politique aux origines de l'Empire romain* (Paris: Fayard, 1988), p. 82; and for commentary on Strabo's account of Posidonius see I. G. Kidd, *Posidonius: The Commentary*, 2 vols (Cambridge: Cambridge University Press, 1988), vol. 1, pp. 216–40.

12 For the views of Posidonius on habitation between the tropics the source is Cleomedes, *Caelestia* (*ΜΕΥΕΩΡΑ*), ed. Robert Todd (Leipzig: Teubner, 1990), 1.4.90-146; see Kidd, *Posidonius: The Commentary*, vol. 2, pp. 749–53. Posidonius' belief that the equatorial region was habitable seems to have remained unknown to the Latin west until the fifteenth century, but the Byzantine dignitary Symeon Seth noted it as early as the eleventh century: *Posidonius: The Fragments*, ed. L. Edelstein and I. G. Kidd (Cambridge: Cambridge University Press, 1972), F211, pp. 189–90.

13 According to Diogenes Laertius, Stoics termed the northern habitable zone 'temperate' (eukratos) and the southern 'counter-temperate' (anteukratos): *Lives of Eminent Philosophers*, 7.156.

14 *Collectanea Alexandrina: Reliquiae minores Poetarum Graecorum*, ed. J. Powell (Oxford: Clarendon Press, 1925), 16 (pp. 62–3); cf. *Commentariorum in Aratum reliquiae*, ed. E. Maass (Berlin: Weidmann, 1958), 29 (pp. 63–4). The poem was preserved by the third-century AD author Achilles Tatius, who included it in his commentary on the poet Aratus, the only surviving part of a larger work entitled 'On the Sphere'. Although standard, Achilles' own description of the zones noted uncertainty about the habitability of the torrid zone.

15 Romm suggests the use of antipodes as a proper noun may have evolved from its initial deployment as an adjective: *Edges of the Earth*, pp. 129 *n*17, 131.

16 According to Suetonius, Crates broke his leg in a sewer hole while on embassy to Rome and, as a result of lectures given during his convalescence, was responsible for changing the way Roman poets were read and commented upon: *De Grammaticis et Rhetoribus*, ed. and trans. Robert A. Kaster (Oxford: Clarendon Press, 1995), 2.1–2, pp. 4/5. Sextus Empiricus noted that Crates styled himself a 'kritikos', expert in a broadly conceived science of speech, and superior to a 'grammatikos', who was able to explain technical matters only: *Adversus Mathematicos* 1, trans. D. L. Blank (Oxford: Clarendon Press, 1998), 1.79, p. 18.

17 On the compatibility of this theory with Stoic principles see Nicolet, *L'inventaire du monde*, p. 78.

18 Geminus, *Introduction aux phénomènes*, ed. Germaine Aujac (Paris: Les Belles Lettres, 1975), 16.1.

19 Aristotle, *De Caelo*, 2.13 [293a.24]. See Plutarch's association of this term with Pythagorean theory in *De placitis philosophorum*, 891F, 895C and E, and *On the Generation of the Soul in the Timaeus* 1028B: *Œuvres morales, vol. 12.2: Opinions des philosophes*, ed. and trans. Guy Lachenaud (Paris: Les Belles Lettres, 1993); *Moralia* vol. 13.1, trans. Harold Cherniss (London: Heinemann, 1976). In *Concerning the Face which appears in the Orb of the Moon*, Plutarch identifies the side of the moon that faces away from heaven as 'House of counter-terrestrial Persephone' (Φερσεφόνης οἶκος ἀντίχθονος): *Moralia* vol. 12, ed. and trans. Harold Cherniss and William C. Helmbold (London: Heinemann, 1984), 944C. The fifth oration of the emperor Julian, 'Hymn to the Mother of the Gods' (362 AD), contains a reference to zonal theory combined with the concept of the antikthones. In a description of the times of the year at which religious rites are performed, the sun is said to depart 'towards the antikthonal zone' (πρὸς τὴν ἀντίχθονα ζώνην): Julian the Emperor, *Œuvres complètes*, ed. and trans. Gabriel Rochefort et al., 2 vols (Paris: Les Belles Lettres, 1924-1963) vol. 2, 173C (p. 122). Pomponius Mela used the term 'anticthones' to describe inhabitants of the southern temperate zone, 'unknown due to the heat of the intervening region': 'anticthones alteram, nos alteram incolimus. illius situs ob ardorem intercedentis plagae incognitus, huius dicendus est': *De chorographia libri tres*, ed. Piergiorgio Parroni (Rome: Edizioni di storia e letteratura, 1984), 1.4. According to Pliny the Elder, writing shortly after Mela, Taprobana (modern-day Sri Lanka) was for a long time regarded as another world and described by use of the term 'antichthones', until Alexander the Great demonstrated its insular nature: *Naturalis historiae libri xxxvii*, ed. C. Mayhoff, 6 vols (Leipzig: Teubner, 1892–1909), 6.81.

20 The term *antoikoumenē*, used to signify a region in the southern hemisphere of corresponding size to the *oikoumenē*, appears in Claudius Ptolemy's *Geography*, in the course of Ptolemy's critique of the measurements of his predecessor Marinos of Tyre. The inaccuracy of Marinos' calculations of latitude, according to Ptolemy, is so great that they would locate regions known to be in Africa in the frigid region of the *antoikoumenē*: *Ptolemy's Geography: An Annotated Translation of the Theoretical Chapters*, trans. J. Lennart Berggren and Alexander Jones (Princeton: Princeton University Press, 2000), 1. 8.

21 Strabo, *Geography*, 2.2.3; Cleomedes, *Caelestia*, 1.4.132–46; see Kidd, *Posidonius: The Commentary*, vol. 1, pp. 229–30, and, on the lack of success of Posidonius' terminology, Germaine Aujac, 'Poseidonios et les zones terrestres, les raisons d'un échec', *Bulletin de l'Association Guillaume Budé* 1 (1976), 74–8. Strabo attributed to Posidonius belief in two further zones 'related to human geography', located directly underneath the tropics: parched, sandy and infertile, they were populated by curly-haired inhabitants with protruberant lips and flat noses.

22 See Geminus, *Introduction aux phénomènes*, 16.19.

23 *De rerum natura*, ed. Cyril Bailey, 3 vols (Oxford: Clarendon Press, 1947), 1. 1058–67, and commentary in vol. 2, pp. 782–7.

24 *De rerum natura*, 1. 1068–82. This section is unfortunately defective.

25 Plutarch, *Moralia* vol. 12, 924A.

26 Cleomedes, *Caelestia*, 1.1.215–28. The term 'antomoi', i.e. those adjacent to, is used (only by Cleomedes, it seems) as an equivalent of 'antoikoi' to denote those in the southern hemisphere but on the same longitude as those in the known world.

27 Cleomedes, *Caelestia*, 1.1.228–34; translation from *Cleomedes' Lectures on Astronomy: A Translation of* The Heavens, trans. Alan C. Bowen and Robert B. Todd (Berkeley: University of California Press, 2004), p. 35. The date of this work is uncertain, and several possibilities have been suggested. See the discussion by Bowen and Todd, who suggest c. 200 AD as the most likely date: *Cleomedes' Lectures on Astronomy*, pp. 2–4.

28 Cleomedes, *Caelestia*, 1.1.232–4. I have not here followed the translation of Bowen and Todd, who translate the

technical terms and use 'convert' for ἀντιστρέφουσι: *Cleomedes' Lectures on Astronomy*, p. 36. They point out that Cleomedes posits a relationship of 'strict reciprocity', different from Aristotle, who allowed the concept of reciprocity to cover the relation between slave and master: *Categories* 6b28–7a5.

29 Cleomedes, *Caelestia*, 1.1.235–61.

30 Cicero, *De re publica*, ed. K. Ziegler (Leipzig: Teubner, 1964), 6.11. Here and throughout this volume all translations are mine unless otherwise indicated.

31 Cicero, *De re publica*, 6.20.

32 Cicero, *De re publica*, 6.20.

33 Cicero, *De re publica*, 6.21.

34 See further Romm's analysis of the 'Somnium Scipionis' in *Edges of the Earth*, pp. 134–40, where he contrasts Cicero's 'warning that the limits of power must not be transcended' with accounts of the antipodal ambitions of Alexander the Great and Messalla Corvinus popular in late republican Rome and the early principate.

35 Cicero, *De re publica*, 6.16.

36 Cicero, *De re publica*, 3.41.

37 Cicero, *De re publica*, 6.11–13.

38 Cicero, *De re publica*, 6.9, 11.

39 On Cicero's cosmographical reading, which seems to have included Eratosthenes, Polybius, and perhaps Posidonius, see Nicolet, *L'inventaire du monde*, pp. 78–83. Karl Büchner argues that the concept of zones was known in Rome before Cicero, and that he need not have relied on Eratosthenes or Posidonius: *De re publica: Kommentar* (Heidelberg: Winter, 1984), pp. 487–8. Pierre Boyancé, *Études sur le Songe de Scipion* (Bordeaux: Feret, 1936), pp. 151–60, questioned the need for a direct Greek source.

40 Do the 'vastae solitudines' refer to uninhabited regions within the known world, or the spaces between the known world and three other quarters of the earth, or the spaces between the northern and southern temperate zones? Clearly the word 'solitudo', with its connotations of loneliness, is chosen to act in opposition to 'celebritas' and 'gloria'. According to Büchner, Cicero refers not to deserts but to the torrid zone: *Kommentar*, p. 486. The lack of specificity seems deliberate, but Cicero appears to refer to human habitation in its entirety, i.e. including *antoecumenical* as well as ecumenical peoples, and the 'solitudines' may refer to spaces within and without the known world, and therefore include expanses of ocean. On the semantic range of 'gloria' in Cicero's writing see Giancarlo Mazzoli, 'Riflessioni sulla semantica ciceroniana della gloria', in *Cicerone tra antichi e moderni*, ed. Emanuele Narducci (Florence: Le Monnier, 2004), pp. 56–81.

41 Cicero, *De re publica*, 6.21–2.

42 Cicero, *Academicorum priorum*, ed. James S. Reid (London: Macmillan, 1885), 2.123.

43 Cicero, *Tusculan Disputations*, trans. J. E. King (London: Heinemann, 1927), 1.68.

44 Virgil, *Georgics*, ed. Richard F. Thomas, 2 vols (Cambridge: Cambridge University Press, 1988), 1. 233–9.

45 For detailed commentary see *Georgics*, ed. Thomas, pp. 107–8.

46 Thomas explains this as emphasis by omission, and sees it as part of Virgil's interest in 'imbalance' of elements: *Georgics*, p. 109; see also Thomas' *Lands and Peoples in Roman Poetry: The Ethnographical Tradition* (Cambridge: Cambridge Philological Society, 1982), pp. 10–11. The reference to the two 'extremae zonae' seems to draw on a line from the *Chorographia* of the first-century BC poet Varro Atacinus, in which the adjective describes the frigid zones: 'sic terrae extremas inter mediamque coluntur': Richard F. Thomas, 'Virgil's *Georgics* and the Art of Reference', *Harvard Studies in Classical Philology* 90 (1986), 171–98, p. 197; P. Terentius Varro Atacinus, *Chorographia*, in *Fragmenta Poetarum Latinorum Epicorum et Lyricorum*, 3rd edn, ed. Jürgen Blänsdorf (Stuttgart: Teubner, 1995), pp. 235–36, no. 13.

47 Lucretius, *De rerum natura*, 6.1. For other possible references see *Georgics*, ed. Thomas, p. 108.

48 Monica R. Gale, *Virgil on the Nature of Things: The* Georgics, *Lucretius and the Didactic Tradition* (Cambridge:

Cambridge University Press, 2000), pp. 82–3.

49 Virgil, *Georgics*, 1.242–51.

50 Thomas identifies 'clear and careful' adaptation of Aratus' *Phaenomena* in the lines on the Snake and the Bears (244–6), but supplemented with direct reference to the *Iliad*, and Cicero's translation of Aratus: *Georgics*, ed. Thomas, p. 110; 'Virgil's *Georgics* and the Art of Reference', p. 197. For the lines in Aratus see the commentary in *Phaenomena*, ed. and trans. Douglas Kidd (Cambridge: Cambridge University Press, 1997), pp. 192–5.

51 Servius, *In Vergilii carmina commentarii*, ed. Georgius Thilo and Hermannvs Hagen, 3 vols (Leipzig: Teubner, 1881–7), vol. 3: *In Vergilii bucolica et georgica commentarii*, p. 187; I discuss this matter further in chapter 3.

52 Mynors suggests that the apparent contradiction of 1.233–51 might be resolved if the location of Tartarus is understood to be in the Antarctic region, thereby allowing for the temperate southern zone to remain 'granted to feeble mortals': *Georgics*, ed. R. A. B. Mynors (Oxford: Clarendon Press, 1990), p. 56. Gale sees the location of the underworld within a system of zones as a deliberate contradiction of Lucretius: *Virgil on the Nature of Things*, p. 118.

53 Franz Cumont, *Recherches sur le Symbolisme funéraire des Romains* (Paris: Geuthner, 1942), pp. 54–5 *n*3: 'une imprécision qu'il jugeait poétique'. Cf. *Georgics*, ed. Thomas, p. 109.

54 Virgil, *Georgics*, 1.36–7

55 Virgil, *Aeneid*, in *Opera*, ed. R. A. B. Mynors (Oxford: Clarendon Press, 1969), 6.794–7.

56 Virgil, *Aeneid*, 1.278–9.

57 Virgil, *Aeneid*, 7.222–8.

58 Respectively *Aeneid*, 7. 96–101; 7. 120–7; 7. 219–21.

59 A similar formulation, in which something is so well-known that even the antipodeans have heard of it, can be found in two places in Plutarch's *Moralia*. The first, in 'On the malice of Herodotus', indicates its hyperbolic nature by drawing attention to the hypothetical nature of the antipodes: 'If there are antipodean peoples [ἀντίποδες], as some say, who dwell on the under side of the world [περιοικοῦντες], I imagine that even they have heard of Themistocles and the Themistoclean plan': *Moralia* vol. 11, trans. Lionel Pearson and F. H. Sandbach (London: Heinemann, 1965), 869C. The second, in 'On Stoic self-contradictions', similarly hyperbolic, concerns philosophy not history: 'Now, that the universal nature and the universal reason of nature are destiny and providence and Zeus, of this not even the antipodes are unaware, for the Stoics keep harping on this everywhere': *Moralia*, vol. 13.2, trans. Harold Cherniss (London: Heinemann, 1976), 1050B.

60 Romm, *Edges of the Earth*, pp. 133–40.

61 *Appendix Tibulliana*, ed. Hermann Tränkle (Berlin: de Gruyter, 1990), 3.7.149–76. On this poem see Romm, *Edges of the Earth*, pp. 136–7. The date of the poem is disputed: a reference to Messalla's installation as consul in 31 BC as a recent event suggests composition in that year or soon after; however, possible imitation of Tibullus' elegy on Messalla's triumph in 27 BC, and allusions to that event, support a later date.

62 Manilius, *Astronomica*, ed. G. P. Goold (Stuttgart: Teubner, 1998), 1.384–6. On the textual difficulties of line 386, which according to manuscript tradition should read 'Caesar nunc terris post caelo maximus auctor', see *Astronomicon*, ed. A. E. Housman (London: Richards, 1903), p. 39, and *contra* Enrico Flores, 'Augusto nella visione astrologica di Manilio ed il problema della cronologia degli *Astronomicon libri*', *Annali della Facoltà di Lettere e Filosofia dell'Università di Napoli* 9 (1960–1), 5-66: pp. 21–4.

63 *Metamorphoses*, ed. R. J. Tarrant (Oxford: Clarendon Press, 2004), 1.45–51.

64 *Metamorphoses*, 1.32–48.

65 *Metamorphoses*, 2.126–37.

66 *Metamorphoses*, 2.202–3.

67 *Metamorphoses*, 2.201–300.

68 *Metamorphoses*, 2.319–39. Obviously, a lot depends on the translation of 'excidit'; the more literal meaning of 'excidere' ('to fall out or from') has the line teetering on the mock-heroic: 'he fell off with great daring'.

69 Thomas, 'Virgil's *Georgics* and the Art of Reference', p. 195.

70 Lucan, *De bello civili*, ed. D. R. Shackleton Bailey, 2nd edn (Stuttgart: Teubner, 1997), 10.36–45. 'Hammon' was the sanctuary of Ammon (i.e. the Egyptian deity Amun) in the region of the Syrtes, where Alexander was allegedly proclaimed son of Zeus. On the debate over line 43 (A. E. Housman, following Richard Bentley, emended 'qua totum ceperat orbem' to 'quo totum etc.') see *Bellum civile liber X*, ed. Emanuele Berti (Florence: Le Monnier, 2000), p. 86, whose reading I follow here.

71 *Arnulfi Aurelianensis Glosule super Lucanum*, ed. Berthe M. Marti (Rome: American Academy in Rome, 1958), p. 497. The line has been interpreted as referring to an expedition along the southern perimeter of the known world only: see Romm, *Edges of the Earth*, p. 138. But the interpretation of latitudinal and longitudinal circumnavigation remains most convincing: *Bellum civile liber X*, ed. Berti, pp. 84–5.

72 *De bello civili*, 8.159–64; 9.876–8.

73 *De bello civili*, 10.20, 42, 70, 279, 333. The adjective is used in connection with Alexander on three occasions in the works of Seneca: *Bellum civile liber X*, ed. Berti, p. 74.

74 *De bello civili*, 10.272–85. The nature of the Nile, Acoreus goes on to explain, is unique: the only river to rise at midsummer, 'only it is permitted to wander through both hemispheres. In this [i.e. northern] one its source is sought; in that one its end' (solique vagari/ Concessum per utrosque polos. Hic quaeritur ortus,/ Illic finis aquae): 10.300–2.

75 Pliny, *Naturalis historia*, 2.170. On the significance of this passage in the context of Pliny's representation of land and sea in the *Naturalis historia* see Mary Beagon, *Roman Nature: The Thought of Pliny the Elder* (Oxford: Clarendon Press, 1992), pp. 159–201.

76 *Naturalium quaestionum libri*, ed. Harry M. Hine (Leipzig: Teubner, 1996), Praef. 8–11: '"hoc est illud punctum quod inter tot gentes ferro et igne diuiditur!" O quam ridiculi sunt mortalium termini!'

77 Pliny, *Naturalis historia*, 2.172.

78 Pliny, *Naturalis historia*, 2.174. I have emended 'gloria' to 'gloriae', following the edition of Jean Beaujeu (Paris: Les Belles Lettres, 1950).

79 On this paradox of Pliny's posture of *contemptus mundi* and his acts of describing the world (the 'problem of totality'), see Sorcha Carey, *Pliny's Catalogue of Culture: Art And Empire in the* Natural History (Oxford: Oxford University Press, 2003), esp. pp. 99–101, 179–83. Trevor Murphy characterizes Pliny's geography in books three to six of the *Naturalis historia* as straightforwardly imperial: *Pliny the Elder's* Natural History*: the Empire in the Encyclopedia* (Oxford: Oxford University Press, 2004), pp. 129–64.

80 As argued by Moretti, *Gli antipodi*, chapter 1.

81 Seneca, *Epistulae morales*, ed. L. D. Reynolds, 2 vols (Oxford: Clarendon Press, 1965), 122.3. Cato's saying is preserved only in this letter.

82 Discussed in more detail by Moretti, *Gli antipodi*, pp. 33–4, who emphasizes the a-technical mention of the antipodes and the significance of their appearance in the course of a diatribe.

83 Lucian, 'The Ship or the Wishes' in *Opera*, 8 vols (London: Heinemann, 1913–67), vol. 6, trans. K. Kilburn, p. 483.

84 See Moretti's comments on this text: *Gli antipodi*, pp. 37–8.

85 Moretti, *Gli antipodi*, p. 37.

86 Lucian, 'A True Story II', in *Opera*, vol. 1, trans. A. M. Harmon, p. 355.

87 See Silvia Mattiacci, *I carmi e frammenti di Tiberiano* (Florence: Olschki, 1990), pp. 61, 201–6; Moretti, *Gli antipodi*, pp. 46–9.

88 Aristotle, *Meteorologica*, 2.5.

89 Moretti, *Gli antipodi*, p. 47.

Chapter Three

Realignment: the Antipodes between Classicism and Christianity

Three crucial moments in the transmission of antipodal theory to the medieval west occurred during the first half of the fifth century. The first, Augustine of Hippo's denial of human habitation of the antipodes in *De civitate dei* (On the City of God) (413–26), was indubitably Christian. The other two moments can be characterized with reasonable, although not complete, security as non-Christian: in Rome around 430 Macrobius Ambrosius Theodosius composed a commentary on Cicero's 'Somnium Scipionis', in which he elaborated upon and illustrated Scipio's cosmic and terrestrial vision; later in the fifth century, in North Africa, Martianus Capella composed an allegorical encyclopedia entitled *De nuptiis Mercurii et Philologiae* (On the Marriage of Mercury and Philology), including a brief summary of antipodal theory in the chapter dedicated to geometry. All three texts, in different ways, testify to a realignment of debate about the antipodes, in which new contexts – philosophical, pedagogical, and exegetical – were beginning to take shape. All three reveal the antipodes at the fault-line between Christian and non-Christian belief.

The theory of the antipodes was rejected by Christian authors such as Augustine, not because it had to be – other options were possible – but because the rejection of the idea of people living beyond the word of God consolidated the space of Christianity. As I have suggested in the previous chapter, theorists of Roman imperial space used the vision of peoples beyond the known world to emphasize or to urge the integrity of dominion. Augustine, and Christian commentators like him, denied the possibility of those peoples because the presence of races unreachable by the word of God threatened not only the integrity of the apostolic dispersal of Christianity to the ends of the earth, but the integrity of scripture itself, for it was there that the descent of humanity from Adam, and the spread of salvation throughout humanity, was made manifest. By contrast, for non-Christian authors such as Macrobius and his contemporary Servius, the compiler of seminal commentaries on the works of Virgil, the antipodes were an integral part of the classical culture they wished to reconstitute. To understand the works of Cicero and Virgil fully it was necessary to understand the antipodes, just as, for Martianus Capella, antipodal theory was a non-negotiable part of classical geometrical learning.

Yet it is possible to exaggerate the breach between Christian and non-Christian cultures: while there are only suggestions of dialogue across this fault-line over the matter of the antipodes, the discussions of the topic by Augustine, Macrobius, and Martianus can all be seen to exist within the context of interaction between classical and Christian literary cultures. In defending the theory of the antipodes, Macrobius may well have had in mind the Christian apologist Lactantius' earlier denunciation of the concept. By locating the antipodes under the sign of geometry, the virtuosic encyclopedism of Martianus Capella provided the basis for future discussion of the topic within the tradition of the seven liberal arts. And even Augustine's rejection of the antipodes stemmed from his profound engagement with Neoplatonist literature.

The central role of commentary in the reception and transmission of classical knowledge brings out a broader, more theoretical, issue concerning tradition. The ways in which the works of classical authors such as Cicero and Virgil were copied and annotated suggests, firstly, that commentary was necessary to enable even learned fifth-century readers to comprehend classical literature. Secondly, when viewed from the perspective of the Middle Ages, it is noticeable that the annotations of commentators such as Macrobius and Servius formed a nearly indelible apparatus; conjoined within a manuscript tradition, it became very hard to read Cicero without Macrobius, Virgil without Servius. The question that arises from these observations concerns the formation of tradition: quite simply, can tradition exist without commentary? Is commentary a necessary component of tradition-making, and, more profoundly, is it possible to perceive tradition *as tradition* outside of the embrace of commentary? In the case of the antipodes, it is possible to identify two models of interaction and interdependence between commentary and tradition: reception as reshaping and supplementation; and reception as intervention and rejection. Both models were transmitted to the Middle Ages, and both, in their different ways, determined the course of the classical traditions of *terrae incognitae*.

Servius: hell and other peoples

For both Servius and Macrobius, in a Rome in which external political and religious forces had seriously disrupted the foundations of traditional authority, the explication, even excavation, of earlier Roman literary culture appeared a vital task.[1] The objective of such a movement was not merely to explain and reconstruct, but arguably to reanimate works such as Virgil's poems and Cicero's 'Somnium Scipionis'. The desire for cultural restoration meant that classical authors could not be seen simply as literary authorities; instead their works had to take on a quasi-encyclopedic function. The works of Virgil became for Servius vast, heterogeneous, 'a quasi *summa* of classical culture';[2] for related reasons Macrobius' commentary has struck more than one reader as a perversely long-winded response to Ciceronian elegance.[3] However, if seen not as products of *otium*, a leisurely appreciation of past masters, but rather as concerted attempts to preserve and grasp a culture that appeared under threat – or, less dramatically, in transformation – as a result of the dual impact of the Christian faith and barbarian invasions, the heavy freight placed on the works of Cicero and Virgil by their late antique commentators becomes both more understandable and more interesting. In one corner of that effort to scrutinize the legacy of

Rome, to interrogate the meaning of its literature and the value of its traditions, lay the question of the antipodes.

Although little is known about Servius, other than that he worked as a grammarian in Rome, he appears as a precociously learned and modest young man in Macrobius' *Saturnalia*, a series of fictive dialogues between actual people set c. 384, but probably written several decades later.[4] The precise nature of the relationship between these two authors is unclear, and the chronology of their works is uncertain, but it seems likely that Servius' commentaries on Virgil were compiled around the turn of the fifth century,[5] and that they predated Macrobius' commentary on the 'Somnium Scipionis', now tentatively dated to c. 430.[6] The similarities between the two authors should not be pressed: there is no solid evidence that Macrobius actually used Servius' commentaries even if he evidently knew of his reputation as a commentator on Virgil, and a case can be made for significant differences in their work and approaches, particularly if it is accepted that Macrobius belonged to the generation after Servius.[7] Nevertheless, the appearance of Servius in the *Saturnalia* testifies not only to Macrobius' knowledge of him and to the former's eminence, but to their shared investment in what has been described as the 'desire for the cultural restoration of the antique'.[8]

The function of commentary was not, however, simply to restore meaning and prestige to works of ancient literature. As Servius' explication of Virgil's lines in *Aeneid* 7 on the spread of the news of the Trojan war shows, commentary supplemented meaning, adding to and subtly realigning the original text. Virgil's lines 7.227–8, '[audiit] et si quem extenta plagarum/ quattuor in medio dirimit plaga solis iniqui' ([these things were heard] even by those cut off by the region of the oppressive sun, stretched in the middle of the four), are glossed by Servius:

> audierunt etiam illi qui separantur zona ea, quae est in medio quattuor, id est feruens. significat autem antipodas.[9]
>
> even they heard who are separated by that zone, which is in the middle of four, that is the burning zone. He [i.e. Virgil] means the antipodeans.

Servius supplements Virgil's invocation of zonal theory with the use of technical terms: 'zona' for Virgil's 'plaga', and 'antipodas' for the anonymous 'quem'. This act gives the lines a much greater specificity of meaning. But specificity works two ways: as well as fixing the meaning of the lines, Servius opens up the text to the many associations that the word 'antipodes' carries.[10] Some of these associations are discussed in greater detail in his commentary on the *Georgics*. There, to explain the lines beginning 'quam circum extremae dextra laevaque trahuntur' (1.235: around it [gloomy] extremities are stretched to the right and left), Servius describes the temperate zones in terms of habitation, and their location between heat and cold:

> unam nos habitamus, alteram antipodes, ad quos hinc torrente zona, hinc frigidis ire prohibemur. antipodes autem dicuntur, quod contra nos positi sunt contrariis vestigiis.[11]
>
> we inhabit one, the antipodeans the other, to whom we are prevented from going by the burning zone on one side, and the frigid zone on the other. They are called antipodeans because they are positioned with their footsteps against ours.

This is a quite straightforward, literal, explanation of the meaning of 'antipodes', using terms standard to classical antipodal tradition: access to the antipodeans is denied by both torrid and frigid zones, and their location defined by their stance 'contra nos'.

A more complex issue was Virgil's conflation of the Tartarean regions with the world 'beneath our feet'. For Servius, Virgil's purpose was to contrast the 'various opinions of philosophers' (variae philosophorum opiniones). In particular he wished to play off the Stoic doctrine that the sun passed through both upper and lower hemispheres, bringing alternate night and day to each, against the Epicurean notion that the lower hemisphere remained in darkness:

> nam alii dicunt a nobis abscedentem solem ire ad antipodas, alii negant et volunt illic tenebras esse perpetuas. mire autem ait quasi de inferis 'Styx atra videt manesque profundi', ut ostenderet illud quod dicunt philosophi, recedentes hinc animas illic alia corpora sortiri: unde et Lucanus ait regit idem spiritus artus orbe alio[12]
>
> For some say that the sun goes to the antipodeans when it leaves us, others deny this and want there to be perpetual darkness there. He says, as if about the lower regions, 'black Styx sees and the infernal shades', so that he might show that which philosophers say, that souls leaving from here obtain another body there: whence Lucan says the same spirit rules the body in another world.

The theory made sense, according to Servius, since, in order to be purged, souls had to pass through fire, wind, or water, acts that could easily be accomplished by crossing to the underworld by way of the frigid or torrid zones. Servius notes that Virgil had inserted such 'philosophia' 'out of poetic licence' (per poeticam licentiam).[13] Yet the potential conflation of antipodal and infernal regions posed certain challenges to interpretation. In his commentary on book 6 of the *Aeneid*, Servius confronts this issue in two places. In response to line 127 ('noctes atque dies patet atri ianua Ditis': the gate of black Dis stands open day and night), Servius notes that Lucretius denies the existence of infernal regions, and he asks where the lower regions can be located, if the antipodes are said to be 'beneath the earth' (sub terris).[14] Subsequently he considers the question asked of Aeneas in the underworld by the shade of the murdered Trojan prince Deiphobus: 'what brings you, a living man, here? Do you come propelled by wanderings on ocean or by some divine admonition?' (pelagine uenis erroribus actus/ an monitu diuum).[15] Did Deiphobus' reference to oceanic wanderings carry the implication that Aeneas might have reached the underworld by travelling across the seas? Servius says no – but then explains why one might say yes:

> non ad inferos, sed ad locum, in quo inferorum descensus est, id est ad Avernum, si intra terram sunt inferi. alii altius intellegunt: qui sub terra esse inferos volunt secundum chorographos et geometras, qui dicunt terram *σφαιροειδη* esse, quae aqua et aere sustentatur. quod si est, ad antipodes potest navigatione perveniri, qui quantum ad nos spectat, inferi sunt, sicut nos illis. hinc est quod terram esse inferos dicimus, quamquam illud sit, quia novem cingitur circulis. Tiberianus etiam inducit epistolam vento allatam ab antipodibus, quae habet superi inferis salutem.[16]
>
> [you come] not to the lower regions, but to the place in which the descent to the lower regions is located, that is to Lake Avernus, if the lower regions are within the earth. Others understand it differently: they want the lower regions to be beneath the earth following the theories of chorographers and geometers, who say that the earth is spherical, held up by water and air. If it is,

> it is possible to reach by navigation the antipodeans, those who from our perspective are lower, just as we are 'lower' to them. We say that the earth is 'lower', however that may be, because it is surrounded by nine circles. Tiberianus even alleges a letter carried across on the wind from the antipodes, which says 'those above greet those below'.

Two points emerge from these lines with particular clarity. First, that commentary of this kind opens up lines of interpretation rather than closing them down: an apparently decisive statement is followed by its alternative, one significantly backed by the authority of 'chorographers and geometers', marked by the usage of Greek terminology, and then augmented by the invocation of Tiberianus' poem. Commentary, in other words, operates on a logic of accretion, such that the meaning of the primary text is not determined but rather spread out and transmitted as a range of possibilities through conference with a number of other surrounding texts, both named (Lucan, Lucretius, Tiberianus) and unnamed, as well as other passages within the primary text itself. Servius' staging of the debate (alii dicunt … alii negunt) constructs the antipodes as a topic of controversy; significantly, though, it is a topic not restricted to philosophers and other scientists, but one in which poetry may play an active and even exemplary role. Second, Servius consolidated three equally tantalizing possibilities that were to resonate for the medieval reception of antipodal theory: that antipodal regions are dark and/or infernal; that the underside of the earth may be a place of purgation, the destiny of souls after death; and that if, on the other hand, habitable antipodal regions are located on the other side of the earth, 'it is possible to reach them *by navigation*'.

The ambiguous connection between hell and the *antoecumene* that concerned Servius was given visual form in a rather mysterious map, apparently Egyptian in origin, that survives in several late medieval manuscripts, but which has been tentatively dated to the second to third century AD (fig. 4).[17] The map shows the earth framed by an outer circle containing the two poles and ten winds, and divided by five horizontal lines: the Arctic and Antarctic circles, the summer and winter tropics, and the equator. A vertical line extends from pole to pole. Geographical representation in the northern hemisphere is confined to four features: a rectangular cartouche contains the toponyms 'Lower Egypt' (Katô chôras), 'Heptanomia' (a Roman administrative district of Middle Egypt), the cities of 'Syene' and 'Hierasycaminus', and the 'marsh of Meroë'; a semi-circle represents 'The Persian Gulf of the Erythrean [i.e. Arabian] Sea'. Libya, the 'Ethiopian Ocean' and the 'Ethiopian Sea' are marked in the far west, just above the equator, and opposite them in the east appears the 'Indian Ocean'.[18] Some manuscript copies of the map also contain the toponym 'Persia'. The inhabited parts of the southern hemisphere are divided from the inhabited parts of the north by an equatorial ocean, termed 'the sea of the *antoikoumenē*' or, in some versions, 'fiery unnavigable sea'. Habitation of the *antoikoumenē* is marked between the winter tropic and Antarctic circle, an expanse somewhat incoherently referred to as 'in latitude 40 stades'.[19]

The most remarkable feature of the map is the infernal topography that forms a kind of circle around the intersection of the polar axis with the Antarctic circle: The River of Pyriphlegethon; The River of Lethe; and Marsh of Acheron. There is no Styx, the most obvious of the four rivers of hell, and therefore no direct connection with Virgil, but without question this is a

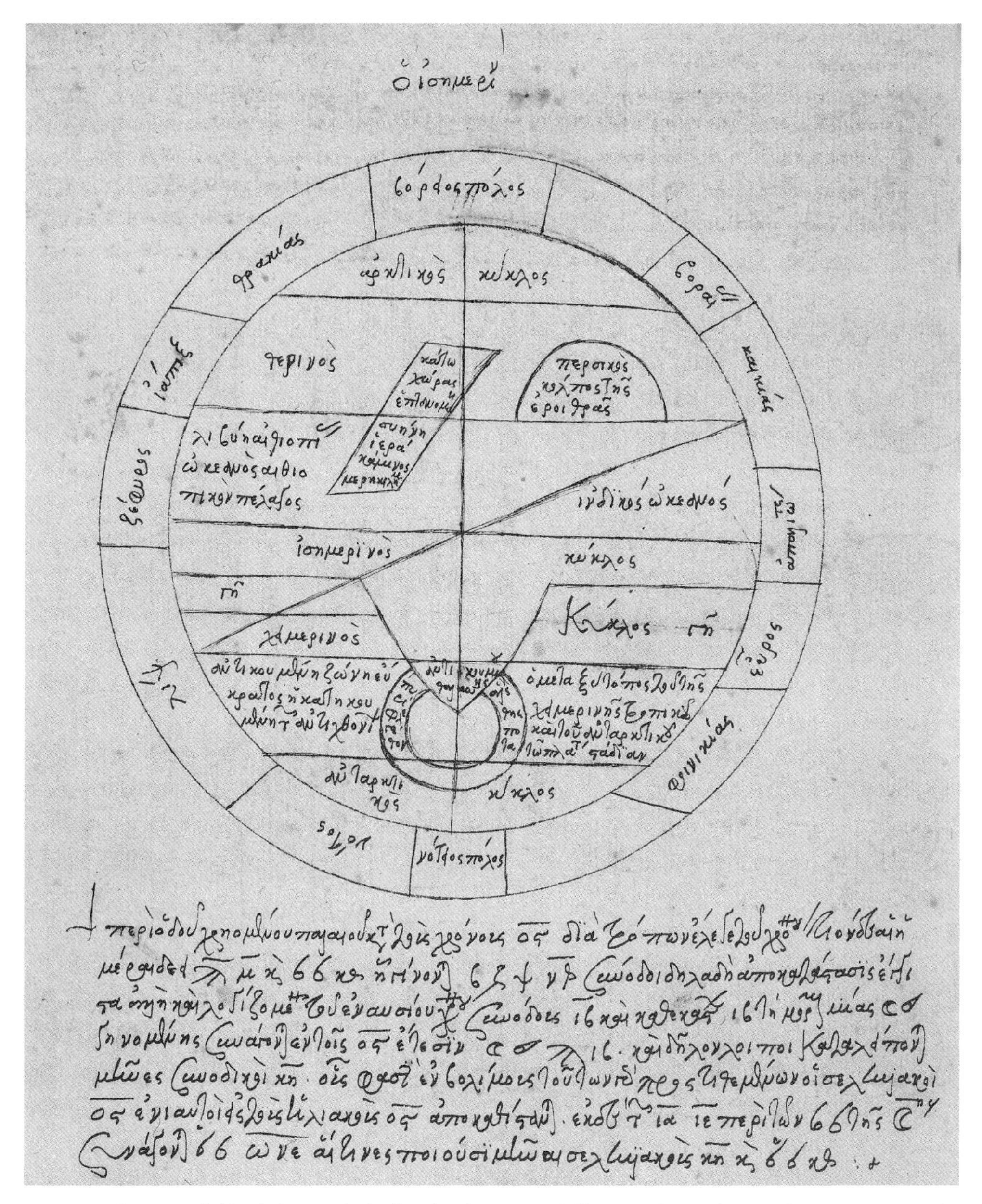

Fig. 4. Berlin, Staatsbibliothek, Handschriftenabteilung, MS Phill. 1479, f. 28v (sixteenth-century copy of a second- or third-century original). Egyptian world map showing hell in the southern hemisphere. The earth is framed by an outer circle containing the two poles and ten winds, and divided by the Arctic and Antarctic circles, the summer and winter tropics, and the equator. In the northern hemisphere, the toponyms 'Lower Egypt' (Katô chôras), 'Heptanomia', the cities of 'Syene' and 'Hierasycaminus', and the 'marsh of Meroë' appear in the rectangular cartouche. 'The Persian Gulf of the Erythrean [i.e. Arabian] Sea' is represented in a semi-circle. Libya, the 'Ethiopian Ocean' and the 'Ethiopian Sea' are marked in the far west, just above the equator; the 'Indian Ocean' appears in the east. The equatorial ocean is termed 'the sea of the *antoikoumenē*'. In the intersection of the polar axis and the Antarctic circle the infernal rivers of Pyriphlegethon and Lethe appear, along with the Marsh of Acheron.

representation of the underworld in, or at least alongside, the *antoikoumenē*. The context of the image – it is found amongst anonymous scholia to Theon of Alexandria's commentary on Ptolemy's *Handy Tables*, and also in an anonymous astrological miscellany – gives little clue as to what might have prompted such a juxtaposition.[20] The underworld and inhabited *antoikoumennē* are certainly marked off from one another, and it is possible that the rivers of hell may be conceived of as subterranean (there is also a suggestion that they may flow into or from the sea of the *antoikoumenē*). Later medieval maps showed the terrestrial paradise either in or near to the ecumene, but conceptually distinct,[21] and the same might be said for the relationship between the two types of underworld here. But there is another feature of significance. The map is obviously concerned with the relationship of land to water: in the northern hemisphere, ocean is marked in the far east and west, along with a gulf and a marsh; in the southern hemisphere the mapmaker has been careful to mark off land from sea and infernal waters (rivers and marsh), simply by using the word '*γῆ*' (land) as well as more elaborate inscriptions. Passage between the northern and southern landmasses is explicitly ruled out by the designation of the equatorial sea as 'fiery unnavigable' (διάπυρον πέλαγος ἄπλωτον), but the map nevertheless suggests the possibility of maritime pathways between known and unknown worlds for those whose ambition may encourage them to challenge the elements and reach beyond the ecumene.

As the map also suggests, acts of commentary could be given a visual counterpart. Just as Servius was able to hold several spaces – real, unknown, and supernatural – within his commentary as possible referents of Virgil's poetry, so the 'astrologer's map' is able to depict worlds known, hypothesized and mythologized, within its frame of zonal representation. The map is not itself, in this instance, an act of commentary, but, as the work of Macrobius was to show, the two could be aligned in order to reconfigure classical tradition.

Macrobius: figuring the antipodes

In Rome some two decades after the sack of 410, where a now Christian aristocracy nevertheless maintained non-Christian traditions, Macrobius composed his philosophical commentary on Cicero's 'Somnium Scipionis'. Macrobius was not, it appears, Christian,[22] but it has been suggested that his paganism was 'essentially nostalgic and literary', one that looked back to a vibrant but overturned culture, of which his predecessor Servius had been an important figure.[23] However accurate such characterizations may be, Macrobius makes clear from the start of the *Commentarii in Somnium Scipionis* his intent to interpret Cicero by reference to the thought of Plato and his followers.[24] The significance attributed by Macrobius to the 'Somnium Scipionis' was great: he held that it contained in distilled form all three varieties of philosophy in the tradition of Plato – moral, physical, and rational. In the course of his discussion, Macrobius attempted to explicate the theory behind Cicero's reference to *antoecumenical* spaces and peoples, and its broader significance in terms of philosophy both physical, because it derives from cosmological and geographical theory, and moral, because the *antoecumene* signals the need for patriotism and the renunciation of glory. Three aspects of Macrobius' discussion of the *antoecumene* were of particular importance for the subsequent medieval reception of the *Commentarii*: his self-positioning as interpreter of, and mediator between, Cicero and Virgil; his insistence on the possibility of habitation in unknown, antipodal, parts of the world; and his use of images to supplement classical theory.

The *Commentarii* go well beyond paraphrase, including lengthy expositions of matters only tangentially related to the 'Somnium Scipionis'. Macrobius introduces his work by comparing the dream of Scipio to the vision of Er at the end of Plato's *Republic*, then replies to Epicurean critiques of the validity of myth, legend, and fiction within philosophy, and outlines five principal varieties of dream. Proceeding through selected passages from the 'Somnium Scipionis', he provides a detailed account of number theory, including geometric and cosmological applications, before moving on to the human life cycle, divination, the four philosophical virtues (prudence, temperance, courage, and justice), the origin and nature of the soul and its relationship to the body, astronomy (the order of the celestial and planetary spheres, stars, the sun, the zodiac, and the size of the earth), the production of sound, and the music of the celestial spheres. Chapters 5–9 of the second book of the *Commentarii* are devoted to the section of the 'Somnium Scipionis' in which Scipio looks down upon the earth; they consist of an exposition of the terrestrial zones including their dimensions, their correspondence with the celestial zones, and the situation and function of Ocean. The final chapters of the *Commentarii* contain further discussion of astronomical phenomena and the immortality of the soul.

By contrast with Servius' representation of debate, one of the explicit objectives of the geographical section of Macrobius' *Commentarii* was to achieve a concordance between the world images of 'the two parents of Roman eloquence', Cicero and Virgil. A particular problem, Macrobius thought, was that Cicero's use of the term 'cinguli' might lead the reader to suppose that he professed a doctrine contrary to that of Virgil, who used the Greek term 'zonae' to describe the five celestial bands.[25] Virgil's description is, Macrobius emphasizes, of the heavens, whereas Cicero was writing about the earth's surface. To show their essential compatibility, though, Macrobius goes on to quote Virgil's line about the two zones allotted by the gods 'to feeble mortals' (*Georgics* 1.237–8). This sustained and rather elaborate play of concordance reveals the complex textual politics of the commentary tradition. Macrobius is able to use the apparent discrepancies between the two authors to establish his own relationship to Roman eloquence. He is both its descendant and its interpreter: the 'parents' give authority to his text, but he is necessary for them to be properly understood. It is also the case that, to continue the familial metaphor, Macrobius does not want the authority of one parent to be set against the other.

The consequence of Macrobius' stance as heir of classical eloquence is that he ignores significant differences between the representation of the *antoecumene* in the works of Cicero and Virgil. He makes no mention of Virgil's association of antipodal regions with the underworld, and instead, like Servius, emphasizes Virgil's use of philosophical material in a poetic context, and his imitation of Homer. As modern commentators have pointed out, the objective here is to establish two parallel axes between Greek and Roman antiquity: the poetic axis of Homer-Virgil complements the philosophical axis of Plato-Cicero.[26] The former emerges specifically in the context of cosmological theory. Macrobius finds Virgil's description in the *Georgics* of the path of the zodiac 'per ambas' (through both, i.e. the two zones granted to mortals) to be imprecise: Virgil surely did not mean that the sun's path extends beyond either tropic as far as the northern and southern frigid zones. Instead Macrobius posits three alternatives: Virgil may have meant that human habitation existed beneath the tropics (hence that the sun, when at the tropic, passed through inhabited regions); he may have meant 'per ambas' to be understood as 'sub ambas' (underneath both); or he may have meant it to be understood as 'inter ambas' (between both).[27] All three possibilities are explained by genre: if it is the first case, Virgil was speaking 'out of poetic exaggeration, which always exalts all things to the highest' (per poeticam tubam, quae omnia semper in maius extollit);[28] if the second or third, he was using poetic licence ('poetica licentia'), imitating Homer in substituting one particle (per) for another (sub, inter).[29] This tidying of Virgil may appear excessive, but it illustrates both the enormous significance that was attributed to classical authors by late antique commentators, and the ways in which their words could be insistently moulded to form a consensus of authority.

If Macrobius was concerned to smooth over differences between Virgil and Cicero, he was moved to refute directly arguments that the existence of antipodeans was impossible, indeed ridiculous. At an early point in his elaboration of the scientific basis of Cicero's use of the antipodes in the 'Somnium Scipionis', Macrobius addresses directly the belief that inhabitants of the other side of the world might 'fall into the sky':

> si enim nobis, quod adserere genus ioci est, iusum habetur ubi est terra et susum ubi caelum, illis quoque susum erit quod de inferiore suspicient, nec aliquando in superna casuri sunt. adfirmaverim quoque et apud illos minus rerum peritos hoc aestimare de nobis, nec credere posse nos in quo sumus loco degere, sed opinari, siquis sub pedibus eorum temptaret stare, casurum.[30]
>
> For if for us – to assert it is a kind of joke – 'below' is where the earth is, and 'above' where the sky is, for them also 'above' will be what they look at from below, nor will they ever fall into the upper regions. I affirm also that amongst them those less educated believe the same about us: they do not believe that we can inhabit this place, and they opine that if anyone tried to stand beneath their feet, he would fall.

This passage doubtless testifies to Macrobius' recognition of the currency of the notion of upside-down peoples in satiric and polemical literature; it is likely to respond to Epicurean views about the impossibility of antipodal habitation transmitted by Lucretius, and may also be aimed at early Christian polemicists such as Lactantius, who ridiculed the notion of antipodeans.[31] Significantly, Macrobius moves from the 'genus ioci', the absurdity of the need to affirm that the earth is down and the sky up, to assert the antipodal perspective, to affirm what *they* are saying. Following his symmetrical model, what they say is precisely what we say, just as their topography contains features similar to ours: and if we do not fall upwards, neither will they. Their jokes, moreover, are our jokes; the unlearned among them cannot believe in us, or in our place. The logic of antipodal symmetry encouraged Macrobius to imagine the imagination of antipodal others: an imagination in which, momentarily, the ecumenical *we* are laughed out of existence.

Macrobius' explanation of antipodal habitation allowed for the multiplication as well as the reversal of perspective. As well as its division into frigid, torrid, and temperate zones, Macrobius described the division of a spherical earth into four habitable portions of land, separated from each other by a central Ocean running along the path of the equator, and a vertical Ocean running from pole to pole.[32] While the known part of the northern zone 'is inhabited by every type of people we are able to know, Romans and Greeks or barbarians of whatever nation' (incolitur ab omni quale scire possumus hominum genere, Romani Graecive sint vel barbari cuiusque nationis),[33] Macrobius stated that it could only be deduced by reason ('sola ratione intellegitur') that the other three parts of the world were inhabited. Importantly, however, he maintained the possibility of not just one race of men in the *antoecumene*, but at least three. Macrobius describes them in terms that echo Cleomedes' notion of reciprocity between the four parts of the earth, but significantly he uses only one technical term to do so, and even there he is careful to draw attention to the fact that it is Greek:

> hi quos separat a nobis perusta, quos Graeci αντοίχους vocant, similiter ab illis qui inferiorem zonae suae incolunt partem, interiecta australi gelida separantur; rursus illos ab antoecis suis, id est per nostri cinguli inferiora viventibus, interiectio ardentis sequestrat, et illi a nobis septentrionalis extremitatis rigore removentur.[34]

> Those who are divided from us by the torrid zone, whom the Greeks call antoikoi, are similarly separated from those who inhabit the lower part of their zone by the intervening southern frigid zone; they in turn are separated from their antecians, that is from those living on the lower part of our band, by the intervention of the torrid zone, and the latter are kept apart from us by the cold of the northern extremity.

The term antoikoi – significantly Latinized and assimilated after its initial appearance in Greek letters – is used in this passage not to denote the dwellers of any particular one of the four parts, but in a relational sense, to indicate people dwelling across the torrid zone, though in the same (upper or lower) part. The result is that all four peoples are conceived of as antoikoi to one of the others. Macrobius emphasizes the symmetricality of the model, but offers no definitive nomenclature for Cicero's others. He does not use the term antipodes,[35] despite his use of it in book one of the *Commentarii*, where he promises that his explication of the attraction of weights to the earth will be used in book two 'in the examination of that passage in which [Cicero] refers to the antipodeans' (ad illius loci disputationem quo antipodas esse commemorat), and in the *Saturnalia*, where he associates the 'lower circle of the earth and the antipodes' with Proserpina.[36] Indeed he is rather vague in his explanation of Cicero's use of the terms oblique, transverse, and adverse to describe the stance of peoples in the unknown parts of the world. The 'transversi' Macrobius identifies as the inhabitants of the 'lower part of our (i.e. northern) zone', but his description of the dwellers of the southern hemisphere is less clear: the 'adversi' 'dwell in the part of the sphere which is opposite us' (in parte sphaerae quae contra nos est morantur),[37] while the 'obliqui' are 'those who were allotted the downward slopes of the southern band' (eos qui australis cinguli devexa sortiti sunt).[38] Macrobius' reasons for this unwillingness to name unknown peoples are themselves oblique, but he undoubtedly shows a marked reluctance to talk in terms of the direction of feet (antipodes, sub pedibus), and a preference for using the notion of upper and lower parts of each zone along with the torrid and frigid zones as reference points. Perhaps by not conceiving of antipodal inhabitants as dwellers 'standing beneath the feet' of those in the known world, Macrobius' critique of those who ridiculed the notion of people living 'with their feet above their heads' became easier. At any rate, the significant effect of commentary in the matter of antipodal habitation is to scatter the alignment of vision that characterized the classical models. Macrobius' explication takes us away from the perspective of Scipio, and behind him his grandfather, and behind him Cicero – and behind them Plato. Instead we view 'alios' from the perspective of 'alii', antoikoi from the position of antoikoi, *as well as* the familiar reversal of we for they.

Here and throughout his explication of the world image in the 'Somnium Scipionis', Macrobius was drawn to consider the means and the very legitimacy of representation of unknown lands and their peoples. Hypothesis is differentiated from speculation and assumption, the latter excluded from schemes of representation. Consequently the existence of an equivalent topography, for example the presence of a sea identical to the Mediterranean in the southern hemisphere, can be supposed but not asserted. Precisely because of this lack of experience, such a topography could not be *represented*: 'this should not be drawn on the basis of our presumption, [because] the location of these things remains unknown to us' (describi hoc nostra attestatione non debuit, cuius situs nobis incognitus perseverat).[39] Inaccessible, the 'situs incognitus' can be known by hypothesis alone.

The need to represent hypothesis (without assertion) led to Macrobius' crucial act of commentary: the provision of visual reference points for the theories he discussed. The text of the *Commentarii* gives instructions for the construction of four diagrams: the earth and seven planets within the zodiac (1.21.3–5); rainfall on the earth (designed to demonstrate that weights are drawn to the earth) (1.22.11–12); the zonal division of the earth (2.5.13–17) (fig. 5), and the zonal division of the sky in relation to the earth (2.7) (fig. 6); the presence of a fifth diagram, showing the relationship between Ocean and land, is implied (2.9.7–8) (fig. 2). The illustration of the *Commentarii* with diagrams explicitly acknowledges the pedagogic value of image over word: the explanation given for drawing the first (terrestrial) zonal diagram is that 'reason sinks into the mind more easily when expressed by drawing than by speech' (animo facilius inlabitur concepta ratio descriptione quam sermone).[40] *Sermo* nevertheless is required to form the image: the diagram is to be constructed by inscribing letters at particular points on a circle and drawing lines between specified letters to construct divisions between the zones. For example, Macrobius begins his instructions for the first zonal diagram (2.5.13–17): 'draw the earth as a circle on which the letters ABCD are inscribed; and around A, let N and L be written; around B, the letters M and K; and around C, G and I; and around D, E and F ...' (esto orbis terrae cui adscripta sunt ABCD; et circa A, adscribantur N et L; circa B autem M et K; et circa C, G et I; et circa D, E et F). He goes on to make clear the meaning of the spaces between the letters, while referring to Cicero's more elliptical description: 'the spaces thus opposite each other, that is the one from C up to the line which is drawn from I, the other from D up to the line which is drawn from F, should be understood to be "frozen with perpetual winter", for the upper one is the furthest northern zone, the lower the furthest southern' (spatia igitur duo adversa sibi, id est unum a C usque ad lineam quae in I ducta est, alterum a D usque ad lineam quae in F ducta est, intellegantur pruina obriguisse perpetua – est enim superior septentrionalis, inferior australis

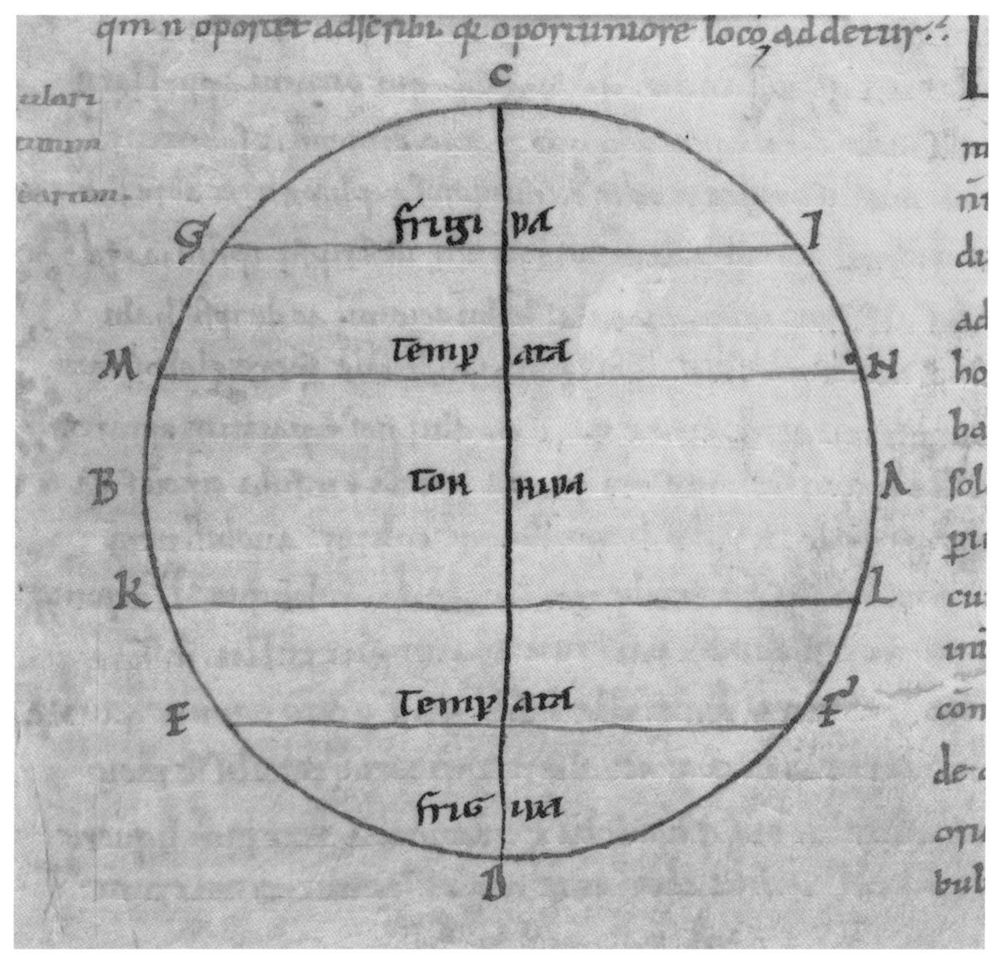

Fig. 5. London, BL, MS Harley 2772, f. 67v. Zonal diagram from an eleventh-century manuscript of Macrobius, *Commentarii in Somnium Scipionis* (Commentary on the Dream of Scipio). This simple image shows the division of the earth into five zones: two outer frigid zones ('frigida'), a central torrid zone ('torrida'), and two habitable zones in the northern and southern hemispheres ('temperata'). The letters correspond to the text of the *Commentarii,* and are intended to allow the reader to understand Macrobius' explanation of zonal theory.

extremitas).[41] The text ensures the maintenance of figural and interpretive order: it generates the diagram, indeed its instructions are meaningless without visualization, but the diagram is held within the interpretive framework of the written word.

The first four diagrams that appear in the *Commentarii* are deeply embedded: the instructions for drawing them and the explanation of their purpose are an unavoidable feature of the complete text. A much less specific prompt is given for the fifth diagram, essentially a world map that will show Ocean, the ecumene, and the southern temperate zone. Unlike all four previous diagrams the copyist/reader is given no instructions for construction. Instead of using the imperative, Macrobius switches to the second person to tell the reader what he or she can expect to find in the image:

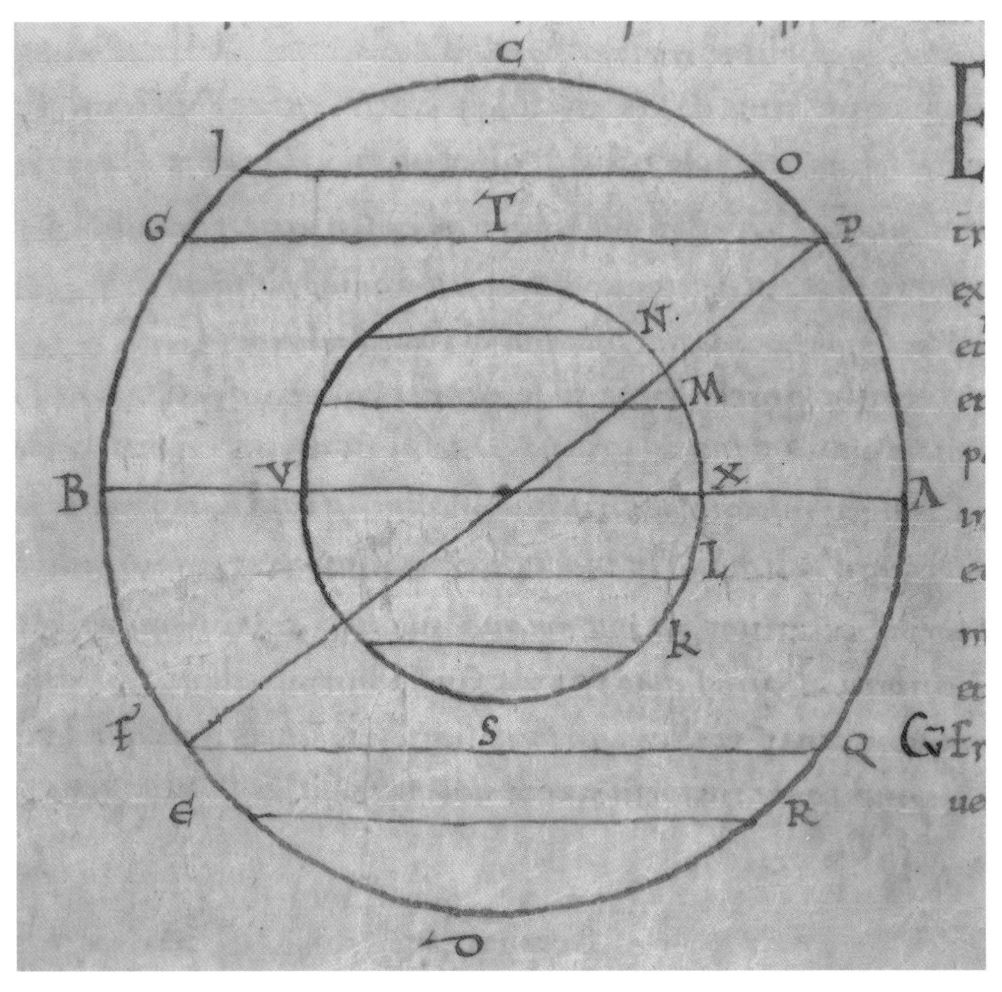

Fig. 6. London, BL, MS Harley 2772, f. 69r. Terrestrial-celestial zonal diagram from an eleventh-century manuscript of Macrobius, *Commentarii in Somnium Scipionis* (Commentary on the Dream of Scipio). This image shows the relation between the terrestrial zones (inner circle) and the celestial zones (outer circle). The diagonal line of the ecliptic (inaccurately drawn in this example) intersects both sets of zones. The letters correspond to the text of the *Commentarii*, and are intended to allow the reader to understand Macrobius' explanation of zonal theory.

> omnia haec ante oculos locare potest descriptio substituta, ex qua et nostri maris originem, quae totius una est, et Rubri atque Indici ortum videbis, Caspiumque mare unde oriatur invenies: licet non ignorem esse non nullos qui ei de Oceano ingressum negent.[42]

> all these things can be found before the eyes in the inserted diagram: from it you will see both the origin of our sea, which is one part of the whole Ocean, and the origins of the Red and the Indian seas; you will also find the source of the Caspian Sea: I am aware of some who deny that it enters [the ecumene] from Ocean.

The diagram, Macrobius continues, will also illustrate Cicero's statement that our part of the world is 'narrow at the top and broad at the sides' (angustam verticibus, lateribus latiorem);

indeed it will reveal what the ancients meant by the remark (conveyed most notably by Strabo) that our habitable region is shaped 'like a chlamys' ('chlamydi similem').[43] Macrobius terms the image a 'descriptio', a word that could imply visual or verbal representation. Seen in the context of the entire passage, however, it is clear that when he uses the verb 'describere' to denote what should not be done (sed describi ... non debuit) Macrobius surely intended a precise technical meaning, such as 'draw', or as in William Stahl's translation, 'mark off'.[44] This passage is, on one level, a set of instructions to the copyist, in which the limits of representation are stated, as well as the content of the image; at the same time, it explicates the content and significance of the image for the reader.

Macrobius' introduction and description of the world image indicate that his text was designed to be accompanied from the first with a visualization of the known world, featuring major oceans and seas. It is possible to imagine this illustration as a map of the known world alone. However, this is certainly not how it was interpreted by the vast majority of subsequent copyists, and there is an implicit suggestion in Macrobius' remarks that the image will show the known world in the context of the theory of zonal divisions. His comment on the certainty of a southern-hemisphere Mediterranean, representation of which would be inappropriate, might reasonably lead a copyist to the assumption that the diagram should show the temperate zone of the southern hemisphere, but that it should not be marked with any presumed maritime features. The limits of the diagram are the boundaries of attestation, that to which the author bears witness. Yet the image alters theory even as it represents it: by not showing the symmetry of topographical relations posited by Macrobius, it emphasizes the contrast between the known, albeit in some cases (such as the Caspian) controversial, features of the northern hemisphere, and the unknown space of the southern hemisphere. In this regard it offers a rather different picture to that of the first zonal diagram in the *Commentarii*. There, no topographical features are shown, but the size and relation of the 'other' temperate zone to the torrid and frigid zones is shown to be identical to 'our' zone; in the case of the world map, greater specificity of geographical detail means that northern and southern hemispheres have a very different appearance.

What is the relationship of image to word here? By embedding images in his writing, Macrobius ensured that they (or at least the potential for their representation) were transmitted as an integral part of the whole. He established, in other words, an iconographic tradition interior to the text. The diagrams are more than visual aids that accompany the text; they are explicitly a mode of understanding with a status equal to words. The significance of this incorporation of image for the representation of unknown land is tremendous. From the commentary emerged the diagrams of the zones and the image of the hemispheres that were to become standard for the next millennium. These diagrams ensured that the southern temperate zone was visualized, the space of the 'antoikoi' given the clarity and the legitimacy of image.

Macrobius makes it clear to the reader towards the end of the cosmographical section of the *Commentarii* that the invocation of the world image in Cicero served a moral purpose: the theories of zonal division and ocean flows emphasize the small size of the earth and the folly of pursuing political glory. 'Virtutis fructum sapiens in conscientia ponit, minus perfectus in gloria': the wise man finds the fruit of his virtue in his conscience; the lesser man in glory.[45] If the image of the world demonstrates Cicero's point about the spatial limits of fame on the 'point' of the earth's surface, it also, according to Macrobius, shows fame's temporal limits. For, the reader is informed, the central band of ocean in the torrid zone, visible on the world map, is perpetually burned up by the heat of the sun and then replenished with water.[46] This alternation of fire and flood signifies the catastrophes that periodically beset civilizations, wiping out culture, memory – and renown. The visual image of antipodal space was intended to signify the limits of glory, and thereby to act as a supplement both to Cicero's text and Macrobius' explication of it. Yet the map of Macrobius ran the risk of countering the purpose of the vision it supplemented: it made an integral feature of the *Commentarii* the very sight that distracted Scipio from the mysteries of the cosmos and non-secular existence. Moreover, as I will discuss in the following chapter, the map of Macrobius was adapted over the course of its transmission from the ninth to the fifteenth century. It came to show not the world as described by Cicero, or even Macrobius, but the world as it was known to the Middle Ages, a geography in which classical foundations had been realigned. Within, yet beyond, that world remained the space of the *antoecumene*, subject to vision and above all to speculation, but increasingly distant from its Ciceronian frame.

Martianus Capella: Geometry and her confusions

Martianus Capella's *De nuptiis Philologiae et Mercurii*, like the commentaries of Servius and Macrobius, received and reshaped classical tradition, but unlike those two works its compilation of classical learning was organized within an allegorical structure. *De nuptiis* consists of an initial two books describing the betrothal and marriage of the unlikely pair, followed by one book each devoted to the speeches of personifications of the seven liberal arts, guests at the wedding. The figure of Geometria announces herself at the beginning of the sixth book of *De nuptiis*, making explicit the importance of measurement of the earth's surface within the many applications of her art:

Geometria dicor, quod permeatam crebro admensamque tellurem eiusque figuram, magnitudinem, locum, partes et stadia possim cum suis rationibus explicare, neque ulla sit in totius terrae diversitate partitio, quam non memoris cursu descriptionis absolvam.[47]

I am called Geometry, for I could set forth, with their explanations, the size, position, parts and distances of the frequently traversed and measured earth as well as its figure; nor, in the course of description from memory, is there any portion in the breadth of the entire earth for which I may not account.

Encouraged by the wedding guests – including Jupiter, desirous of learning all hiding places on the earth so that no girl might escape him – Geometria proceeds to a proof of the sphericity of the earth, its size, its position at the middle and bottom of the universe, its division into zones and into 'upper' and 'lower' hemispheres. She further explains the division of the world between 'us' and three other groups: the 'antoikoi' to the south; those in the lower hemisphere who are 'obverse to us', called 'antipodes' (i.e. on the underside of the northern hemisphere), and the 'antikthones', opposite the 'antoikoi'.[48] Geometria goes on to elaborate this territorial division by means of a series of differences between the four regions and their inhabitants. In the case of the antikthones these differences are based upon the binary oppositions 'we' and 'they', and 'here' and 'there':

> nam cum aestate torremur, illi frigore contrahuntur; [nam] cum hic ver pubescit florentibus pratis, illic edomita aestas teporibus autumnascit; hic bruma, solstitium illic apparet; nobis Arctoa lumina spectare permissum, illis penitus denegatum.[49]

> for when we are burned with summer heat, they are stiff with cold; when spring matures here with flowering meadows, there summer, subdued by lukewarm temperatures, turns to autumn; when midwinter appears here, summer solstice appears there; to us it is permitted to see the lights of the Great Bear, to them it is completely denied.

These temporal differences, of course, are structured around a shared experience between the separate parts of the world, since the identical nature of the seasons in the *antoecumene* is not in doubt. On the other hand, the model is not entirely symmetrical: the inability of those in the southern hemisphere to observe particular constellations constitutes significant difference. Similarly, the rather garbled passage that follows appears to assert that the (northern hemisphere) 'antipodes' have seasons at the same time as inhabitants of the known world but experience days and nights of inverse length, e.g. long nights in summer and long days in winter; in the southern hemisphere, the 'antikthones' share seasons with the 'antoikoi', as well as the ability to look towards the south pole.[50] To complicate matters still further, despite having previously denied the habitability of parts of the earth in the torrid zone, Geometria proceeds to describe the experience of dwellers at the equator: all stars are visible to them; days and nights are of equal length for them and for 'their antipodeans'; but unlike dwellers in non-equatorial regions, they experience two winters and two summers per year, since the sun passes them twice – once on its way to the winter solstice, and once on its return to the northern solstice.[51] Confused? Martianus' medieval readers certainly were. But his incoherence here – the product of his attempt to fuse many centuries' worth of diverse sources and the diverse traditions therein –[52] shows the logic of similarity and difference under strain. Once the relationship of known world to unknown moved beyond binary opposition between 'we' and 'they' to encompass multiple unknown peoples, error (in its literal sense) could infiltrate and disrupt analogy. Wandering from one group of unknown people to the next, the desire to explain the totality of terrestrial habitation caused a proliferation of antipodeans: ours, those of the antoikoi, those of the equatorial dwellers – each the other's antipodeans. Description here serves to blur and break down differences, to undo, rather than reproduce and reinforce, categorization.

Martianus' deployment of antipodal theory differed from Macrobius' in a number of ways. As part of an encyclopedia of the liberal arts rather than a commentary on an earlier author, the cosmographical and geographical section in Martianus is a relatively insignificant part of Geometria's speech, and of the work as a whole. The section on inhabitants beyond the known world is the prelude to a description of the ecumene, which itself is just one aspect of Geometria's teachings. Further, Martianus provides, however confusingly, a nomenclature for the three unknown peoples of the unknown regions, adopting both the geometrical terminology used by Cicero (obverse, opposite) and Greek-derived terminology (antoikoi, antikthones, as well as antipodes). Showing a lyrical quality in his writing rarely found in Macrobius (for instance, the flourishing meadows of spring and summer 'conquered by tepid warmths'), Martianus also discusses in more detail than Macrobius the different experience of the seasons, visibility of stars, and length of nights and days in the *antoecumene,* subject matter that later commentators would seize upon. Crucially, Martianus does not include any statement to indicate that his writing should be illustrated. Certainly no 'map of Martianus Capella' survives, and it seems clear that, while images could and occasionally did accompany the text, they were not integral to it.[53]

Both Macrobius and Martianus responded to a Platonism evident in other works of the fourth and fifth centuries; their works, in turn, became an important means of transmitting Stoic and Platonist ideas and theories, of which the vision of an earth of divided habitations was an important part. The influence of both fifth-century authors was swift – as witnessed by Boethius' use of Macrobius in *De consolatione Philosophiae,* and references to Martianus' work by Cassiodorus[54] – and it was sustained: though the popularity of each rose and fell (usually around the same time), these texts continued to be copied into the fifteenth century and beyond.[55] Along with Servius' commentaries, the works of Macrobius and Martianus Capella therefore provided a significant channel for the transmission of antipodal theory and the zonal division of earth and sky to the Middle Ages. Yet each conduit preserved subtle differences. The commentaries of Servius and Macrobius supplemented original texts, serving to open them out to several different interpretive possibilities rather than fixing meaning. By including images within his *Commentarii* – a practice that, I will suggest, bled into the medieval manuscript transmission of Servius' works – Macrobius took supplementation further by embedding a visual element in the representation of spaces beyond the known world, with the result that the image itself could function as a mode of commentary, or alternatively stand outside of commentary as an independent text. The formation and codification of tradition was not a moment of ossification: Martianus' *De nuptiis* located discussion of the antipodes within a scheme of knowledge that became foundational for medieval education, but in a way that served to encourage a breakdown of categorization through the proliferation of *antoecumenical* others. For all three authors, theories of the antipodes were crucial to the understanding of antique poetic and philosophical culture, and integral to a much broader body of knowledge that had been transmitted from antiquity. All three, however, reshaped the theories – distorting, amplifying, and visualizing them – as they transmitted them. Christian reception of classical theories of the antipodes, by contrast, revealed a model different to that of reshaping and supplementation, one in which classical theories were read in terms of their compatibility with Christian belief. This was reception as intervention, opposition, and rejection.

Sons of Adam

As the case of the antipodes makes clear, the attitude of early Christian authors to classical scientific theories was by no means uniform. At one end of the spectrum was polemic, perhaps most vividly represented by Lactantius' early fourth-century denunciation of 'men whose feet are above their heads'. At the other end, however, it is possible to discern the acceptance and adaptation of antipodal theory amongst authors such as Irenaeus, Tertullian, and Origen. In these instances the antipodes appear to occupy a relatively uncontroversial, if marginal, position within Christian belief. Between those two poles was the position expressed by Augustine in *De civitate dei*. Ultimately, Augustine declared a clear reluctance to allow for the possibility of human habitation beyond the reach of the word of God, but even within his denial there was ambivalence about the other side of the earth, uncertainties about whether it consisted of land or ocean, and whether it was completely uninhabited or inhabited by non-human life forms. The range of early responses reveals the position of the antipodes between two fundamental impulses in Christian thought: evangelization and exegesis. The dissemination of the word of God to the ends of the earth demanded some consideration of precisely what and where those ends were; equally, scriptural authority – as Augustine pointed out – affirmed that such evangelization had already taken place as the result of the proselytizing activities of the apostles. Antipodeans, according to classical scientific theory, were beyond the reach of peoples in the ecumene, but no humans, according to scripture, were beyond the reach of Christianity, hence the impossibility of antipodal habitation. If, in other words, evangelization naturally encouraged the idea of the expansion of Christian space to include the inhabitants of the antipodes, scripture, on at least one reading, prohibited faith in their existence, and demanded a consolidated vision of Christian space within the known world.

At least three different responses to the questions of antipodal habitation – could antipodeans exist beyond the reach of God, and were such peoples in fact human? – can be identified in early Christian writing: assimilation to the category of monstrous race; inclusion within the mysteries known only to God; denial and ridicule. The logic of the catalogue enabled Tertullian (c. 160–220) to invoke the idea of the antipodeans as one of the monstrous races, linked with creatures such as dog-headed men (Cynocephali) or the 'Sciapodes', a race of one-legged men each with a gigantic foot, which enabled them to move with great rapidity and, in supine moments, to obtain shade from the sun. The assimilation of antipodeans to the monstrous races enabled a striking act of mock identification. In defending Christians from being called a 'third race' (after the Romans and the Jews), Tertullian suggests that such nomenclature associated them with monstrosities:

> Plane, tertium genus dicimur. Cynopennae aliqui uel Sciapodes uel aliqui de subterraneo Antipodes? Si qua istic apud uos saltem ratio est, edatis uelim primum et secundum genus, ut ita de tertio constet.[56]

> Assuredly, we are called the third race. Like Cynopennae or Sciapodes, or those antipodeans from beneath the earth? If at least on this point you are capable of reason, I would like you to give out the first and second race, so that we may establish the third.

Christians as antipodeans? Tertullian's response is that Christians are not one nation but many: the name 'Christian' extends to all peoples, and cannot simply be added to a list of the earth's inhabitants; moreover Christians do not constitute a 'third people' after Romans and Jews – they represent a new order that seeks to displace both.[57] It is the extremity of antipodal people, their position 'de subterraneo', as well as the obvious verbal connection between sciapodes (shadow-feet) and antipodes (opposite feet) that enables their alignment with the monstrous races.

The categorization of the antipodes as phenomena subject only to divine knowledge appears first in a reference of Irenaeus of Lyon (c. 130–202) to worlds situated beyond Ocean, and therefore beyond the knowledge of Christians. In this instance, the antipodes are cited as just one of a number of things, including the source of the Nile and the ebb and flow of ocean tides, that only God could truly know.[58] Like Irenaeus, Origen (c. 185–255) used the antipodes as a sign of divine knowledge and human ignorance when he brought up the 'antikthones' in his *De principiis* during a digression on the meaning of the term 'mundus' and its scriptural significance. Having outlined several possible meanings of 'mundus' ('cosmos', 'ornament', the known world, the universe consisting of heaven and earth),[59] Origen cited a passage in Clement of Rome's letter to the Corinthians to argue that the word might refer to peoples beyond 'our world with its inhabitants'.

> Meminit sane Clemens, apostolorum discipulus, etiam eorum, quos ἀντίχθονας Graeci nominarunt, atque alias partes orbis terrae, ad quas neque nostrorum quisquam accedere potest, neque ex illis, qui ibi sunt, quisquam transire ad nos, quos et ipsos mundos appellauit, cum ait: 'Oceanus intransmeabilis est hominibus et hi, qui trans ipsum sunt mundi, qui his eisdem dominatoris dei dispositionibus gubernantur'.[60]

> When he said 'the Ocean cannot be crosssed by men, nor can those worlds [be crossed] which are beyond it, which are governed by the very same dispositions of the lord God', Clement, the disciple of the apostles, doubtless had in mind those whom the Greeks call *antikthonas*, and other parts of the earth to which none of us ever was able to approach, and from which none of those, who are there, was ever able to cross over to us; he called these places 'worlds'.

According to Origen, in speaking of worlds plural Clement indicated that all these concepts of *mundus* could be assimilated into the idea of the universe, the 'one and perfect' world, celestial and supercelestial, earthly and infernal, 'within which or by which others, if they are there, can be thought to be contained' (intra quem uel a quo ceteri, si qui illi sunt, putandi sunt contineri).[61] The status of Clement as apostolic disciple enabled Origen to assert impressive authority for his synthesis of Greek cosmology with Christian doctrine. In this synthesis, other worlds and other peoples are allowed to exist beyond the knowledge of men – but not beyond the governance of God. Such thinking allowed for the presence of antipodal peoples within a Christian science; it was, however, difficult to reconcile with the notions of Christian universalism inherent in Tertullian's understanding of Christianity as a name and a force above and beyond that of race.

At the heart of the tension between universalism and the acceptance of the existence of spaces beyond human knowledge was an uncertainty about the nature of antipodal habitation: were antipodeans to be conceived of as humans mirroring those in the known world, or as monsters, not descended from Adam? Or should they be conceived of at all? From the point of view of the polemic to which Lactantius subjected antipodal theory in book three of his *Divinae Institutiones* (composed 304–11), the notion that men might exist on the other side of the earth 'standing opposite to' those in the known world was the product of corrupt and disreputable philosophizing. Drawing on Cicero's *Academica*, Lactantius attacks the natural philosopher

Xenophanes for suggesting the habitation of the moon by 'lunatici homines', as well as the theory, attributed by Seneca to certain Stoics, that the sun might have its own peoples. He then turns to the possibility that there might be 'those who think there are antipodeans with their footsteps opposite to ours' (illi qui esse contrarios vestigiis nostris antipodas putant).[62] Lactantius' opposition to the notion of the antipodes enjoyed a considerable post-medieval notoriety,[63] but it has to be seen in the context of a more general attack on Stoic theories of nature, and specifically on the process of hypothesizing false positions from false suppositions. There is also, as Moretti has suggested, a continuum between Lactantius' views and earlier satirical and polemical treatments of the antipodes.[64] The thrust of Lactantius' anti-antipodal argument is that the theory of the antipodes rests on the notion of an 'upside-down' world:

> aut est quisquam tam ineptus qui credat esse homines quorum uestigia sint superiora quam capita? aut ibi quae aput nos iacent, inuersa pendere, fruges et arbores deorsum uersus crescere, pluuias et niues et grandines sursum uersus cadere in terram? et miratur aliquis hortos pensiles inter septem mira narrari, cum philosophi et agros et urbes et maria et montes pensiles faciant?[65]

> But is there anyone so foolish as to believe that there are men whose feet are above their heads? Or that those things that lie amongst us, in that place hang upside-down, (that) fruits and trees grow downwards, while rain and snow and hail fall upwards to the earth? Indeed one might wonder that the hanging gardens are reckoned amongst the seven wonders, when philosophers construct hanging fields and towns, and even seas and mountains.

Clearly Lactantius' particular target is the philosophical reasoning that has created 'these hanging antipodeans'. He explains the theory even as he ridicules it: it is the consequence of observations of the course of the stars, and the sun and the moon, from which arose the notions of the earth as a ball ('sicut pila'), uniformly covered with land, and uniformly inhabited by men as well as animals.[66] Defenders of the 'hanging men' theory claim that weights tend towards the centre, like the spokes of a wheel. Such theories are, Lactantius suggests, either the result of stupidity, play, or mendacious showing-off:

> quid dicam de his nescio, qui cum semel aberrauerint, constanter in stultitia perseuerant et uanis uana defendunt, nisi quod eos interdum puto aut ioci causa philosophari aut prudentes et scios mendacia defendenda suscipere, quasi ut ingenia sua in malis rebus exerceant uel ostendant.[67]

> I do not know what to say about these people who, when once they have lost their way, persist in unremitting stupidity and defend inanities with inanities. Unless, I suppose, they now and then philosophize as a form of jest, or, practised and knowing, defend lies for the sake of it, as if to exercise or show off their talents for evil things.

Lactantius' hostility towards deceitful or playful philosophizing, and in particular his reference to philosophizing as 'a form of jest' (ioci causa), may have prompted Macrobius' comment that it is a 'kind of joke' (genus ioci) to assert that the sky is up for us and the earth down.[68] Lactantius' jokers are those who propagate the notion of an upside-down race, while for Macrobius the joke is the assertion of the obvious: that antipodeans will not 'fall into the sky'. In the absence of any explicit reference it is hard to be sure of a connection between these texts, but certainly Macrobius' excursus on why those on the other side of the earth will not fall into the sky addresses the very problem with antipodal theory raised by Lactantius. This verbal echo suggests the existence of a dialogue, however strained, on questions of natural science between Christian apologists and those engaging with antique philosophical traditions from a non-Christian perspective. Above all, it is important not to lose sight of the knowledge of scientific theories displayed by Lactantius, in spite of his resistance to them, nor that, in parodying the discourse of inversion common to antipodal literature, he perpetuated its fantastical possibilities. Lactantius imagined fruits and trees, rain and hail, fields, towns, seas and mountains in the antipodes, all the while condemning them as fictions. Even – perhaps especially – within rejection, there remained space to represent the manifold possibilities of places and peoples on the other side of the earth.

The question of the identity or difference of peoples 'on the underside of the earth' with those in the known world was at the crux of Augustine's rejection of belief in the existence of inhabitants of antipodal regions in book 16 of *De civitate dei*. Such beliefs were based, he argued, on conjecture rather than historical knowledge (no-one had ever been there, let alone met an antipodean), and they could not be reconciled with scripture. If these unknown lands could not be reached by us, any inhabitants could not be descended from Adam, and had not been redeemed by Christ. Augustine's denial of an antipodes inhabited by humans (he leaves open the possibility that they might be inhabited by other creatures) leads him to restrict the search for the city of God to the known world:

> Quod uero et antipodas esse fabulantur, id est homines a contraria parte terrae, ubi sol oritur, quando occidit nobis, aduersa pedibus nostris calcare uestigia: nulla ratione credendum est … Quoniam nullo modo scriptura ista mentitur, quae narratis praeteritis facit fidem eo, quod eius praedicta conplentur, nimisque absurdum est, ut dicatur aliquos homines ex hac in illam partem, Oceani immensitate traeiecta, nauigare ac peruenire potuisse, ut etiam illic ex uno illo primo homine genus institueretur humanum. Quapropter inter illos tunc hominum populos, qui per septuaginta duas gentes et totidem linguas colliguntur fuisse diuisi, quaeramus, si possumus inuenire, illam in terris peregrinantem ciuitatem Dei …[69]

> And that there are supposed to be antipodeans, that is men at the opposite part of the earth, where the sun rises when it falls for us, who tread their footsteps opposite to our feet: there is no reason for belief in them … Since scripture itself in no way lies (the credibility of its historical narratives is attested by the fulfilment of its predictions), and it would be too absurd to say that certain men could have sailed and penetrated from this part to that, across the immensity of Ocean, so that there too from that one first man the human race might have been founded. On account of this let us see if we can find that city of God on pilgrimage on the earth amongst those races of humanity who are inferred to have been divided into seventy two peoples and the same number of tongues.

In Augustine it is the ocean, rather than intervening torrid or frigid zones, that renders impossible knowledge and contact with 'that part of the world', suggesting a familiarity with the Cratesian theory of two bands of encircling ocean. On the whole, however, Augustine's description of antipodal peoples shows an awareness of the idea, rather than a thoroughgoing interest in the theory. Augustine's consideration of the antipodes certainly seems to be indebted to Lactantius in several ways. His description of antipodeans standing with footsteps 'opposite to ours' echoes the wording used by Lactantius ('contrarios vestigiis nostris antipodas'), although the ultimate source of the expression seems to be Cicero's mention of the antipodes in the *Academica*: 'those, whom you [Lucullus] call antipodeans, who on the opposite side of the earth stand with footsteps directly against ours' (in *contraria parte terrae*, qui *aduersis uestigiis* stent *contra nostra* uestigia, quos *antipodas* uocatis).[70] Like Lactantius, Augustine also gives a brief summary of the reasoning that produced the theory of antipodal habitation: the earth, suspended within the heavenly spheres, occupies both the lowest and the most central position, and therefore it is opined that the part of the earth underneath must contain human habitation. Nevertheless, the comments on the antipodes in *De civitate dei* actually represent a significant modification and moderation of the Lactantius' polemical attack. Whereas for Augustine scripture is the primary objection to the existence of antipodeans, for Lactantius the notion expressed perfectly the vanities of human philosophizing. Lactantius does not articulate the same theological concerns as Augustine: he does not identify the antipodes as a contradiction of Genesis, nor does he raise the problem of the impossibility of converting inaccessible antipodal peoples to Christianity.

The key issue for Augustine is that of habitation, since he entertains the possibility of the existence of *uninhabited* land beyond the known world. Even allowing for a spherical earth, land on the other side of the world may be covered by water (following Genesis 1.10: 'ab aquarum congerie'), or exposed ('nuda') but not inhabited. The notion of a land populated with humans unknown to the sons of Adam was openly contradicted by the account of the flood in Genesis 7–10, according to which post-diluvian humanity is descended from the three sons of Noah. Moreover, the presence of unknown humanity raised the problem of explaining the statements in the Gospels and Acts concerning the mission of the apostles to spread Christianity to the ends of the earth ('ad ultimum terrae') (Acts 1:8), and to every nation (Matthew 28:19, Luke 24:47), and the firm statement in Acts 17.26 that God made the world to be inhabited by one race of men 'over the entire surface of the earth' (super universam faciem terrae). The prelude to Augustine's remarks on the antipodes is his discussion of accounts of monstrous races and deformities (16.8). The overriding issue addressed in these sections of *De civitate dei* is the question of what constitutes the 'genus humanum': according to the logic applied by Augustine, monstrous races, if they really exist, may or may not be human – but if they are human, they are descended from Adam.

The problem of human habitation beyond the ecumene, beyond the word of God, and therefore in contradiction of scripture, did not go away. It continued to be debated throughout the Middle Ages, and it gained particular urgency at the moment of the discovery of the New World, when the presence of vast numbers of peoples hitherto unknown to Christendom seemed to cast doubt on the theory of human monogenesis from Adam. Augustine's denial of antipodal habitation insisted that, in the absence of discovery, it was necessary to limit theoretical speculation. His statements in *De civitate dei* should not, however, be seen as a closing of vision. For one thing, they show a willingness to think about the other side of the earth – to envisage it, to consider the possibility of habitable land there, even while arguing against the possibility of human habitation. As I will argue in the following chapters, denial on the basis of scriptural authority acted to stimulate more than to restrict debate about the antipodes. Part of the reason for this is that, in addition to Augustine's response to, and reconfiguration of, classical traditions of the antipodes, literate medieval culture received the pagan exposition of antipodal theory provided by authors such as Servius, Macrobius and Martianus Capella. The transmission of both exposition and denial meant that it was possible to read Augustine alongside Macrobius, and vice-versa, with one authority contradicting but not silencing the other. Of equal importance was the dissemination of an image of the world that represented land beyond the known world. The world image contained in Macrobius' *Commentarii* became a standard means of figuring torrid, frigid, and temperate climatic zones, as well as the distribution of land and sea; it also became a standard reference point for discussion of antipodal habitation. The image of unknown land persisted, and with it, in spite of denials, the idea of its inhabitants.

Notes

1 See Andrea Pellizzari, *Servio: Storia, cultura e istituzioni nell'opera di un grammatico tardoantico* (Florence: Olschki, 2003), pp. 15–31.

2 Pellizzari, *Servio*, p. 26.

3 See for example Macrobius, *Commentary on the Dream of Scipio*, trans. William Harris Stahl (New York: Columbia University Press, 1952), p. 12.

4 Macrobius, *Saturnalia*, ed. J. Willis (Leipzig: Teubner, 1963), 1.2.15. Alan Cameron, 'The Date and Identity of Macrobius', *Journal of Roman Studies* 56 (1966), 25–38, argued that the *Saturnalia* was written in the years immediately after 431; this constituted a revision of earlier scholarship, which, based on references to a Macrobius in the Theodosian Code, had variously dated Macrobius' *Commentarii* to c. 390–410, and had tended to suggest dates in the last decade of the fourth century for the *Saturnalia*. A valuable summary of the debate about the date of Macrobius and his interrelationship with Servius is provided in Paolo De Paolis, 'Macrobio 1934–1984', *Lustrum* 28–9 (1986–7), 107–249: pp. 113–25.

5 The dates of Servius' life and hence his commentaries are uncertain. For prosopography see the summary in Robert A. Kaster, *Guardians of Language: The Grammarian and Society in Late Antiquity* (Berkeley: University of California Press, 1988), pp. 356–9.

6 For a brief summary of recent arguments about the date see Macrobius, *Commentaire au Songe de Scipion*, ed. Mireille Armisen-Marchetti, 2 vols (Paris: Les Belles Lettres, 2001–3), vol. 1, pp. xvi–xviii.

7 See Robert Kaster, 'Macrobius and Servius', *Harvard Studies in Classical Philology* 84 (1980), 219–62: esp. 255–62.

8 Pellizzari, *Servio*, p. 18, comparing Servius' *Commentaries* with the *Saturnalia*, both 'ispirati dalla medesima volontà di restaurazione culturale dell'antico', although belonging to different genres.

9 Servius, *Commento al libro VII dell'Eneide di Virgilio*, ed. Giuseppe Ramires (Bologna: Pàtron, 2003), p. 36.

10 Other commentaries adopted a different vocabulary, using the Greek 'ἀντίχθονες' (antikthones), and 'antoecumene': see 'Probi qui dicitur in Vergilii Bucolica et Georgica commentarius', in Servius, *In Vergilii bucolica et georgica commentarii*, pp. 361, 363 (in Verg. *Georg.* 1.233). In what appears to be an example of a commentator distorting the meaning of his author, the commentary on Statius' *Thebaid* made by Lactantius Placidus – a work of uncertain date but possibly roughly contemporary with Servius' commentaries on Virgil – interprets the phrase 'hidden world' (mundus latens) as a reference to the antipodes ('antipodas'): *Lactantii Placidi in Statii Thebaida Commentvm*, vol. 1, ed. Robert Dale Sweeney (Stuttgart: Teubner, 1997), 6.363–4.

11 Servius, *In Vergilii bucolica et georgica commentarii*, p. 186. Servius attributes his understanding of the zones to the explanations of a certain 'Metrodorus philosophus' (possibly Metrodorus of Scepsis), who composed at least five books on the subject and defended Virgil against the charge of being ignorant of astrology: *In Vergilii bucolica et georgica commentarii*, p. 185.

12 Servius, *In Vergilii bucolica et georgica commentarii*, p. 187. The reference is to Lucan's description of Druidical beliefs in *De bello civili*, 1.454–8: 'vobis auctoribus umbrae/ non tacitas Erebi sedes Ditisque profundi/ pallida regna petunt: regit idem spiritus artus/ orbe alio; longae, canitis si cognita, vitae/ mors media est.' (according to you the shades do not seek the silent seats of Erebus and sunless kingdoms of deepest Dis: the same spirit rules the body in another world; if your songs are right, death is but the mid-point of a long life). 'Orbe alio' has been taken to refer to another part of the world – e.g. the other hemisphere, or the Isles of the Blessed – or to another world entirely, or to Pythagorean theories of the transmigration of the soul. Its meaning was by no means certain to Arnulf of Orléans in the twelfth century: 'VOBIS AUCTORIBUS that is, you say souls do not descend to infernal regions, but are reincarnated in the world of the other hemisphere, or in another part of the world remote from you' (VOBIS AUCTORIBUS id est sicut uos dicitis anime ad inferos non descendunt, sed in orbe alterius hemisperii incorporantur iterum uel in aliqua parte orbis a uobis remota): *Arnulfi Aurelianensis Glosule super Lucanum*, p. 59.

13 Servius, *In Vergilii bucolica et georgica commentarii*, p. 188.

14 Servius, *In Vergilii carmina commentarii*, vol. 2, pp. 27–8: 'nam Lucretius ex maiore parte et alii integre docuent inferorum regna ne posse quidem esse: nam locum ipsorum quem possumus dicere, cum sub terris esse dicantur antipodes?' For discussion of this passage in terms of Servius' Neoplatonism see Pellizzari, *Servio*, pp. 160–1.

15 Virgil, *Aeneid*, 6.531–3.

16 Servius, *In Vergilii carmina commentarii*, vol. 2, pp. 75–6

17 Evelyn Edson and Emilie Savage-Smith, 'An Astrologer's Map: A Relic of Late Antiquity', *Imago Mundi* 52 (2000), 7–29.

18 For the toponyms see Edson and Savage-Smith, 'Astrologer's Map', pp. 27–8.

19 Discussed by Edson and Savage-Smith, 'Astrologer's Map', p. 14, who suggest that it is a corrupt reference to Ptolemy's calculation of the total latitude of the inhabited world.

20 The mention of the Roman colony of Hierasycaminus (abandoned in 298) along with other Egyptian toponyms has suggested the map's provenance in Roman Egypt: Edson and Savage-Smith, 'Astrologer's Map', pp. 15–21.

21 Scafi, *Mapping Paradise*, p. 84.

22 Although it has been suggested that Macrobius may have been a Christian, the absence of any explicit reference to Christianity in his work, its incompatibility with Christian doctrine, and his apparent location within non-Christian circles in Rome continue to convince most scholars of his paganism: see De Paolis, 'Macrobio 1934–1984', pp. 125–32. Macrobius states in the *Saturnalia* that he was born 'beneath another sky' (sub alio caelo), possibly a reference to Africa.

23 For this argument see Alan Cameron, 'Paganism and literature in Late Fourth-Century Rome', in *Christianisme et formes littéraires de l'antiquité tardive en Occident* (Geneva: Foundation Hardt, 1977), pp. 1–30: 25–6.

24 See Jacques Flamant, *Macrobe et le néo-Platonisme latin, à la fin du IVe siècle* (Leiden: Brill, 1977). For a summary of Platonic and Neoplatonic influence on the commentary, as well as other possible sources, see Armisen-Marchetti, *Commentaire*, vol. 1, pp. liv–lxvi.

25 'quas Graeco nomine zonas uocat': Macrobius, *Commentaire au Songe de Scipion*, ed. Armisen-Marchetti, 2.5.7.

26 Observed by Mario Regali: Macrobio, *Commento al Somnium Scipionis*, ed. and trans. Mario Regali, 2 vols (Pisa: Giardini, 1990), vol. 2, p. 170.

27 Macrobius, *Commentarii in Somnium Scipionis*, ed. J. Willis (Leipzig: Teubner, 1963), 2.8.2–8. The equivalence of 'per' with 'inter' is suggested in a scholia contained in the expanded version of Servius' commentary printed by Pierre Daniel in 1600 and known as 'Servius auctus': *In Vergilii bucolica et georgica commentarii*, p. 188. It is quite possible that Macrobius drew upon scholia of this kind (i.e. not composed by Servius); the equation of 'per' with 'sub' is surprising, however: Armisen-Marchetti, *Commentaire*, vol. 2, p. 144 *n*172.

28 Macrobius, *Commentarii*, 2.8.4.

29 Macrobius, *Commentarii*, 2.8.5.

30 Macrobius, *Commentarii*, 2.5.25–6.

31 See further Moretti, *Gli antipodi*, pp. 41–3; Flamant, *Macrobe et le néo-Platonisme latin*, pp. 469–74; and the comments of Regali: Macrobio, *Commento al Somnium Scipionis*, vol. 2, pp. 158–9. The possibility of a response to Lactantius appears to me more likely than a response to Augustine.

32 Macrobius, *Commentarii*, 2.5; 2.9. It has been argued that Macrobius was using a commentary on Virgil that explained his work in Cratesian terms, emphasizing physical rather than astrological factors. Hans Mette noted the parallel between Macrobius and comments of Pseudo Probo on *Georgics* 1.233, where the earth is described in the form of the Greek letter Θ: *Sphairopoiia: Untersuchungen zur Kosmologie des Krates von Pergamon* (Munich: Beck'sche, 1936), 78 *n*1; see 'Probi qui dicitur in Vergilii Bucolica et Georgica commentarius', in Servius, *In Vergilii bucolica et georgica commentarii*, p. 364. The parallel is hardly an exact one, though, since Macrobius does not mention the letter.

33 Macrobius, *Commentarii*, 2.5.16.

34 Macrobius, *Commentarii*, 2.5.33.

35 Stahl's translation is misleading on this point, since he inserts 'antipodes' in rendering 'similiter ab illis qui inferiorem zonae suae incolunt partem, interiecta australi gelida separantur' as 'next, those who live on the underside of the southern hemisphere, the Antipodes, separated from the Antoeci by the south frigid zone':

Commentary, p. 206.

36 Macrobius, *Saturnalia*, 1.21.3: 'Proserpina ... quam numen terrae inferioris circuli et antipodum diximus'; Macrobius, *Commentarii*, 1.22.13.

37 Macrobius, *Commentarii*, 2.5.35.

38 Macrobius, *Commentarii*, 2.5.36.

39 Macrobius, *Commentarii*, 2.9.7.

40 Macrobius, *Commentarii*, 2.5.13. See also the introduction of the first diagram: *Commentarii*, 1.21.3: 'because the path to the intellect is easier through the eyes, that which language describes should be given visual form' (quia facilior ad intellectum per oculos uia est, id quod sermo descripsit uisus adsignet). On Macrobius' use of diagrams in the *Commentarii* see Barbara Obrist, *La cosmologie médiévale. Textes et images*, vol. 1: *Les fondements antiques* (Florence: Edizioni del Galluzzo, 2004), pp. 180–94. Obrist sees the function of the diagrams as 'more illustrative and persuasive than demonstrative' of particular hypotheses.

41 Macrobius, *Commentarii*, 2.5.14

42 Macrobius, *Commentarii*, 2.9.7. The debate about the open or land-locked nature of the Caspian was long-standing: see Armisen-Marchetti, *Commentaire*, vol. 2, p. 147 *n*182.

43 Macrobius, *Commentarii*, 2.9.8. The term refers to a type of Macedonian coat. Regali notes Macrobius' adaptation of Cicero here: the introduction of the simile, probably derived from Eratosthenes rather than Strabo, provides an image absent from Cicero's text: *Commento al Somnium Scipionis*, vol 2, p. 173.

44 Macrobius, *Commentary*, trans. Stahl, p. 215.

45 Macrobius, *Commentarii*, 2.10.2.

46 Macrobius, *Commentarii*, 2.10.4–13. For the lineage of this fairly standard feature of Stoic cosmology from Aristotle onwards see Armisen-Marchetti, *Commentaire*, vol. 2, p. 159 *n*210.

47 *De nuptiis Mercurii et Philologiae*, ed. James Willis (Leipzig: Teubner, 1983), 6.588.

48 Martianus Capella, *De nuptiis*, 6.604–5.

49 Martianus Capella, *De nuptiis*, 6.605.

50 Martianus Capella, *De nuptiis*, 6.606.

51 Martianus Capella, *De nuptiis*, 6.607.

52 As noted in Ramalli's commentary: *Le nozze di Filologia e Mercurio*, ed. Ilaria Ramelli (Milan: Bompiani, 2001), p. 916.

53 The theories advanced by Richard Uhden on this matter are not supported by the almost total absence of maps from manuscripts of *De nuptiis*: Richard Uhden, 'Die Weltkarte des Martianus Capella', *Mnemosyne*, 3rd ser. 3 (1935–6), 97–124. Leonardi discusses the zonal map and T-O map that illustrate *De nuptiis* in Florence, Biblioteca Laurenziana, MS S. Marco 190, a manuscript from the end of the tenth or early eleventh century: Claudio Leonardi, 'Illustrazioni e glosse in un codice di Marziano Capella', *Bulletino dell' Archivio Palaeografico Italiano* n.s. 2–3 (1956–7), 39–60, esp. pp. 45–50. The zonal map in this manuscript is accompanied by the quotation 'quinque tenent celum zone' (*Georgics* 1.233), which may be an allusion to Macrobius as much as to Virgil.

54 Boethius, *De consolatione philosophiae*, ed. Claudio Moreschini (Munich: Saur, 2000), 2.7.2–12. On the influence of Macrobius on Boethius, see Pierre Courcelle, 'La postérité chrétienne du *Songe de Scipion*', *Revue des Études Latines* 36 (1958), 205–34, pp. 215–23; but see also the tempering comments of Joachim Gruber, *Kommentar zu Boethius de consolatione Philosophiae* (Berlin: de Gruyter, 1978), p. 216: 'Boethius beide Texte [i.e. Cicero's and Macrobius'] kannte und die Argumente nach seinem Bedarf verwendete'. *Cassiodori Senatoris Institutiones*, ed. R. A. B. Mynors (Oxford: Clarendon Press, 1937), 2.2.17, 2.3.20. The references suggest that Cassiodorus had heard of Martianus' work but had not read it: Bernard Ribémont, *Les origines des encyclopédies médiévales: D'Isidore de Séville aux Carolingiens* (Paris: Champion, 2001), p. 23.

55 C. R. Ligota, 'L'influence de Macrobe pendant la Renaissance', in *Le soleil e la Renaissance ... Colloque international*

(Brussels: Presses universitaires de Bruxelles, 1965), pp. 465–82; Courcelle, 'La postérité chrétienne du *Songe de Scipion*', pp. 205–34; Claudio Leonardi, 'Nota introduttiva per un'indagine sulla fortuna di Marziano Capella nel Medioevo', *Bullettino dell'Istituto Storico Italiano per il Medio Evo e Archivio Muratoriano* 67 (1955), 265–88; Leonardi, 'I codici di Marziano Capella', *Aevum* 33 (1959), 443–89, and *Aevum* 34 (1960), 1–99, 411–524. For a summary of evidence for Martianus' reception see Cora E. Lutz, 'Martianus Capella', in *Catalogus translationum et commentariorum: Medieval and Renaissance Latin Translators and Commentaries*, ed. P. O. Kristeller and F. Edward Cranz, vol. 2 (Washington: Catholic University of America Press, 1971), pp. 367–81; also William H. Stahl, 'To a better understanding of Martianus Capella', *Speculum* 40 (1965), 102–15, pp. 106–15; and the informative discussion in Natalia Lozovsky, *'The Earth Is Our Book': Geographical Knowledge in the Latin West ca. 400–1000* (Ann Arbor: University of Michigan Press, 2000), pp. 113–38.

56 Tertullian, *Ad nationes*, ed. J. G. Ph. Borleffs, in *Tertulliani Opera*, Corpus Christianorum, Series Latina, 2 vols (Turnhout: Brepols, 1954), 1.8.1. 'Cynopennae' or 'cynophanes' were used by Tertullian to refer to the creatures usually known as 'Cynocephali': *Le premier livre Ad Nationes de Tertullien*, trans. André Schneider (Neuchâtel: Institut Suisse de Rome, 1968), pp. 191–2.

57 *Ad nationes*, 1.8.1–13.

58 Irenaeus of Lyon, *Adversus haereses/Contre les hérésies*, ed. Adelin Rousseau and Louis Doutreleau, 10 vols (Paris: Éditions du Cerf, 1965–82), 2.28.2.

59 Origen, *De principiis/Traité des principes*, ed. and trans. Henri Crouzel and Manlio Simonetti, 3 vols (Paris: Éditions du Cerf, 1978–84), 2.3.6.

60 *De principiis*, 2.3.6. For the reference to Clement see *Épître aux Corinthiens*, ed. and trans. Annie Jaubert (Paris: Éditions du Cerf, 1971), 20.8. Origen's citation of Clement's letter is a variant reading: his 'impassable' (intransmeabilis or ἀπέρατος), a reading also found in the Syriac version, appears instead of ἀπέραντος ('without limit'): *Épître aux Corinthiens*, p. 135 *n*4. Clement's reference to worlds beyond the ocean is vague; Irenaeus' remark in *Adversus haereses* may be an echo of it: Robert M. Grant, *Irenaeus of Lyons* (London: Routledge, 1997), p. 197 *n*7.

61 *De principiis*, 2.3.6.

62 Lactantius, *Opera omnia*, ed. Samuel Brandt and Georg Laubmann, 2 vols, vol. 1 (Vienna: Bibliopola Academiae Litterarum Caesareae Vindobonensis, 1890): *Divinae Institutiones*, 3.23. The reference to Xenophanes is directly derived from Cicero's *Academicorum priorum* 2.123, where the theory of lunar habitation is attributed to Xenophanes. Anaxagoras, according to Diogenes Laertius, was the actual proponent of this opinion.

63 See Jeffrey Burton Russell, *Inventing the Flat Earth: Columbus and Modern Historians* (New York: Praeger, 1991), pp. 64–5, 72–4.

64 Moretti, *Gli antipodi*, pp. 43–4, where Lactantius' views on the antipodes are compared with the earlier treatments of Lucretius and Plutarch.

65 *Divinae Institutiones*, 3.24.

66 *Divinae Institutiones*, 3.24.

67 *Divinae Institutiones*, 3.24. The association of the theory of the antipodes with the excesses of philosophical reasoning is equally clear in the epitome of the *Divinae Institutiones*, thought to have been compiled c. 320: 'about the antipodes too it is not possible to hear or to talk without a smile, but it is asserted as if a serious thing that we should believe there to be men who have their footsteps opposite ours. Anaxagoras raved more tolerably, when he said that black was white, making not only opinions but even facts laughable' (de antipodis quoque sine risu nec audiri nec dici potest, adseritur tamen quasi aliquid serium, ut credamus esse homines, qui uestigiis nostris habeant aduersa uestigia. tolerabilius Anaxagoras delirauit, qui nigram niuem dixit. quorundam non modo dicta, sed etiam facta ridenda sunt): Lactantius, *Epitome divinarum institutionum*, ed. Eberhard Heck and Antonie Wlosok (Stuttgart: Teubner, 1994), 34.2–3. The reference to Anaxagoras is again a quotation from Cicero's *Academicorum priorum*, 2.72.

68 Moretti, *Gli antipodi*, p. 43.

69 *De civitate dei*, 16.9.

70 Cicero, *Academicorum priorum*, 2.123. My emphases.

Chapter Four

Representing the Unknown: the Antipodes on the World Image

The antipodes did not become a heretical concept following the dicta of Augustine and Lactantius, and they were neither studiously ignored nor rejected out of hand.[1] However, the incompatibility of the idea of the antipodes with the view of Christian history enunciated by Augustine demanded that it be repositioned within the intellectual framework, the knowledge of the nature of things, elaborated by the most influential of early medieval Christian scholars. What emerged was not a single uniform response to the idea of the antipodes but ambiguous ones: instead of denial alone, uncertainty, compromise – and fictionalization. In the immensely influential quasi-encyclopedias of Isidore of Seville and Bede, the antipodes and specifically antipodeans came to be classed within the category of *fabula.* Subsequently, however, the revitalization of the classical tradition of *antoecumenical* spaces, discernable in continental Europe from the ninth century, challenged and complicated straightforward rejection of antipodal habitation. The new energies directed towards the reconstruction and study of classical and late antique literature at this time saw the emergence of Macrobius' *Commentarii* and Martianus Capella's *De nuptiis* as authoritative texts, with serious attention given to their descriptions of the earth and its inhabitants. Most strikingly – and perhaps paradoxically, given its ambiguous status – from the ninth century the idea of the antipodes was expressed in widely disseminated visual form. The maps and diagrams that gloss and illustrate the works of Macrobius and Virgil, as well as maps produced in the contexts of encyclopedic, computistic, and eschatological works, all register the presence of land beyond the ecumene: land that may be represented, but that cannot be known.

The representation of antipodal space within the world image brings out a further dimension of commentary's construction of tradition. It is possible to argue for a visual tradition of representing antipodal space on world maps extending as far back as Macrobius and as far forward as the seventeenth century. However, the transmission of the Macrobian tradition involved not mere copying, but active reconstruction and reshaping of an image either lost or tenuously preserved between the time of its original construction in the fifth century and its reappearance in the tenth. In the case of Virgil manuscripts, it is clear that, rather than the *Aeneid* or the *Georgics*, it was the commentary of Servius that generated a world image designed to illustrate the theory of the division of the heavens and earth into climatic zones. Yet in some manuscripts of the *Georgics* a zonal map appears not as a marginal gloss, but within the space of the poem, implying the integrity of the image with the poem, and thereby helping to blur lines between original and commentary (see fig. 11). Again, textual tradition emerges as something dependent upon commentary, and therefore continually dismantled and remade.

As the product of long-standing cosmological traditions, zonal maps self-evidently expressed theory rather than direct observation of the earth's surface. But here too it is possible to see traditions as fluent and responsive rather than static. From the eleventh century, increasing amounts of information were poured into representations of the ecumene on *mappaemundi*, and it is possible to discern a corresponding impulse to add material to the *antoecumene* on zonal maps. In place of geographical information, the blank space of antipodal *terrae incognitae* became the repository of expository statements. Paratextual material, writing normally located outside the frame of the world map that explained the theoretical basis not only of the antipodes, but of the map as a whole, came to be located in the *antoecumene*. The function of the antipodes on the world image began to emerge, I will suggest, as that of meta-cartographical space – as the part that speaks the whole, expresses its function and rationale. In this regard, it can be argued that the *antoecumene* acted simultaneously as *anti-* and *ante*-ecumene: opposite to the known world, it also preceded it, signifying land itself, the fundamental basis of habitation, and the precondition for cartographic representation.

Representing the unknown

As a component of the world image, the antipodes, and unknown lands generally, were part of medieval visual culture. In terms of geographical representation, that culture was in no way separate from the written word: the world image illustrated, and derived from, verbal description. Nevertheless images could establish their own lines of transmission, and perhaps more significantly they posed a particular set of representational problems. Reasonably clear indications existed, for example, for the representation of the ecumene in the world map in Macrobius' *Commentarii*; but little to nothing was said about what should be shown in *antoecumenical* regions, except that they should not mirror features of the known world. The zonal map invited labelling, the distinction of one zone from another, including hypothesized zones in the southern hemisphere. But labelling raised more questions: should, for example, the southern temperate zone be labelled in terms of its climate ('temperata'), its tropic ('hiemalis' or 'brumalis', i.e. winter), its putative peoples ('temperata antipodum'), or left blank? Each of these possibilities implied opinions about the fundamental questions of the antipodes (did habitable land exist; was it inhabited?), and the validity of such representation. The problem of representation posed by the putative habitation of land opposite to the known world certainly contributed to the vehement rejection of the antipodes by some Christian authors, in part because visual representation enhanced rather than diminished the credibility of the idea. In the course of ridiculing the possibility of the antipodes in his polemical *Christian Topography*, the

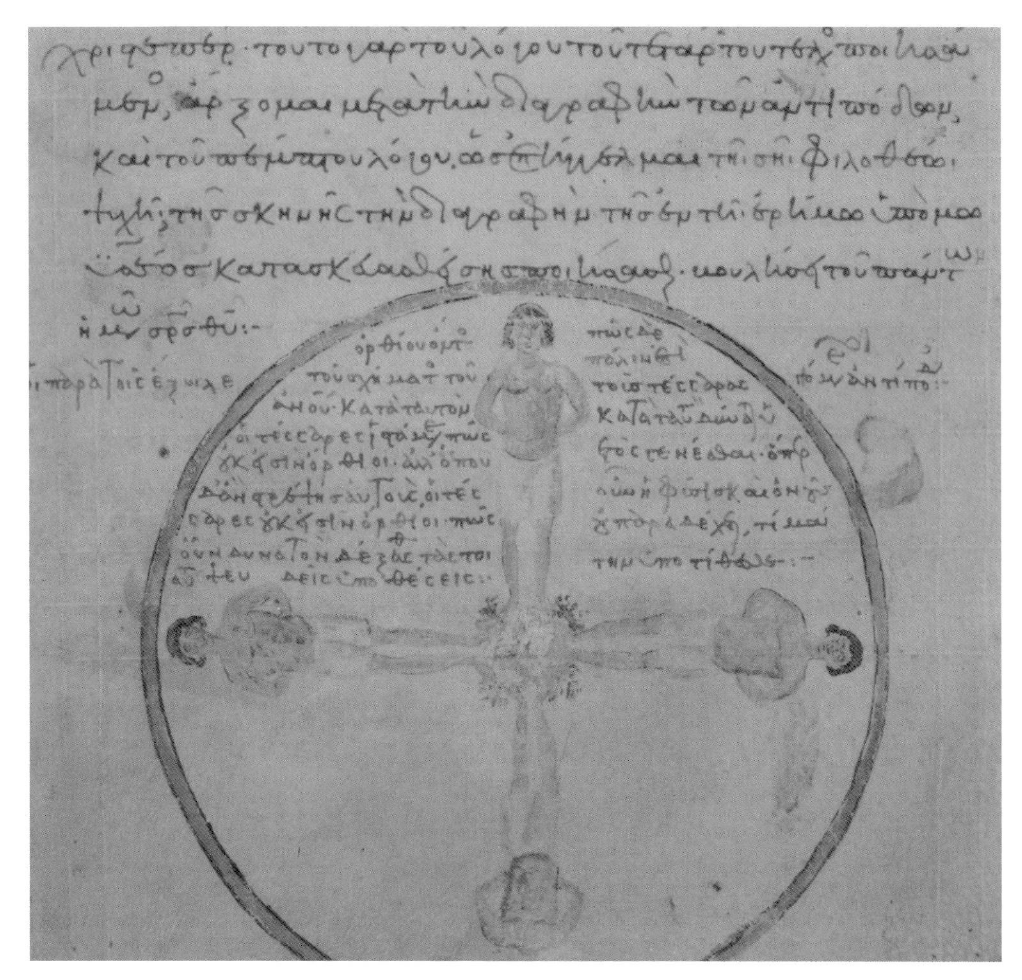

Fig. 7. Florence, Biblioteca Laurenziana, MS Plut. IX.28, f. 98v. Image demonstrating the concept of the antipodes in an eleventh-century copy of Cosmas Indicopleustes, *Christian Topography*. Cosmas mocked the idea of the antipodes, but he appears nevertheless to have illustrated it with a diagram which shows four men standing at the four cardinal points of the globe. Compare with figure 17.

sixth-century Alexandrian merchant Cosmas Indicopleustes illustrated the concept with a diagram which shows four men standing at the four cardinal points of the globe (fig. 7).[2] Cosmas, who denied the sphericity of the earth and instead argued on scriptural authority that it had the form of a tabernacle, intended the image of the antipodes to represent absurdity: the notion of an upside-down man on what he termed the 'antiface' of the earth.[3] However, the image bears similarities to other illustrations of antipodal theory used in texts that supported, rather than opposed, the idea, such as that of the thirteenth-century author Gossouin of Metz (see fig. 17). It is also nearly identical to diagrams that showed personified winds – or occasionally personified seasons – standing at the vertical and horizontal axes of the earth.[4] What may, in one context, have been intended as critique could, in another context, represent a rational exposition of the sphericity of the earth's surface and the distribution of its peoples, winds, and seasons.

Cosmas' work appears to have exerted no influence on writers in the Latin west, where the tradition of the zonal map formed the basis for most visual representations of the antipodes. Here too the variety of contexts in which the zonal map appeared ensured that neither its form nor function were static. *Antoecumenical* regions appear on the images that illustrated Macrobius' *Commentarii in Somnium Scipionis*, in accordance with the instructions given by him for drawing a zonal diagram (2.5.13–17), a diagram of the terrestrial and celestial zones (2.7.4–12), and a world map showing the division and flows of Ocean (2.9.7–9). Prior to the twelfth century, zonal diagrams of varying levels of detail also appeared as glosses to Virgil's *Georgics*, in manuscripts of Calcidius' translation and commentary on Plato's *Timaeus*,[5] in one instance only in a Martianus Capella manuscript,[6] and largely independent of a written text in contexts associated with the *computus*, a set of tables used for the calculation of the Christian calendar.[7] Clear distinctions can be made between what may be termed Macrobian, Virgilian, and Isidorean strands of *antoecumenical* representation. Influence unquestionably exists between each of these strands. Macrobius' *Commentarii* included an attempt to harmonize Virgil and Cicero, and was simultaneously a response to Servius' commentary; Isidore's sources included both Servius and Macrobius; in turn, elements of Isidorean commentary could be used to gloss the world image in Macrobius. On the other hand, maps in Virgil manuscripts gloss a literary text, and thereby acquired a particular relationship with poetic structure and content, whereas in the Macrobian tradition the world map is integral to a philosophical-political treatise. The Isidorean tradition is more diffuse, but two contexts can be readily identified: scientific and encyclopedic material, and Christian commentary. It is in maps of this variety that the potential for unknown parts of the world to be used as meta-cartographical space first began to be realized.

The map of Macrobius

The rise in significance and popularity of Macrobius' *Commentarii* from the ninth to the twelfth centuries can be gauged from the tally of surviving manuscripts of the text. Prior to the ninth century the *Commentarii* appear to have been copied in the British isles, and from there re-introduced to the continent – a process perhaps visible in an eighth-century manuscript from the monastery of Bobbio which contains extracts from Macrobius written in Insular script.[8] Seven manuscripts and fragments can be reasonably securely dated to the ninth century, most – perhaps all – copied in French monasteries. Only eight date from the tenth century, although the geographical spread is broader, extending to the monasteries in the south of Germany and modern-day Switzerland. A surge of interest in the text seems to have occurred in the eleventh century, with thirty-one surviving exemplars, and the peak of production was reached in the twelfth century, from which 106 manuscripts survive. Thereafter the numbers tail off: twenty-eight manuscripts of the *Commentarii* from the thirteenth century, and just eleven from the fourteenth, before humanist interest in classical texts prompted a rise to forty from the fifteenth century.[9]

The world map that illustrates Macrobius' *Commentarii* exists in many different states, in thirty-five manuscripts prior to 1100, and around 150 manuscripts overall, up to and including the fifteenth century.[10] It is perhaps impossible categorically to identify the earliest state of the image; however, certain observations can be made about the earliest surviving exemplars, and it is possible to identify something resembling a standard form of the map. The basis of the image is a relatively schematic representation of the relationship of ocean to land in the northern hemisphere, as well as representation of the five zones (northern and southern frigid, northern and southern temperate, central torrid or 'perusta', i.e. 'burned up'), an equatorial ocean, and indication of the direction of ocean flows ('refusiones') from the equator to the poles by means of four inscriptions around the outside of the map (see fig. 2). The vast majority of pre-twelfth-century Macrobius maps contain some representation of the Red, Indian, and Caspian seas, the latter usually represented as an inlet from the outer, encircling Ocean. A small number of toponyms – usually just 'Italia' and 'Orcades' – are marked in the northern hemisphere. Intriguingly, some maps (represented by at least four pre-1100 exemplars) include the toponyms Syene and Meroë, respectively at the northern tropic and in the torrid zone above the equator.[11] These cities marked the first two of the ancient *climata*, divisions of the northern hemisphere into seven latitudinal bands. As the locations of these cities are discussed by Macrobius in the surrounding text (2.8.3), their appearance on the map could be an original feature. Copyists of the world map in manuscripts of Macrobius' *Commentarii* after 1100 tended to increase the amount of geographical information in the northern hemisphere. One twelfth-century form of the map dispenses with representation of the seas, adding Spain, the Alps, Egypt and Asia to the earlier version's 'Italia', as well as a selection of Mediterranean islands: 'Gades', 'Balearia', Sardinia, and Sicily.[12] Others followed the innovations made by adaptors such as William of Conches: in these versions the westernmost extent of Europe and Africa is represented by the toponyms 'Calpes' (one of the fabled columns of Hercules) and 'Atlas' (i.e. Mount Atlas), and the standard aquatic divisions between the three parts of the known world – the Tanais, Nile, and Mediterranean – also become more frequent.[13] By contrast, at first glance very few adaptations appear to have occurred to *antoecumenical* space on the image over the course of its transmission. Whether a large or small number of toponyms were used to locate the European reader in 'our temperate' region, the temperate zone of the southern hemisphere appeared strikingly monolithic, unmarked by any representation of seas or land formations, in apposition to the ecumene's geographical detail. A close study of *terrae incognitae* on the maps of Macrobius reveals three ways in which the *antoecumene* could, nevertheless, be subject to subtle changes of emphasis and meaning: through the variety of nomenclature used to describe the unknown inhabitants of spaces beyond the known world; through the function of the image as a means of invoking and representing debate about such habitation; and through the technical innovation of certain copyists faced with the difficulty of representing places and peoples on the other side of the earth.

The most frequently used description of the southern temperate zone on Macrobius maps prior to 1100 is 'temperata antecorum' (temperate [zone] of the antoikoi), or some variant thereof (antoecorum, antechorum, antetorum); more rarely the zone is marked 'temperata antipodum' (temperate zone of the antipodeans),[14] in one instance 'temperata antyrtorum',[15] and in another 'temperata antiktorum' (temperate zone of the antikthones; 'antyrtorum' is possibly a corruption of 'antiktorum' or 'antetorum').[16] In eight instances in surviving pre-1100 manuscripts the southern temperate zone is referred to not in terms of its inhabitants, but simply designated 'temperata', 'temperata zona', 'temperata australis', or 'habitabilis'.[17] In only one pre-twelfth-century exemplar is the entire southern hemisphere left blank.[18] In no cases are topographical features marked in the southern temperate zone, a reticence in accordance with Macrobius' instruction to refrain from depicting hypothesized but unproven details in the southern hemisphere.[19]

To what extent did medieval scribes make conscious choices in their selection of *antoecumenical* terminology? Two examples from manuscripts of the *Commentarii* suggest something of the confusion, or at least fluidity, within scribal copying practices. On the world map in the eleventh-century manuscript BL, Harley 2772, the temperate southern zone of Macrobius' map is marked 'temperata antetorvm' (i.e. temperate zone of the antoikoi) (fig. 8).[20] Two folios previously, the scribe, trying to preserve Greek letters and possibly flummoxed by the text's terminology, initially rendered Macrobius' 'rursus illos ab antoecis suis' (2.5.33) as 'rursus illos ab ARCTICTIC suis'. However, 'ARCTICTIC' has been corrected (in superscript) to 'antoetis' (f. 68v). Similarly, a marginal correction has replaced 'antiktos' (in Macrobius' 'quos greci αντοίκους uocant': 'whom the Greeks call *antoikoi*') with 'antoetos'. Precisely the opposite correction was made in a late tenth- or early eleventh-century south German manuscript of the *Commentarii* (not apparently related to Harley 2772): there 'quos greci anteecos uocant' has been corrected to 'antiktos', and the inscription on the southern temperate zone of its map reads 'temperata antiktorvm' (fig. 2).[21] The significance of the substitution of the term 'antoikoi' with 'antikthones' to form 'temperata antiktorum' is that the latter is usually used to describe those in the southern and lower hemisphere (i.e. diametrically opposed to inhabitants of the known world), whereas 'antoikoi' was usually supposed to refer to inhabitants in the southern and 'upper' hemisphere (on the same longitude as inhabitants of the ecumene). Clearly many scribes were simply copying exemplars; others, comparing one state of the text with another, attempted to correct and to rationalize. But still others were aware that different terms for antipodal inhabitants signified distinct, if related, concepts.

The issue of *antoecumenical* nomenclature was addressed directly in the ninth-century glosses on Martianus Capella's *De nuptiis* attributed to the philosopher John Scottus Eriugena. Eriugena's glosses were subsequently augmented by Remigius of Auxerre (841–c. 908), a member of the 'school of Auxerre' and therefore one of the key figures in the dissemination of both Macrobius' *Commentarii* and Martianus' *De nuptiis*.[22] The comments of Eriugena and Remigius on Geometria's speech in book six of Martianus' work include the fullest discussion of the terminology of antipodal spaces prior to the twelfth century. Eriugena and Remigius were particularly concerned to explain the terminology of Greek origin used by Martianus, from 'geometria' itself to 'antikthones'. In so doing they reinforced the classical idea of multiple *antoecumenical* habitation: 'Antikthones are those who possess the opposite part of the earth, for cthonos means of the earth. Therefore there are four, that is, we and our antipodeans and the antoikoi and the

Fig. 8. London, BL, MS Harley 2772, f. 70v. World map in an eleventh-century manuscript of Macrobius, *Commentarii in Somnium Scipionis* (Commentary on the Dream of Scipio). North at the top. The frigid zones are marked uninhabitable ('inhabitabilis'). The northern hemisphere shows a schematic representation of the ecumene ('temperata nostra'), with the Orcades and Italia marked, and representation of the Mediterranean, Adriatic, and an umbrella-shaped Palus Maeotis-Tanais (Sea of Azov/Don river) confluence. The Red ('rvbrvm') and Indian ('indicvm') seas are shown as indentations from the equatorial ocean, while the Caspian sea ('mare caspium') appears at the far north-east of the ecumene as an inlet of the outer Ocean. The temperate southern zone is marked 'temperata antetorvm', instead of the more usual 'temperata antecorum' (i.e. 'of the antoikoi', inhabitants beneath the *oikoumenē*).

antikthones' (ΑΝΤΙΧΘΟΝΕΣ dicuntur qui contrariam terre partem possident. χθονόσ enim terrae. Ergo iiii sunt, id est nos et nostri antipodes et antoikoi et antikthones).[23] Eriugena's commentary attempted to clarify further, although not always with success, Martianus' rather brisk assignation of unknown parts of the world. Eriugena, and Remigius after him, located the 'antipodes' 'against us beneath the earth' (contra nos sub terra), i.e. in the 'underside' of the northern hemisphere, and 'antoikoi' and 'antikthones' in the southern hemisphere. The antoikoi dwell on the same longitude as those in the ecumene, but are separated by the torrid zone, while the antikthones inhabit the 'lower' hemisphere (i.e. they are in the southern hemisphere on the same longitude as the northern-hemisphere antipodeans).[24] In addition, Eriugena and Remigius clarified Martianus' potentially confusing discussion of dwellers at the equator, and tackled the difficult question of the length of days and nights enjoyed by people living in unknown places.[25]

The combination of different sources of authority, and the textual layering of the commentary tradition (in which commentary explains commentary) is evident in a late tenth- or early eleventh-century manuscript of Macrobius' *Commentarii* written at Echternach. Here one scribe used Martianus' *De nuptiis*, via Remigius' commentary on it, to gloss the terms 'antoikoi', 'antipodes' and 'antikthones':

> Ex Martiano. Sicut nos antipodes illos qui sunt in brumali circulo habemus subtus nos positos illi ita alios antipodes subtus terram constitutos habent. Antipodes dicuntur nobis qui pedes suos pedibus nostris obuersos habent. ANTYKOI, id est: contra nos habitantes. Nam OYKOC domus dicitur, inde 'antoykoc' id est: contraria domus, id est: contrarii habitatores contra nos qui sunt terrestres. IKTONOC terra dicitur, inde 'antiktones' dicuntur qui contrariam terrae partem habitant, id est: contra terranes.[26]

> From Martianus. Just as we have the antipodeans who are in the southern circle beneath us, they in turn have other antipodeans located beneath the earth. Our antipodeans are so called because they have feet obverse to our feet. ANTYKOI, that is: living against us. For OYKOC means home, thus 'antoycoc' means opposite to the home, that is contrary dwellers on the earth who are opposite to us. IKTONOC means earth, thus 'antiktones' means those who inhabit the contrary part of the earth, that is: counter-earthly.

Etymologies reduce the issue of *antoecumenical* peoples to the basics of the known world: home, feet, land. Yet in this case that etymological movement to the origin meets its antithesis: etymologies lead home, but these unknown peoples live opposite to home, opposite to earth. And simultaneously their home is a reversal of our home, our feet, our land. Crucially, the nature of the relationship between inhabitants of known and unknown worlds expressed by such etymologies could also be given the clarity of visual expression on the map. Equally crucially, representation of unknown and unknowable peoples courted controversy, because images that adverted to antipodal populations could not be viewed outside of the context of the medieval debate about the acceptability of a doctrine that posited habitation beyond the reach of Christianity. For this reason the map of Macrobius came to act not simply as a means of explicating the words of Cicero, but as the target of disagreement and even polemic.

The clearest expression of opposition to the Macrobian description of the world and its inhabitants was provided in the late eleventh-century *Liber contra Wolfelmum* of Manegold of

Lautenbach, a work that attempts to reveal the wickedness and heresy contained in pagan philosophy in general, and more particularly in Neoplatonism.[27] Addressed to Wolfelmus of Cologne (Wolfhelm, Abbot of Brauweiler from 1065 to 1091), whom Manegold characterizes as a proponent of Neoplatonism, the *Liber contra Wolfelmum* identifies as 'dangerous' faith in the Macrobian division of the world into four habitable regions – Manegold uses the Ciceronian word 'maculae', spots or patches – and the consequent existence of 'antipodeans or antoecians'.[28] In opposition to this theory Manegold asks two questions: how could Christ have come to earth to save the entire human race if three races of men were excluded? And how could the faith be disseminated to the ends of the earth, as scripture (Is. 52.10) demanded, 'if the other ends of the earth are inhabited by men to whom nature forbids the voices of our prophets and apostles to pass, by means of impassable lengths of waters, or cold or hot spaces' (si aliqui fines terre sunt ab hominibus inhabitati, ad quos sonus prophetarum et apostolorum nostrorum prohibente natura per inaccessibiles aquarum, frigorum calorumve distantias transire nequivit)?[29] Objections of this kind were made from time to time in manuscripts of the *Commentarii*, with authorities cited in direct contradiction of Macrobius' views. In a south German manuscript written around 1000, for example, Macrobius' passage defending the possibility of habitation in the southern temperate zone at 2.5.22 of the *Commentarii* is accompanied by the gloss: 'Augustine speaks against this in book 16, *De civitate dei*' (Contra id loquitur Augustinus in VIX liber de ciuitate dei).[30] Similarly, the diagram of celestial and terrestrial zones in the twelfth-century manuscript Paris, BNF Nal. 923, echoes Augustine (and Isidore) in describing the habitable zone in the southern hemisphere as 'temperate circle of land which southerners are mythically supposed to inhabit' (temperatus circulus terre quem australes inhabitare fabulose iactantur).[31] Such juxtaposition indicates the ways in which manuscript culture allowed for the confrontation of divergent world views. On the pages of manuscripts, authorities could be played off against each other; as a result of these interpolated comments, Macrobius was read through the frame of Augustine.

Precisely the same gesture of contradiction – but this time on the map itself – was enacted in Paris, BNF lat. 6622, a twelfth-century manuscript of the *Commentarii* produced in Germany or north-east France. Macrobius' call for a world map at 2.9.7–8 was illustrated in this instance by a relatively standard copy of the world map in which the temperate zone of the southern hemisphere was designated 'temperata antoecorum' (temperate zone of the antoikoi).[32] But to these words a later hand, probably of the fifteenth century, has added the inscription: 'quod hic sint homines habitantes reprobat Augustinus 16 libro de ciuitate dei' (that there may be men living here Augustine rejects, Book 16, *De civitate dei*).[33] This type of intervention on the map is highly unusual, but it exemplifies the endemic nature of controversy about habitation of antipodal regions, and suggests the often sensitive nature of visual representation. The fifteenth-century annotator here does not object to the representation of a temperate zone *per se*, but specifically to the word 'antoecorum': it is the designation of the zone as inhabited that draws the citation of Augustine.

The juxtaposition of Augustine alongside Macrobius responded to the theological challenge posed by the representation of *terra incognita* beyond the reach of humanity. Yet there was another, more practical, difficulty faced by copyists of zonal diagrams: how could one portray on a flat surface zones that theoretically encircle the entire world? A small but significant number of manuscripts of Macrobius' *Commentarii* preserve interesting and innovative attempts to escape this difficulty. These manuscripts contain a zonal diagram that shows antipodal habitation in the western, rather than southern, hemisphere (fig. 9). On this image a north to south zonal division still pertains, with the two temperate zones marked ('temperata') on either side of the central torrid zone ('perusta'). However, a vertical line divides the sphere in two: on the right, two inscriptions identify, firstly, the 'upper' surface of the earth ('hec est superior superficies terre'), and secondly the ecumene, marked 'hic habitamus' (here we live). To the left of the vertical axis is antipodal land: the long vertical inscription identifies the 'lower' surface of the earth ('hec est inferior superficies terre'), and the region inhabited by antipodeans ('hic antipodes'). Symmetry has been maintained: habitation is only marked in the temperate zone of the northern hemisphere; the vertical inscriptions have simply added the upper-lower binary to the north-south divide. In Paris, BNF lat. 18421, a twelfth-century manuscript probably of French provenance, a gloss in the right margin explains the schema of the diagram, perhaps in the knowledge that readers used to the standard form of the zonal diagram may be confused: 'do not think that this shows only the upper ambit of the earth, but rather the whole sphere both above and below, for otherwise it cannot be shown on a plane surface' (Non intellige superiorem tantum ambitum terre hic esse depictum sed totam speram terre et supra et infra sed aliter non potuit in plano depingi).[34] Another version of the diagram manages to indicate all four inhabited regions: an eleventh-century Italian manuscript (BAV MS Ottob. lat. 1939) illustrates the *Commentarii* at 2.5.12 with a diagram similarly divided into eastern and western as well as northern and southern hemispheres. The polar axis separates 'Antipodum habitabilis' (the habitable region of the antipodeans) in the northern hemisphere from 'nostra habitabilis' (our habitable); in the southern hemisphere this disposition is mirrored by 'Antipodum' (of the antipodeans) in the west and 'nostra australis' (our southern) in the east.[35]

The appearance of these diagrams in manuscripts of the *Commentarii* undoubtedly constituted an innovation within the corpus of diagrams that accompanied Macrobius' text. They acted as a substitute not for the world map (i.e. the illustration of 2.9.7–8 of the *Commentarii*), but for the zonal diagram about which Macrobius gives instructions at 2.5.13–14. In these instructions, Macrobius does distinguish between 'upper' (superior) and 'lower' (inferior) regions, but only to distinguish northern from southern temperate zones.[36] In applying this division to eastern and western hemispheres, the revised image did not contradict antipodal theory, which posited just such an antipodal region in the northern hemisphere, along with the two regions of the southern-hemisphere *antoecumene*. But it does testify to an element of fluidity in the interpretation of Macrobius' instructions. It seems to have arisen from dissatisfaction with the traditional zonal map's inability to show the lower or western hemisphere, and therefore its tendency to posit antipodal habitation only in the southern hemisphere. Such attempts to readjust and infiltrate the binary between a northern 'we' and a southern race of unknown men were not confined to Macrobius manuscripts.

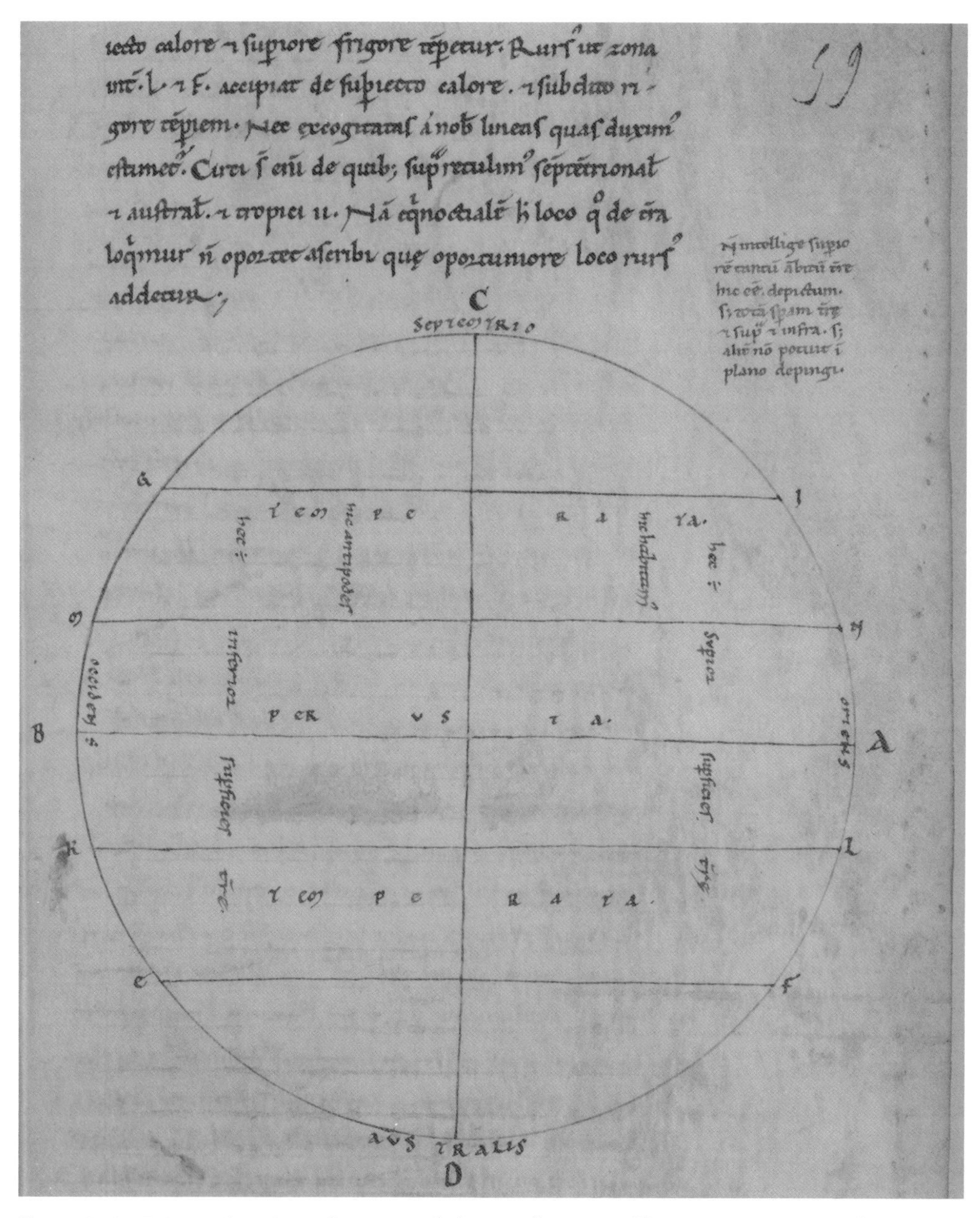

Fig. 9. Paris, BNF, MS lat. 18421, f. 59r. Zonal diagram from a twelfth-century manuscript of Macrobius, *Commentarii in Somnium Scipionis* (Commentary on the Dream of Scipio), showing eastern and western hemispheres. North at the top. The cardinal points are marked, as is the division between temperate zones in the north and south ('temperata') and the central torrid zone ('perusta'). The scribe has attempted to indicate a quadripartite division of the earth by marking an eastern and a western hemisphere on either side of a vertical axis. On the right, two inscriptions identify, firstly, the 'upper' surface of the earth ('hec est superior superficies terre'), and secondly the ecumene, marked 'hic habitamus' (here we live). To the left of the vertical axis is antipodal land: the long vertical inscription identifies the 'lower' surface of the earth ('hec est inferior superficies terre'), and the region inhabited by antipodeans ('hic antipodes'). A gloss in the right margin assures the reader that the image shows 'the whole sphere both above and below'.

The zonal map between Servius and Virgil: from gloss to centrality

There is no tradition of zonal map illustration evident in late antique manuscripts of Virgil's poetry; the earliest surviving example of the use of a zonal map to gloss *Georgics* 1.233 appears in a ninth-century manuscript, and thereafter the zonal map seems to have become an infrequent but not unusual means of illustrating the poem. It is likely that the commentaries of Servius and others played a pivotal role in initiating this visual tradition. While there is little indication that Servius' commentary originally contained a zonal diagram (unlike Macrobius' *Commentarii* there are no instructions for the construction of images), it seems probable that at some relatively early point in its transmission scribes began to add diagrams to explicate Servius' discussion of zonal theory in the *Georgics*. It is notable that even in manuscripts where they appear independently of Virgil's texts, Servius' commentaries were occasionally illustrated with zonal images.[37] On the other hand, in some manuscripts of the *Georgics* the world image has 'migrated' from the accompanying marginal commentary of Servius to take up a central position on the page: within, not merely alongside, Virgil's poetry. The diagram emerged from Servius' commentary; it was itself an act of commentary; nevertheless, by a curious paradox, in some instances it obtained a greater proximity to the text than the commentary – becoming, in certain manuscripts, conceptually poised between the two.

The function of uniting the celestial and the terrestrial in the zonal diagram is evident in a number of world images in Virgil manuscripts from the eleventh and twelfth centuries. The two zonal diagrams that illustrate *Georgics* 1.233 in Paris, BNF, MS lat. 7930, an eleventh-century manuscript of French origin (fig. 10), express the contiguity between earth and sky. The upper, terrestrial, diagram marks the zones in terms of human habitation, identifies both tropics, and shows the path of the zodiac; the lower, celestial, diagram represents the same scheme, but instead of nomenclature the zones at the extremities and centre are coloured (brown for the frigid zones, orange for the torrid). The temperate zones are filled with the appropriate zodiacal signs: Taurus, Gemini, Cancer, Leo, and Virgo north of equator; Pisces, Aquarius, Capricorn, Aries, and Sagittarius south of equator (Libra and Scorpio are omitted). The diagrams, in this instance, seem to provide a visual gloss to the verbal explanation that surrounds them. The marginal text above the diagrams makes explicit the connection between terrestrial (Ciceronian) and celestial (Virgilian) zonal division: 'these are the five zones which Virgil says are in the sky and Cicero on land' (Has V zonas quas Virgilius dicit esse in celo Ciciro dicit esse in terris) (f. 28v). The gloss goes on to name the five zones (septentrionalis, solstitialis, equinoctialis, brumalis, australis) and to explain that two of these only are habitable:

> Solsticialem namque inhabitamus nos brumalem uero ut dicunt nostri antipodes inhabitant. Ad quos prohibente perusta plaga accedere non possumus. Dicuntur autem antipodes contrariis pedibus nam anti contra pos pes inde antipodes dicuntur eoquidem contra nos uersis uestigiis incedunt.
>
> For we inhabit the zone of the summer solstice, while, as they say, our antipodeans inhabit the zone of the winter solstice. To whom we are not able to approach because the burnt region prevents us. They are called antipodeans, with feet opposed, because anti – contra, pos – pes, so they are called antipodeans because they proceed with footsteps turned against us.

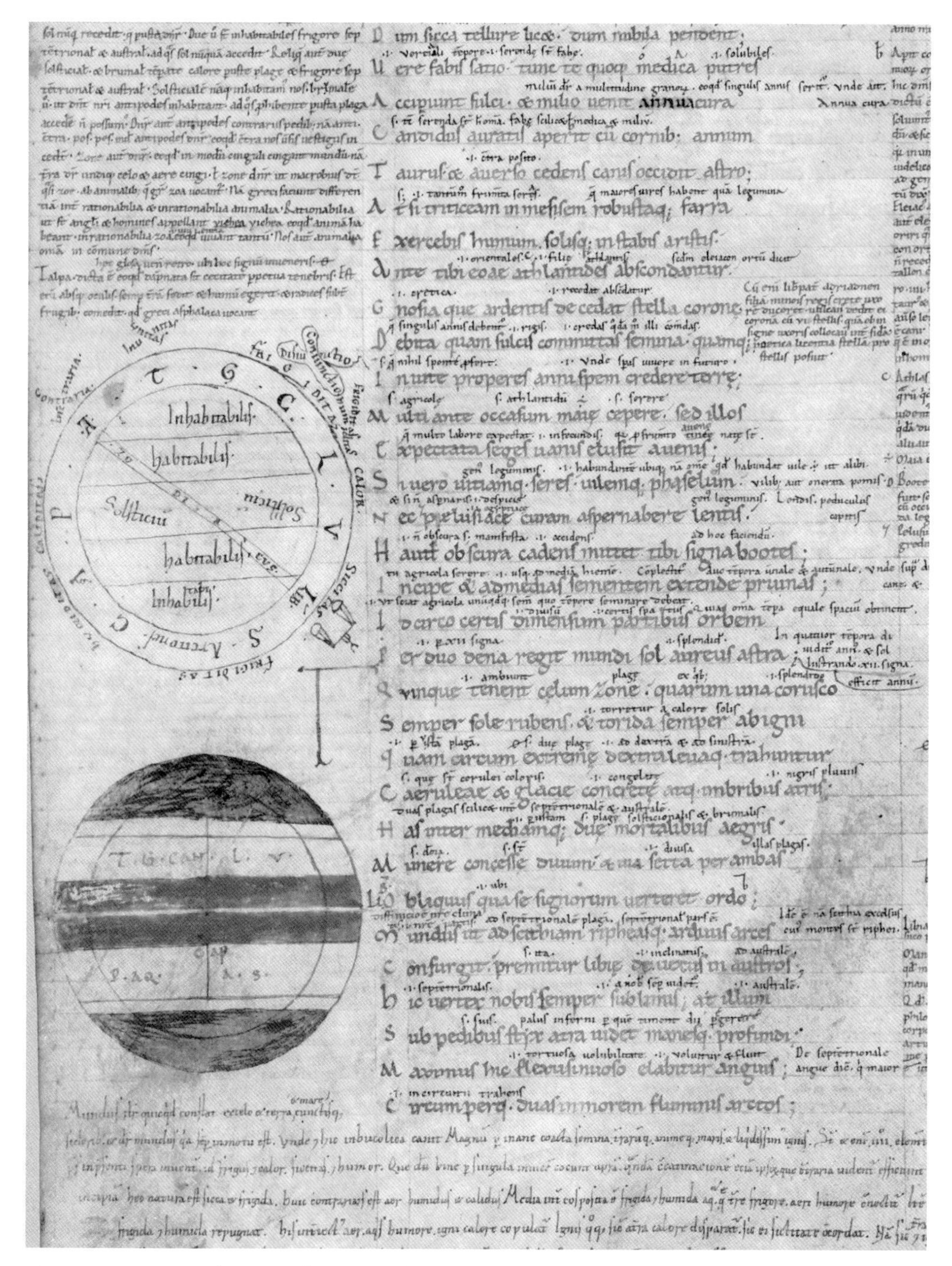

Fig. 10. Paris, BNF, MS lat. 7930, f. 28v. Zonal diagrams illustrating an eleventh-century manuscript of Virgil's *Georgics*. North at the top. In the upper, terrestrial, diagram, the zones are marked in terms of human habitation, both tropics identified, and the path of the zodiac shown; in the lower, celestial, diagram the extremities and centre are coloured (brown for the frigid zones, orange for the torrid). The temperate zones are filled with the zodiacal signs: Taurus, Gemini, Cancer, Leo, and Virgo north of equator; Pisces, Aquarius, Capricorn, Aries, and Sagittarius south of equator (Libra and Scorpio are omitted).

The function of the world image to explicate oblique poetic references is explicitly stated in the rubric above a zonal diagram in another, twelfth-century, Virgil manuscript: 'here, reader, the five zones are drawn for you lest the traditions of Virgil be closed to you' (hic lector zonae depinguntur tibi quinque ne sint clausa tibi tradita Virgilii).[38] However, in the manuscripts where the zonal diagram has moved from margin to centre of the page to occupy the position of the poem rather than the commentary, the visual prominence of the image gives it the impression of instigating rather than responding to Virgil's words. There is no better example of this than the zonal diagram in London, BL, MS Harley 2533. In this twelfth-century manuscript, probably of Italian provenance, a coloured diagram appears directly above *Georgics* 1.233 (f. 18v) (fig. 11). Five zones are marked off: the outer two (inscribed respectively 'Septentrionalis' and 'Australis') are coloured green, the hot central zone red.[39] The initial letters of zodiacal signs are marked in a diagonal band that runs between northern ('Solsticionalis') and the southern ('Brumalis') temperate zones. Alongside the diagram and running down the page appear glosses from Servius' commentary, including his discussion of the antipodes and the location of the Styx. A hand that is possibly that of the principal scribe has written next to the map 'ille sunt zone celi' (these are the zones of the sky), and, next to the southernmost point of the zodiac, 'Quo sidere terram uertere' (beneath which star [] the earth turns), a partial quotation of the first two lines of the poem. The diagram connects poem with commentary, but also one section of the poem with another, mapping heaven, earth, and the *Georgics* itself.

Quite clearly one reason for supplying Virgil's poem with an image of the zones was that the concept of the antipodes demanded visual as well as verbal explication. It was desirable to see the relationship of the zones – temperate, torrid and frigid – to each other, and consequently to give visual meaning to the relationship between inhabitants of the ecumene and antipodeans. In zonal diagrams outside of the Macrobian tradition there is little or no attempt to represent topographical features of the known world. Symmetries are preserved, and the theoretical nature of the image undiluted. In the Macrobian illustrations the combination of ecumenical map, however schematic, with *antoecumenical* space created obvious differences of function even within the world image. Those differences – between experience and theory, reality and hypothesis, passage and barriers – are particularly apparent on maps that owe their representation of the *antoecumene* neither directly to Macrobius nor Virgil, but instead emerge from the traditions of the *computus*, and from Isidore of Seville's description of a 'fourth part of the world'.

Isidore: the antipodes as *fabula*

In order to address the function of the antipodes on maps that derive to at least some degree from the work of Isidore (c. 560–636) it is necessary to consider the semiotic multiplicity of the antipodes in his writings. In his *Etymologiae*, Isidore – bishop, encyclopedist and historian in Visigothic Spain, and therefore heir to a rich inheritance of ancient culture – used the Latin language as the structure from which classical learning could be unfolded. The antipodes find their place at three different points within this structure, as addenda to the fields of ethnography

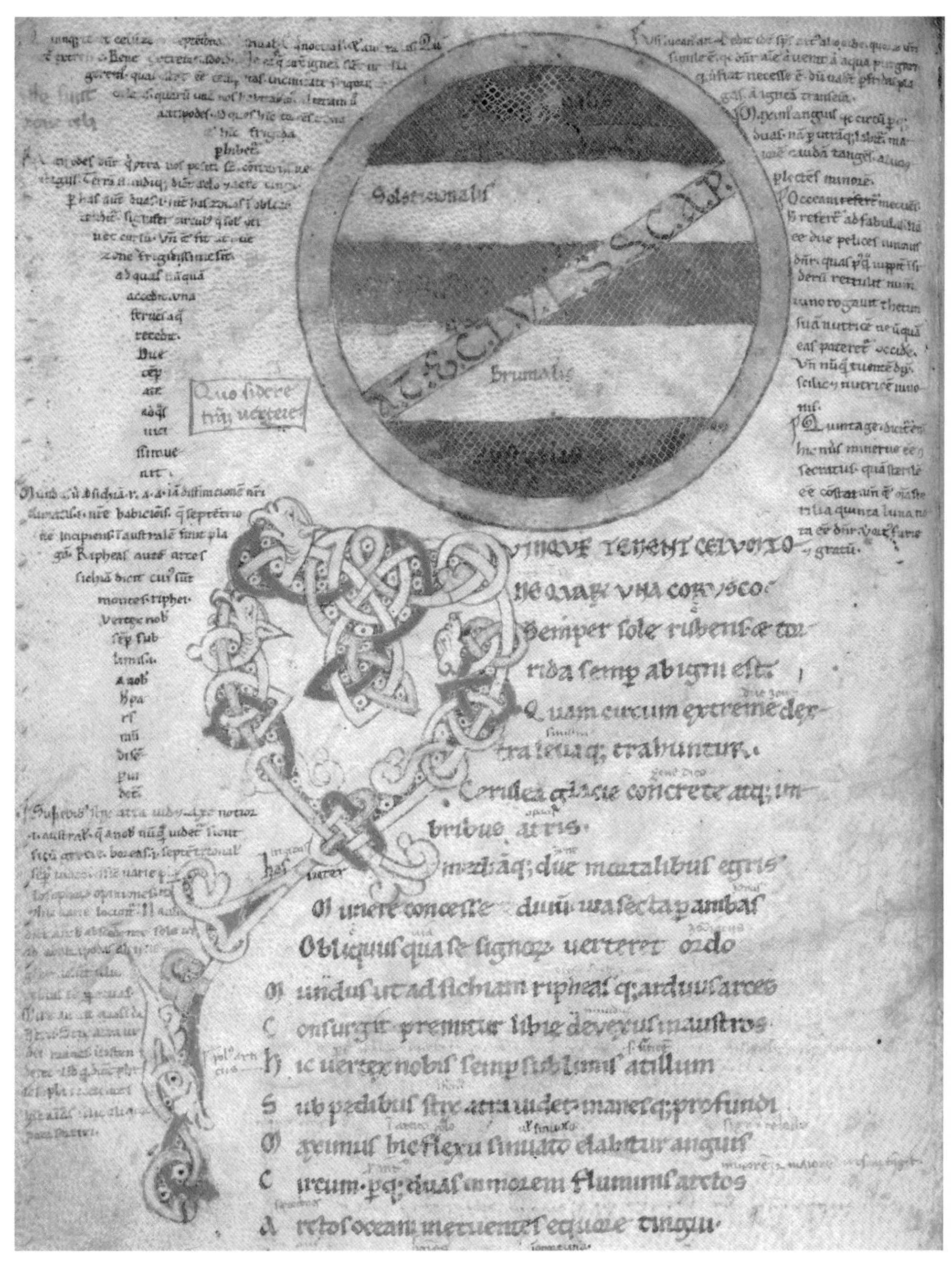

Fig. 11. London, BL, MS Harley 2533, f. 18v. Diagram of the celestial zones illustrating a twelfth-century manuscript of Virgil's *Georgics*. The diagram appears directly above *Georgics* 1.233: 'Quinque tenent celum zone …'. The diagram is surrounded by glosses from Servius' commentary on the *Georgics*. It is divided into five zones (marked [septentrio]nalis; Solsticionalis; [illeg.]; Brumalis; Australis), and the ecliptic (marked with the initial letters of the zodiacal signs). In a rectangular box to the immediate left of the image a scribe has written 'Quo sidere terram uertere' (*Georgics*, 1.1–2); above this inscription, also to the left of the image, appear the words 'ille sunt zone celi' (these are the zones of the sky).

and geography. Isidore begins book 14 of the *Etymologiae*, 'De terra et partibus', with the standard tripartite division of the known world into the continents of Europe, Africa, and Asia. However, in his description of Africa ('Libya') later in the chapter, Isidore mentions a 'fourth part of the world, across Ocean deeper in the south, unknown to us due to the heat of the sun, in the extremities of which the antipodeans are said in fable to live' (Extra tres autem partes orbis quarta pars trans Oceanum interior est in meridie, quae solis ardore incognita nobis est; in cuius finibus Antipodes fabulose inhabitare produntur).[40] The immediate context of this remark is a description of Ethiopia, notable for its proximity to the sun, the blackness of its inhabitants, and its beasts (e.g. rhinoceros, camelopard, dragons). Having identified the location of Ethiopia between the Atlas mountains in the west and Egypt in the east, and between the Nile in the north, and the equatorial Ocean in the south, Isidore adverts to the classical tradition (ultimately derived from Homer's *Odyssey*) according to which there were two Ethiopias, 'one to the east, the other to the west in Mauretania'. At this point he refers to the unknown 'fourth part of the world'. Given the context of this reference in a description of the southernmost extent of Africa, and the connection (made, for example, in the work of the first-century AD geographer Pomponius Mela) between the two Ethiopias and land inhabited by antipodeans or the 'antikthones', it is likely that Isidore conceived of the fourth part of the world and its inhabitants as existing to the south of the equatorial ocean. By designating the antipodeans as fictitious ('said in fable to live') Isidore manages both to retail and to deny fable, acknowledging the potential role of the unknown 'fourth' part of the world as a site for fantasy, and representation of the fantastic, even as he attempts to close down this possibility and insist upon a reiteration of ignorance of the region. The burning sun, which prevents passage to this quarter of the world, also prevents knowledge. Unknowability, poetic fictionality, is aligned with *ardor*, so that the intensity of unbearable heat signifies the absence of human experience: what is known is intensity, the barrier, the fable.

Other references to antipodeans and *antoecumenical* spaces in the writings of Isidore confirm that he was juggling a number of different cosmographic models. In his *De natura rerum*, Isidore accepted the division of the world into zones, illustrating the concept by comparing the five circles or zones to the fingers on his right hand: the thumb is the Arctic circle; the index finger northern, temperate and habitable; the middle finger torrid and uninhabitable; the fourth southern, temperate and habitable; and the little finger the Antarctic circle.[41] Yet in the *Etymologiae* he does not locate the antipodeans in the southern temperate zone (nor in the underside of the northern temperate zone); instead, his description of a 'fourth part of the world' suggests an extension of the three known parts. Neither Macrobius nor Martianus spoke of antipodal regions in terms of a fourth part of the world; nor did they share Isidore's suspicion about the existence and the humanity of antipodal inhabitants. On this matter, though, Isidore left conflicting evidence of his opinions. In book 9 of the *Etymologiae*, in the course of a list of the peoples who inhabit the earth he explicitly rejected the notion of the existence of a race of 'antipodeans' who plant their feet in the opposite direction to those in the known world:

> Iam vero hi qui Antipodae dicuntur, eo quod contrarii esse vestigiis nostris putantur, ut quasi sub terris positi adversa pedibus nostris calcent vestigia, nulla ratione credendum est, quia nec soliditas

> patitur, nec centrum terrae; sed neque hoc ulla historiae cognitione firmatur, sed hoc poetae quasi ratiocinando coniectant.[42]
>
> Now those who are called antipodeans because they are thought to stand opposite to our footsteps, as if placed beneath the earth with footsteps treading against our feet, there is no reason for belief in them. For neither the solidity nor the centre of the earth permits it. And neither is any of this supported by the authority of history: rather, poets, as if by calculation, conjecture it.

The passage is derived from Augustine's '[homines] adversa pedibus nostris calcare vestigia' (men who plant their footsteps opposite ours), with the addition, probably from Servius (*ad Georg.* 1.235), of the more ambiguous 'quasi sub terris positi' (as if placed beneath the earth) in place of 'homines a contraria parte terrae' (men from the opposite part of the earth).[43] Isidore repeats Augustine's judgement that 'nulla ratione credendum est' (there is no reason for belief), although with an additional reference, drawn from Servius' comments on *Aeneid* 6.127, to the physical impossibility of antipodal existence ('nec soliditas patitur, nec centrum'). There is also an echo of Servius' comment that Virgil inserted the matter of philosophy 'by poetic licence'. The theme of ethnographic fictionality is continued in the next entry on Isidore's list in book nine, the Titans, described as people 'whom fables say were created from the enraged earth to wreak vengeance on the gods' (quos ferunt fabulae ab irata contra deos terra ad ejus ultionem creatos).[44] The antipodeans and the Titans follow on from a series of races in the most remote parts of Africa (Trochodytae, Pamphagi [omnivores], Icthyophagi [fish-eaters], Anthropophagi [men-eaters]) and together form a kind of coda to a list of the human *gentes* who inhabit the earth. It was precisely from such a list that Augustine wished to exclude the antipodeans; here, however, they linger as a quasi-ethnographic supplement to the peoples of mankind, unattested by reliable history but nevertheless 'spoken of' – 'dicuntur', 'ferunt fuisse', the product of poets and fables.

If Isidore's reference to the antipodeans in book 14 of the *Etymologiae* followed Augustine in locating them within the category of *fabula*, and his inclusion of them in book 9 complicated this position by opening up the possibility that they could be located on the fringes of ethnographic writing, book 11 of the same work sought to bring them within the ecumene but as part of the multifarious category of 'monstrous men'. In the section of the *Etymologiae* devoted to portents, Isidore describes the 'antipodes' as an African people, whose feet are turned in the opposite direction to their bodies, and who have eight digits on their feet.[45] The description of this race derives from Pliny, via Solinus, although the use of the term antipodes to describe them is unique to Isidore.[46] The continuities between the three contradictory references to antipodeans in the *Etymologiae* are the association with Africa, the amalgamation of information derived from several classical and late antique authors (Virgil, Solinus, Servius, Augustine), and the intellectual positioning of them within or at the edges of fable, a location that corresponds to their geographical positioning either at the ends of the known world, or beyond Ocean and the heat of the sun. The position of the antipodes within fable – a category defined by Isidore, in contradistinction to history and 'argumentum', as events that did not happen, and could not have happened because they are contrary to nature – gave them a particular license.[47] At home only in rumour and poetry, the antipodes could be discussed as something unworthy of belief but nevertheless attached to, and deriving from, belief. They were not separated from items worthy of belief – actual, known, attested peoples and places – but rather added to them almost as a sub-category at the end of the list, where history shades into argument and fable. The antipodes were never excluded, that is, from the organization of knowledge; indeed they functioned as a point of definition that revealed the boundaries of history.

The distinction between *historia* and *fabula* continued to inform the question of antipodal habitation. The Anglo-Saxon historian Bede's extensive work of natural science, *De temporum ratione* (c. 725), condensed and clarified Isidore's formulation:

> Quamvis unam solummodo probare possunt habitatam, neque enim vel antipodarum ullatenus est fabulis accomodandus assensus, vel aliquis refert historicus vidisse vel audisse vel legisse se, qui meridianas in partes solem transierunt hibernum ita ut eo post tergum relicto, trangressis aethiopum fervoribus, temperatas ultra eos hinc calore illinc rigore atque habitabiles mortalium repererint sedes.[48]

> All the same, only one [zone] can be proved to be inhabited, for assent should not be given to fables of any kind concerning the antipodeans: no historian claims to have seen, heard, or read of men who passed beyond the winter sun into parts south, so that, having passed the raging heat of the Ethiopians, they found dwelling places of mortal men made temperate by heat on one side and by cold on the other.

Just as for Augustine 'antipodas esse fabulantur', and in Isidore the antipodeans were said to inhabit the fourth part of the world 'fabulose', so Bede assigns the antipodeans to fiction ('antipodarum … fabulae') and denies any record of them in history, a genre that appears to include travel narrative.[49] Yet in denying the existence of any reliable account of the antipodes Bede himself describes the image of just such a passage, through Ethiopia, across the winter tropic, and into the 'seats of mortals' in the temperate regions beyond. Once again, the path to the antipodes runs through the southernmost part of Africa. And, once again, the idea of the antipodes encourages description in the absence of proof, fantasy in the absence of possibility.

The pronouncements of Isidore and Bede can in no way be said to have settled the question of the antipodes. The eclecticism of early medieval conceptualizations of unknown lands and peoples, and something of the power of fable, is illustrated by the well-known case of Bede's near contemporary Virgil of Salzburg, an Irish cleric resident in Bavaria. Around 748 Archbishop Boniface of Mainz alleged to Pope Zacharias I that Virgil professed that 'there are another world, and other men beneath the earth, and a[nother] sun and a moon' (quod alius mundus et alii homines sub terra sint seu sol et luna).[50] In the matter of men living 'beneath the earth' it is not clear whether Virgil had in mind the kind of theory outlined in Macrobius or Martianus, or some kind of 'cultural synthesis' of scientific and popular traditions. Such a synthesis may have arisen, it has been argued, from a desire to fuse the classical scientific notion of regions beyond the ecumene with folk traditions of a subterranean underworld, one, that is, located literally beneath the ground, or more vaguely on the 'other side' of the known world.[51] Certain early medieval Irish texts, most notably the poem *Saltair na Rann*, contain references to peoples 'on the other side' of the earth; clearly indebted to Latin ecclesiastical literature, such works may also gesture to native lore concerning an Otherworld, 'immanent everywhere, but most often accessible by going underground or underwater'.[52] Ultimately however, in the absence of further information about Virgil's doctrine, the case for a popular-learned synthesis is attractive but also highly speculative. The classical tradition itself contains more than a suggestion of formulations of the antipodes in terms of underground dwellers and dwellings. As an Irish cleric, Virgil would presumably have been familiar with native Irish legends, but he would also have had access to texts such as Servius' commentary on the *Aeneid*, with its discussion of subterranean infernal regions.[53] Perhaps the only certainty is that these accusations did not harm Virgil's career: Zacharias recommended that, if the allegations turned out to be correct, Virgil should be stripped of his priestly status and excommunicated, but by 767 at the latest, and perhaps even as early as 755,[54] Virgil had become bishop of Salzburg, and in the thirteenth century he was canonized.[55]

Despite the apparent ecclesiastical censure, Virgil was certainly not alone in professing some version of antipodal theory: an anonymous Carolingian commentator on pseudo-Augustine's *Decem categoriae* vigorously affirmed 'that [antipodeans] may be able to live beneath the earth is not repugnant to faith, because the nature of the earth is spherical' (quod autem uiuere possint subtus terram non repugnat fidei, quia hoc agit natura terrae quae speroides est).[56] Rather more puzzlingly, the same commentator declares that the antipodeans 'are said to live according to the custom and religion of the Persians' (dicunt uiuere more et cultu per sarum).[57] On the other hand, a set of glosses on Boethius' 'Liber contra Eutichen et Nestorium', once attributed to Eriugena but more likely to have taken shape in Auxerre in the later ninth century, succinctly dismissed the idea of antipodeans living beneath the earth ('subtus terras') on Augustinian grounds: 'if every person [is descended] from Adam alone, whence then the antipodeans' (Si enim omnis homo ex Adam solo, unde ergo antipodas)?[58] Just a few years later, in his commentary on Boethius' poem 'O qui perpetua' (*De consolatione philosophiae*, book 3, metrum 9), Bovo of Corbie similarly hastened to explain to his reader that his discussion of cosmological theory should not be taken to imply that he accepted the 'fables of antipodean peoples' (antipodarum fabulas), which were in every way contrary to the Christian faith.[59]

Dismissal of the idea of antipodeans did not equate to dismissal of the zonal map. It was possible to represent the earth divided into zones without affirming the symmetrical existence of a race or races of men opposite to the known world. Though not a direct or consistent translation of any one work, Ælfric of Eynsham's Old English compilation of computistical material, *De temporibus anni*, explicitly acknowledges Bede, and its description of the zones is consistent with those contained in Bede's *De natura rerum* or *De temporum ratione*. Notably, however, it is even more cautious than its source about the matter of antipodal habitation. Ælfric describes five zones, going to some trouble to introduce and gloss the term: 'fif dælas on middanearde . þa we hatað on leden quinque zonas . þæt sind fif gyrdlas' (five parts on earth, which we call in Latin 'quinque zonas', that is five belts).[60] 'Gyrdlas' is possibly a translation of Cicero's 'cinguli'; the glossing, at any rate, suggests the effort necessary to convey this theory to a non-Latinate audience. The central zone is uninhabitable to 'any earthly man' due to the proximity of the sun, while the two furthermost zones are too cold. Ælfric describes two temperate zones ('naðor ne to hate ne to cealde') but he does not mention the word 'antipodes', and his position on habitation is unambiguous: 'On ðam norðran dæle wunað *eal mancynn* under þam bradan circule þe is gehaten Zodiacus' (in the northern part *all humanity* dwells under the broad circle called Zodiacus).[61]

Ælfric's marked reticence on matters antipodal indicates the potentially controversial nature of this topic. Nevertheless it also suggests that the possibility of habitation in the southern hemisphere was inherent to the zonal image of the world. Representations of the world in zones, whether entirely verbal or diagrammatic, had to have a southern temperate zone, and therefore had to raise the question – even if to respond with a categorical or implicit denial – of the existence of races of men beyond the known world. The question for medieval mapmakers who aspired to show the *imago mundi* in its entirety was how to express in visual and verbal terms an idea whose most fundamental trait was uncertainty. One answer was to represent uncertainty *as* uncertainty, to write 'terra incognita', to make explicit the inclusion of the unknown within the space of the map; another, in the absence of known land, was to make the map itself the object of description.

'There are four parts of the world': the emergence of meta-cartographical space

Isidore's reference to the antipodes as a 'fourth part of the world' in his *Etymologiae* appears to have contributed to a mode of cartographic representation of unknown, non-ecumenical land distinct from the essentially zonal model transmitted by Macrobius. Three examples of maps, dating from the eighth to the eleventh centuries, from contexts eschatological and computistic, illustrate the emerging possibilities for representation of the *antoecumene* beyond those offered by zonal images. The different modes of representing unknown land on these maps indicate the range of expression open to mapmakers, from the acknowledgement of *terra incognita* to the south of the ecumene, to the use of *antoecumenical* space as a site for theoretical and paratextual material. The blank space of unknown land invited statements about the map's content, but also about the theories the map expressed, the function it performed, and even its historical and social context.

The alternative configuration of unknown land prompted by Isidore's *Etymologiae* is found most notably in the world map contained in certain manuscripts of Beatus of Liébana's eighth-century commentary on the *Apocalypse*. The Beatus maps survive in fourteen manuscripts of the commentary that date from the tenth to the thirteenth centuries. The transmission history of this world image is complex, since the extant Beatus maps vary in a number of significant ways from manuscript to manuscript. The maps have been divided in two branches, corresponding to the two branches of the text of the commentary: Branch I is notable for the presence of portrait images of the apostles in the parts of the world to which they had spread the gospel; Branch II consists of maps in manuscripts that emanated from a revival of interest in Beatus' commentary in the tenth century.[62] Branch I (of which only three examples survive) seems closer to the original purpose of the map to illustrate the apostolic mission of extending Christianity throughout Asia, Africa, and Europe prior to the Apocalypse, although most Branch II maps show the places evangelized by the apostles minus the portraits.[63] All the maps, and indeed in the commentary itself, owe a debt to Isidore's *Etymologiae*.[64] This debt seems particularly evident in a feature common to all but one of the surviving Beatus maps: the representation of a strip of land at the far south of the map, beneath a narrow body of water that divides it from the known world (fig. 12). The manner of representation of this *terra incognita* varies from manuscript to manuscript. Four copies of the map inscribe the land with Isidore's statement in the *Etymologiae* that a fourth part of the world lies across Ocean deeper in the south ('trans Oceanum interior est in meridie'), unknown due to the heat of the sun, where in fable antipodeans are said to live (14.5.17).[65] Five copies term the unknown land 'Deserta terra uicina soli ab ardore incognita nobis' (desert/waste land near to the sun, unknown to us due to heat), thereby avoiding reference to a fourth part of the world and to antipodeans. In two, relatively late, manuscripts the strip of unknown land is left blank. Finally, two manuscripts witness the confusion, at some point in the process of transmission, of the antipodeans with the 'sciopodes', monstrous beings located by Isidore in Ethiopia and said to possess one leg and an enormous foot, with which they provide shade for themselves. Hence the placement, on one surviving map, of a sciopod sheltering from the burning sun in *terra incognita*, accompanied by the relevant passage from Isidore.[66]

Several attempts have been made to explain the paradox of the appearance of *terra incognita* on a map apparently designed to show the apostolic mission to the ends of the earth. One school of interpretation of this space has argued that it represents a 'fourth continent' introduced in order to represent the world in its entirety, with unknown depicted alongside known;[67] more recently, it has been suggested that it is intended to show the southernmost part of Africa, rather than an *antoecumenical* space.[68] It is certainly the case that on at least some of the maps the stretch of water that divides the *terra incognita* from the known world is either explicitly or implicitly (by use of colour) designated the Red Sea, rather than an equatorial ocean.[69] Further, there is evident identification of this unknown area with the torrid zone: not only do several exemplars term the land 'deserta' due to its extreme heat, two describe it as 'uninhabitable for us' (nobis inhabitabilis), rather than temperate.[70] Above all, this image underlines the difficulty of interpreting Isidore's reference to the antipodes in *Etymologiae* 14. It is not clear, for example, how Isidore's phrase 'quarta pars trans Oceanum interior est in meridie' should be understood: does this mean that the fourth part of the world is across the Ocean [and] deeper, i.e. further in the south – or inland in the south? Or should 'interior' be understood as a descriptor for 'Oceanum' (the least grammatically satisfying option)?[71] If so, what 'interior ocean' did Isidore have in mind? The different means of inscribing *terra incognita* in the Beatus corpus (quotation, pictorial representation, blank) seem to result from attempts to interpret the contradictory references to the antipodes in Isidore, above all his blurring of ethnographic and geographic reference, and his association of antipodeans with the southern part of Africa. Crucially, the transformation of written description into visual image has led to the representation of *terra incognita* in unusual proximity to the known world.

Contrary to assumptions sometimes made about the 'closed' nature of medieval ecumenical maps,[72] world images produced in the Middle Ages quite often 'open' out into zones of cartographic uncertainty: the known world itself was not represented as entirely closed off, despite the theories of its insularity, and of the impossibility of communication with other

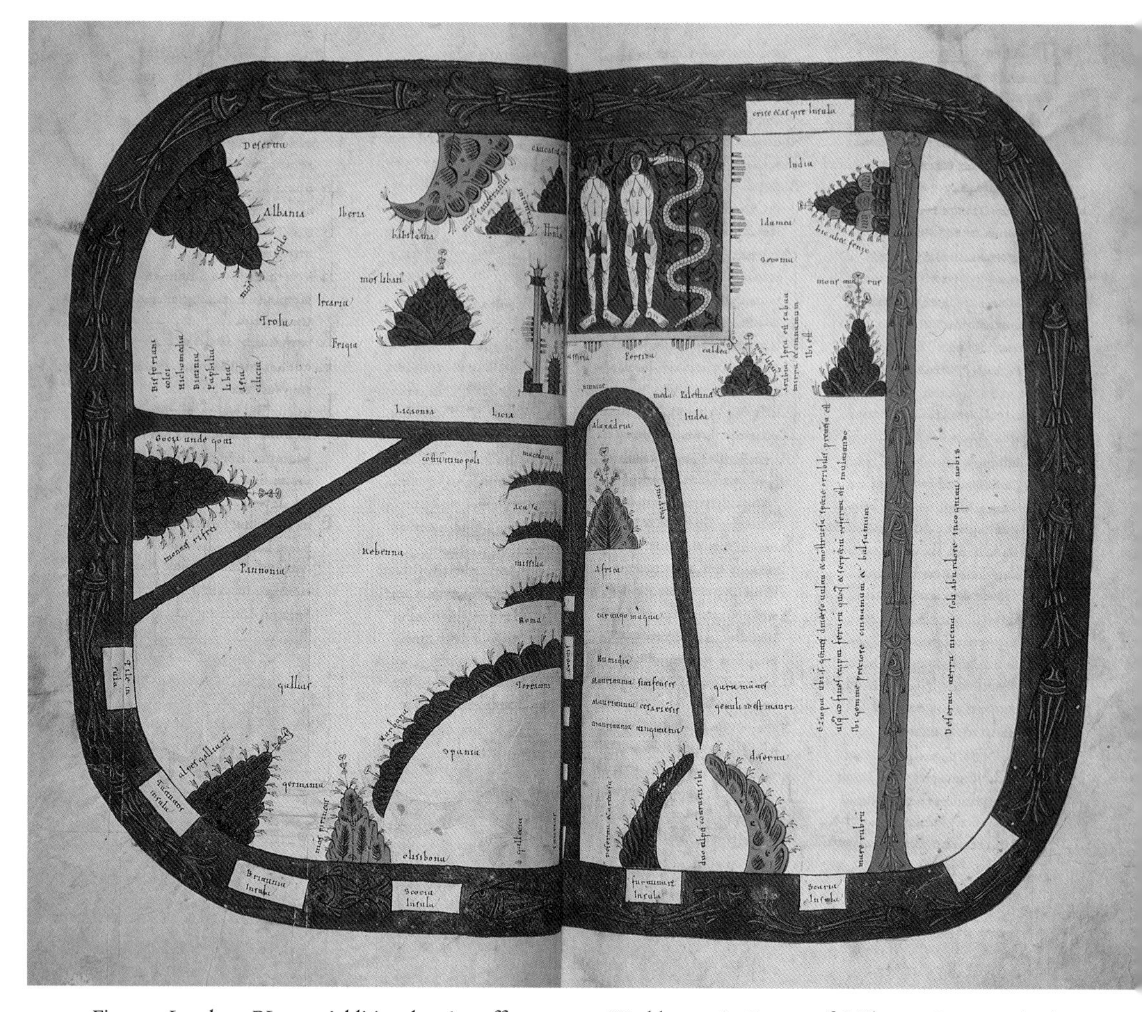

Fig. 12. London, BL, MS Additional 11695, ff. 39v–40r. World map in Beatus of Liébana, *Commentarius in Apocalypsin* (Commentary on the Apocalypse). Monastery of Santo Domingo de Silos, 1109. East at the top. The Beatus world map was designed to show the spread of Christianity throughout the earth, and to look forward to the end of human history. On this example the earthly paradise, complete with Adam, Eve, and serpent, is prominently marked in the far east; the map shows the mission fields of the apostles, but also records aspects of the natural world, such as the phoenix bird in India. Copies of the Beatus map are notable for a strip of land in the far south, often marked, as in this example, 'terra incognita'. The inscription on the land ('Deserta terra uicina soli ab ardore incognita nobis': 'desert/waste land near to the sun, unknown to us due to heat') is a version of a statement in Isidore of Seville's *Etymologiae*, in which land 'across Ocean' and deep in the south is said to be 'inaccessible due to the heat of the sun'.

regions. The three continents of the known world were, in more schematic representations, divided in terms of their geography – with the Mediterranean and the Tanais (Don) river forming the T of the T-O maps – and their demography – the association of the three continents with the three sons of Noah. However, far from being slavishly followed during the medieval period, this system of tri-partite division of the ecumene was, even on many schematic T-O maps, subtly disfigured by the presence of the unknown, or the unassimilated at the margins of the ecumene. The representation of islands within the Ocean that encircled the known world offered a particularly fertile means of 'opening' the world image. Perhaps the most remarkable example of the use of the island to represent unknown land at the edge of the ecumene is found in an eighth-century map traditionally associated with Isidore's *Etymologiae*, but now placed in the context of the medieval *computus*.[73] Here a rather elongated island in the far south-west of the map seems to be marked 'insola incognita o[] sunt iiii partes mundi' (unknown island … there are four parts of the world).[74] This island may have been intended to indicate a Cratesian division of the world between four land-masses, or, as is more likely, to represent Isidore's reference to a 'fourth part of the world, across Ocean deeper in the south' (14.5.17), the same part of the world represented on the world maps of the Beatus tradition.[75] Either way, the 'insola incognita' testifies to a tradition of locating unknown land at the ends of the earth, in a manner more or less symbolic. Detached, unknown, visually similar to several other islands on the map (Taprobana, the oddly-labelled 'mare mortun' and 'oceanus occiduus'), the 'unknown island' enunciates the form of the world: 'sunt iiii partes mundi'. The 'fourth part of the world' is the part that speaks the whole, the site for meta-cartographical observations.

Something of the range of possibilities of meta-cartographical space can be glimpsed in a mid eleventh-century manuscript from Santa María of Ripoll in Catalonia. Contained within this codex is a bifolium in which a world map shows the earth divided into two hemispheres, and two functions: topographic representation of the northern hemisphere on the left-hand side; and cosmographic description of the southern hemisphere on the right (fig. 13).[76] The map was previously regarded as the copy of a much earlier work, one over-optimistically assigned to the Carolingian bishop Theodulf of Orléans; its design in its present state, however, and particularly the disposition of the southern hemisphere, seems more likely to be the product of the monks of Ripoll.[77] The text that extends immediately beneath the equator in majuscule describes the equinoctial circle, its heat, and the impossibility of transit 'from the temperate zone in which we dwell to the other temperate zone' (ab hac temperata in qua inhabitamus zona ad aliam temperatam transitum uetat).[78] The text in minuscule in a cartouche beneath it quotes Bede's list of the five zones or 'circuli' in *De natura rerum*, with Greek terminology added from Isidore's works.[79] A semi-circular line of text that arches below the entire ensemble records Eratosthenes' calculation of the circuit of the globe. The word 'TERA' and an allegorical figure representing the earth, possibly a later interpolation,[80] can be found between two smaller cartouches. These contain a poem composed by Bishop Theodulf, divided in two and supplemented with two lines from another of his poems. The verses in one cartouche describe a work ('opus'), usually presumed to be a table, commissioned by Theodulf with the purpose of showing an image of the world ('orbis imago') that might be contemplated during meal times to encourage spiritual

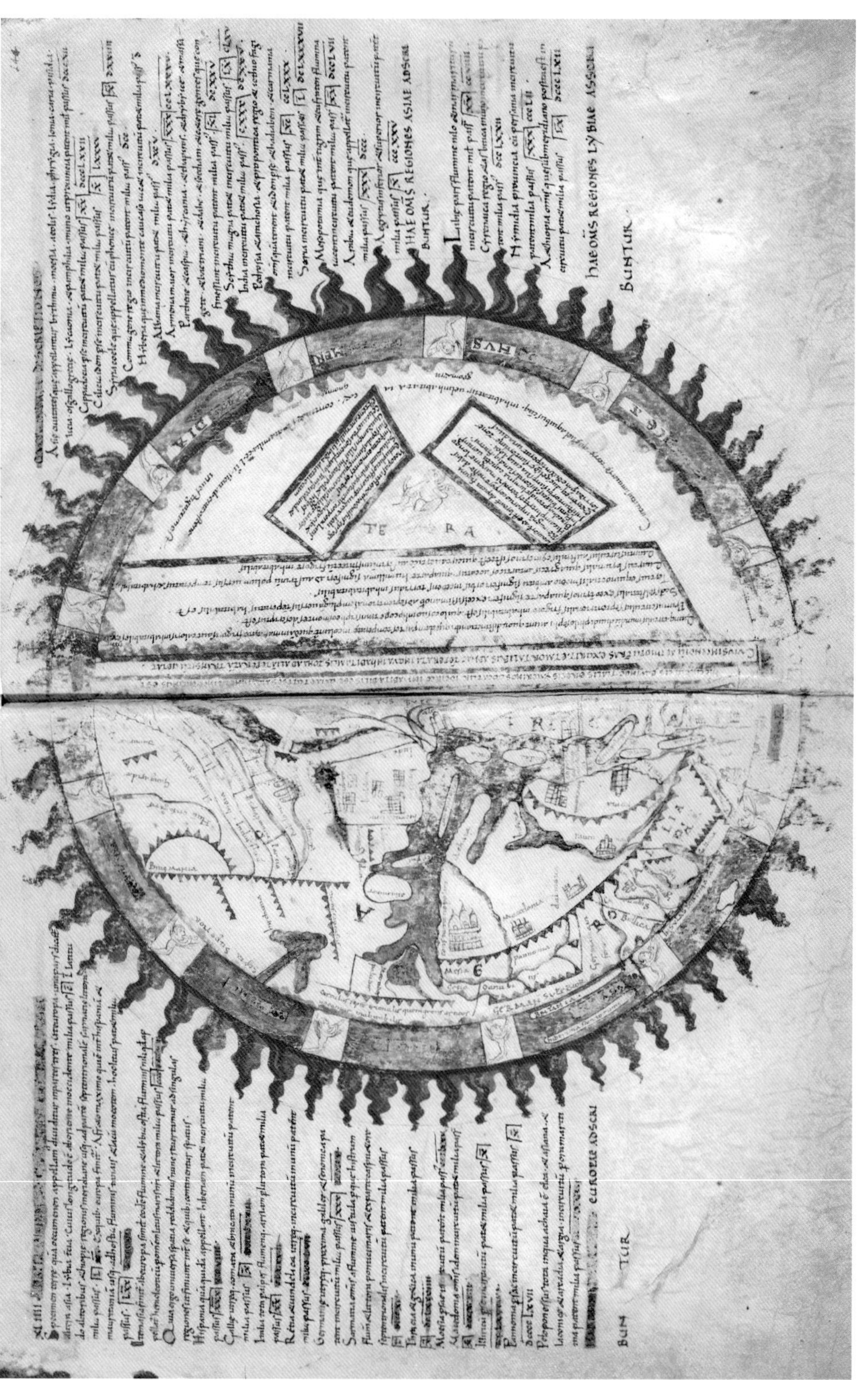

Fig. 13. Vatican City, BAV, MS Regin. lat. 123, ff. 143v–144r. World map from the monastery of Santa María of Ripoll, Catalonia, eleventh century. East at the top. The southern hemisphere of this map (on the right-hand folio) contains an unusual assemblage. The text located immediately beneath the equator describes the equinoctial circle, its heat, and the impossibility of transit between northern and southern temperate zones. The text in a cartouche below it quotes the list of the five zones or 'circuli' from Bede's *De natura rerum.* The two smaller cartouches contain a poem composed by the Carolingian bishop Theodulf of Orléans; between them the word TERA has been written, accompanied by an allegorical figure. A semi-circular line of text arching below the entire ensemble records Eratosthenes' calculation of the circuit of the globe.

rather than carnal consumption. The verses in the other cartouche continue the description: the image is the means by which a great thing might be known in a small space; it shows the ocean, rivers and winds, as well as the earth; the two supplementary lines conclude the verses with a description of the zones.[81] The ensemble located in the southern hemisphere was evidently carefully constructed, comprising words and image, poetry, allegory, and commentary. Why choose this space as its site? To answer this question it is necessary to look outside the map. The text that surrounds it is the 'Divisio orbis terrarum Theodosiana', a late antique description of the ecumene including measures of the circumferences of provinces, associated with the fifth-century emperor Theodosius II.[82] No reference is made to land beyond the ecumene in the 'Divisio', so one function of the map's depiction of earth beyond the known world is to supplement it. In place of the geographical formations in the northern hemisphere the southern offers intellectual *divisio*: the geometrical function of zonal theory (although the southern hemisphere itself is not divided into zones); the rhetorical function of allegory; and a statement – allusive, conflated – of authorship. The poems announce the origin and purpose of the world image. Yet the part of that image in which the verses of Theodulf are located is precisely the part they do not and cannot describe, the part that must be there to make the image whole, but which shows that it is far from complete. A part that can only be *supposed* in one essential element: land.

Meta-cartography might be defined as the act of explaining maps and mapmaking, the expression of the function and mode of execution of a map, an account of how and why a map came into being. The location of such explanation in the *antoecumene* has a logic, since to place meta-cartographical reflection in the known world would be to displace practice with the insertion of theory. When placed in the unknown world, however, meta-cartography does not overwrite geography; instead, it accounts for the image of the world in its entirety, the mapping of the known as well as the representation of the unknown. Accounting could well take place outside of the map, but the ambiguous presence of unknown land – part of the whole, yet of a different order to the ecumene – invites the inclusion of statements about the map, and specifically about the acts of representation involved in making a map, *on the map*. As many examples of medieval zonal maps suggest, antipodal space lent itself to theoretical reflections, to comments, explication, and controversy, because it is itself purely the product of theory. But as the Ripoll map shows, that space was subject to extension and elaboration: there were possibilities other than its simple designation as a 'temperate zone of the antoikoi'. Those possibilities, already emerging by the eleventh century, became more evident in the later Middle Ages as people began to challenge the impermeability of the barriers between known and unknown worlds.

Notes

1 The assertion that belief in the antipodes was 'heretical' during the Middle Ages has been made frequently. See for example John Kirtland Wright, *The Geographical Lore at the Time of the Crusades: A Study in the History of Medieval Science and Tradition in Western Europe* (New York: American Geographical Society, 1925), p. 57; W. G. L. Randles, 'Le Nouveau Monde, l'Autre Monde et la Pluralité des Mondes', in *Actas do Congresso Internacional de História dos Descobrimentos* (Lisbon: Comissão Executiva das Comemorações do V Centenário da Morte do Infante D. Henrique, 1961), vol. 4, pp. 347–82: p. 349; Moretti, *Gli antipodi*, p. 84.

2 The illustration of the antipodes survives only in one, eleventh-century, Byzantine manuscript (Florence, Biblioteca Laurenziana, MS Plut. IX.28, f. 98v). However, modern scholars are inclined to accept the images contained in the three complete manuscripts of the *Christian Topography* as broadly faithful reproductions of the original images: Leslie Brubaker, 'The Relationship of Text and Image in the Byzantine Manuscripts of Cosmas Indicopleustes', *Byzantinische Zeitschrift* 70 (1977), 42–57; Cosmas Indicopleustes, *Topographie chrétienne*, ed. and trans Wanda Wolska-Conus, 3 vols (Paris: Éditions du Cerf, 1968–73), vol. 1, pp. 124–85; and Wolska-Conus, 'La "Topographie Chrétienne" de Cosmas Indicopleustès: hypothèses sur quelques thèmes de son illustration', *Revue des Études Byzantines* 48 (1990), 155–91.

3 *Topographie chrétienne*, vol. 1, 2.107.

4 For discussion and examples see Barbara Obrist, 'Wind Diagrams and Medieval Cosmology', *Speculum* 72 (1997), 33–84, esp. pp. 66–75.

5 Edson, *Mapping Time and Space*, pp. 67–8: Calcidius' diagram was of the heavenly zones, but like the diagrams that illustrated the *Georgics* it could be read as representing terrestrial zones.

6 Florence, Biblioteca Laurenziana, MS S. Marco 190.

7 Edson, *Mapping Time and Space*, pp. 72–96.

8 Bruce Barker-Benfield, 'Macrobius', in *Texts and Transmission: A Survey of the Latin Classics*, ed. L. D. Reynolds (Oxford: Clarendon Press, 1983), pp. 222–35: p. 225.

9 The statistics come from Barker-Benfield, 'Macrobius', p. 224. A helpful finding list of Macrobius manuscripts is provided by Bruce Eastwood, 'Manuscripts of Macrobius, *Commentarii in Somnium Scipionis*, before 1500', *Manuscripta* 38 (1994), 138–55.

10 There is no definitive catalogue of the maps in Macrobius manuscripts. The list of M. Destombes, *Mappemondes A.D. 1200–1500* (Amsterdam: Israel, 1964), pp. 43–5, 88–95 is incomplete and frequently inaccurate, and should be supplemented for BNF manuscripts with Patrick Gautier Dalché, 'Mappae mundi anterieures au XIIIe siècle dans le manuscrits latins de la Bibliothèque nationale de France', *Scriptorium* 52 (1998), 102–62, and in general by the descriptions and reproductions of maps provided in Leonid S. Chekin, *Northern Eurasia in Medieval Cartography: Inventory, Text, Translation, and Commentary* (Turnhout: Brepols, 2006), pp. 95–120. An impressively large number of Macrobius maps (including some no longer extant) were reproduced in Youssouf Kamal, *Monumenta Cartographica Africae et Aegypti*, 5 vols (Cairo, 1926–51). Earlier discussions include Konrad Miller, *Mappaemundi: Die ältesten Weltkarten*, 6 vols (Stuttgart: Jos. Roth'sche Verlagshandlung, 1895–8), vol. 3 (1895), pp. 122–6 and Michael C. Andrews, 'The Study and Classification of Medieval Mappae Mundi', *Archaeologia* 75 (1924–5), 61–76: pp. 68–72.

11 Paris, BNF, MS lat. 6371, f. 20v (eleventh century); Florence, Biblioteca Laurenziana, MS Santa Croce 22 sin. 9, f. 46v (eleventh century); Munich, Bayerische Staatsbibliothek, Clm 6362, f. 74r (eleventh century); Vatican City, BAV, MS Vat. lat. 1546, f. 74v (eleventh century); BNF, MS lat. 15170, f. 125r (twelfth century). Patrick Gautier Dalché regards Syene and Meroë as additions to the original map of Macrobius: 'De la glose à la contemplation. Place et fonction de la carte dans les manuscrits du haut Moyen Âge', *Settimane di studio del Centro italiano di studi sull'alto medioevo* 41 (1994), 693–771, p. 717.

12 For example the twelfth-century manuscripts Paris, BNF MS lat. 16680, f. 60v; Trinity College, Cambridge, MS R.9.23, f. 60v; London, BL, MS Egerton 2976, f. 62v; Vatican City, BAV, MS Vat. lat. 1548, ff. 20v–21r; Naples, Biblioteca nazionale, MS V.A.12, f. 46v. Arentzen notes the infiltration of toponyms on the map, from at least the eleventh century, and in certain instances the use of solely ecumenical maps in place of the map of Macrobius: Jörg-Geerd Arentzen, *Imago Mundi Cartographica* (Munich: Fink, 1984), pp. 69–70. See also Gautier Dalché, 'De la glose

à la contemplation', pp. 718–21.

13 For example Paris, BNF, MS lat. 18421 (twelfth century).

14 Paris, BNF, MS lat. 6371, f. 20v, an eleventh-century manuscript of northern French or German provenance ('temperata antipodvm'); Brussels, Bibliothèque royale, MS 10146, f. 109v, a tenth-century manuscript, possibly French ('temperata australis antipodum'). Gautier Dalché detects the influence of Isidore of Seville's *Etymologiae* 14.5.7 in the substitution of 'antipodum' for 'antoecorum': 'De la glose à la contemplation', pp. 717–18.

15 Metz, Bibliothèque municipale, MS 271 (c. 950–1050), f. 40v. This manuscript was destroyed in 1944 but a photograph of the map is preserved in Kemal, *Monumenta Cartographica Africae et Aegypti*, vol. 3, f. 554.

16 Oxford, Bodleian Library, MS D'Orville 77, f. 100r.

17 Vatican City, BAV, MS Vat. lat. 1546, f. 74v ('temperata'); London, BL, MS Cotton Faustina C.I, f. 87v ('temperata zona'); Oxford, Bodleian Library, MS Auct.F.2.20, f. 53v ('temperata australis'); Munich, Bayerische Staatsbibliothek, Clm 14436, f. 58r; BAV, MS Palat. lat. 1577, f. 79r; Bern, Burgerbibliothek, MS 265, f. 57v; Cologny (Geneva), Fondation Martin Bodmer, MS 111, f. 36v; Zürich, Zentralbibliothek, MS Car.C.122, f. 38v (all 'habitabilis').

18 Munich, Bayerische Staatsbibliothek, Clm 6362, f. 74r. This map is also unusual for the level of ecumenical detail in its northern hemisphere.

19 I am aware of two instances of the intrusion of pictures into the equatorial and southern zones of the map of Macrobius. In a twelfth-century English manuscript, Oxford, Bodleian Library, MS Auct F.2.20, f. 53v, a later hand has drawn a representation of Neptune in the central zone of Macrobius' world map, pouring liquid from an amphora down into the southern hemisphere. The southern continent in the two world maps that illustrate section 2.9.7–8 of the *Commentarii* in Worcester Cathedral, MS F. 68, f. 221r, a fourteenth-century manuscript produced in Avignon, is filled in both instances with generic representations of a church, a hall, and a fortified tower.

20 Edson's description of 'antetorum' as a 'made-up word' is inaccurate: *Mapping Time and Space*, p. 6. The word derives instead from the misreading of 'ant[o]ecis' as 'antetis', and the subsequent formation of a genitive plural, 'antetorum', rather than the more usual 'antecorum'. An earlier copy of this map, also with the term 'antetorum', appears in a late tenth- or early eleventh-century Macrobius manuscript, ex-Malibu, Getty Ludwig XII.4, fol. 22r, now in a private collection.

21 Oxford, MS D'Orville 77, f. 96r, f. 100r. For discussion of the manuscript see Bruce Barker-Benfield, 'The manuscripts of Macrobius' Commentary on the *Somnium Scipionis*', 2 vols (unpublished doctoral dissertation, University of Oxford, 1975–6), vol. 2, pp. 150–9. The map is discussed by von den Brincken, *Fines Terrae*, p. 39.

22 See the discussion in Louis Holtz, 'L'école d'Auxerre', in *L'école carolingienne d'Auxerre de Murethach à Remi 830–908* (Paris: Beauchesne, 1991), pp. 131–46.

23 *Iohannis Scotti Annotationes in Marcianum*, ed. Cora E. Lutz (Cambridge, MA: Medieval Academy of America, 1939), 298.22; compare Remigius of Auxerre, *Commentum in Martianum Capellam Libri III–IX*, ed. Cora E. Lutz (Leiden: Brill, 1965), 298.22. The attribution to Eriugena of this set of glosses on Martianus – particularly those on books 5–7 of *De nuptiis* – is far from settled, although for the purposes of this discussion I assume his authorship. Remigius' commentary, now generally seen as a synthesis of more than one strand of ninth-century scholia, appears to have copied Eriugena's glosses on antipodal matters in book 6 very closely: he adds a few remarks about the relationship between the antipodeans and other *antoecumenical* peoples, expands the gloss on 'oikos', and supplies scholia on words not glossed by Eriugena. For discussion of the attributional questions see Claudio Leonardi, 'Martianus Capella et Jean Scot: nouvelle présentation d'un vieux problème', and Michael Herren, 'The Commentary on Martianus Attributed to John Scottus: its Hiberno-Latin Background', both in *Jean Scot écrivain*, ed. G.-H. Allard (Montreal: Bellarmin, 1986), pp. 187–207, and 265–86, as well as John Marenbon, *From the circle of Alcuin to the school of Auxerre* (Cambridge: Cambridge University Press, 1981), pp. 117–19.

24 *Iohannis Scotti Annotationes in Marcianum*, 298.21–3; Remigius of Auxerre, *Commentum in Martianum Capellam*, 298.21–3.

25 *Iohannis Scotti Annotationes in Marcianum*, 299.6; Remigius of Auxerre, *Commentum in Martianum Capellam*, 299.6.

26 Paris, BNF, MS lat. 10195, f. 31r. See further Alison M. Peden, 'Echternach as a Cultural Entrepôt. The case of Macrobius', in *Willibrord: Apostel der Niederlande, Gründer der Abtei Echternach* (Luxembourg: Editions Saint-Paul,

1989), pp. 166–70: p. 168.

27 Manegold of Lautenbach, *Liber contra Wolfelmum*, ed. Wilfried Hartmann (Weimar: Böhlaus Nachfolger, 1972), pp. 39–40. Manegold also makes a connection between elements of pagan philosophy, such as belief in reincarnation, and supporters of the imperial side in the investiture conflict: *Liber contra Wolfelmum*, pp. 46–7.

28 *Liber contra Wolfelmum*, p. 51.

29 *Liber contra Wolfelmum*, pp. 51–2.

30 Oxford, Bodleian Library, MS Auct T.2.27, f. 40r.

31 f. 39r. Gautier Dalché, 'Mappae mundi ... dans le manuscrits latins de la Bibliothèque nationale', no. 82, pp. 152–3.

32 Paris, BNF, MS lat. 6622, f. 59v.

33 Gautier Dalché, 'Mappae mundi ... dans le manuscrits latins de la Bibliothèque nationale', no. 25, p. 125.

34 BNF, MS lat. 18421, f. 59r. This diagram is also contained in BNF, MS lat. 6570, f. 87r, and Prague, Státní knihovna, MS VIII.H.32 (1650), f. 39r.

35 Similar diagrams are contained in Florence, Biblioteca Laurenziana, MS Plut. 51.14, f. 50v; Vatican City, BAV, Vat. lat. 1546, f. 70v. The inventive design in Paris, BNF, MS lat. 6367, f. 2v, shows a circle divided in four by two intersecting bands of water, the horizontal one marked 'Mediterraneum sub terra perusta' (the sea between the land beneath the burning zone), and the vertical 'Mare athlanticum' and 'Pars australis et ignota' (southern and unknown part).

36 Occasionally the words 'nostra superior' (our upper) and 'inferior sub pedibus' (lower beneath feet) were marked in the northern and southern hemispheres of the diagram used to show rainfall on the earth, located at the end of book 1 of the *Commentarii*. This combination (and variants) is found in London, BL, MS Cotton Faustina C.I, f. 82v (eleventh century): 'Superior terra'/ 'Terra subpedibus'; Florence, Biblioteca Laurenziana, MS Santa Croce 22 sin. 9, f. 36r (eleventh century); Biblioteca Laurenziana, MS Plut. 77.9, f. 24v (twelfth century); Paris, BNF, MS Nal 923, f. 31r (twelfth century): 'nostra superior'/ 'antipodum inferior'; Vatican City, BAV, MS Palat. lat. 274, f. 59r (twelfth century); Munich, Bayerische Staatsbibliothek, Clm 15738, f. 202r (fifteenth century): 'Terra superior'/ 'Inferior sub pedibus'. Most depictions of the rain diagram carry no nomenclature. The dates of the manuscripts above suggest that the addition of nomenclature was an eleventh-century innovation.

37 E.g. Vatican City, BAV, MS Vat. lat. 5993, ff. 6v and 40v; BAV, MS Palat. lat. 1646, f. 54v.

38 Vatican City, BAV, MS Vat. lat. 1575, f. 18r.

39 The use of colours to distinguish different zones is a feature of diagrams in several Virgil manuscripts. An incomplete and therefore revealing diagram in an eleventh-century manuscript of Servius' commentary on the *Aeneid* contains in each zone a set of instructions to an illustrator. From north to south: 'fac uiride' (make this one green); 'uenetum fac' (make this one marine blue); 'minio' (red); 'uenetum' (blue); 'uiride' (green): Vatican City, BAV, MS Vat. lat. 5993, f. 40v.

40 Isidore of Seville, *Etymologiarum sive originum libri xx*, ed. W. M. Lindsay, 2 vols (Oxford: Clarendon Press, 1911), vol 2, 14.5.17. For a discussion of Isidore's antipodal references preoccupied with the question of whether he regarded the earth as a sphere or a disk see William D. McCready, 'Isidore, the Antipodeans, and the Shape of the Earth', *Isis* 87 (1996), 108–27.

41 Isidore of Seville, *De natura rerum (Traité de la nature)*, ed. Jacques Fontaine (Bordeaux: CNRS, 1960), 10.1–2. Cf. Isidore, *Etymologiae*, 13.6.

42 Isidore, *Etymologiae*, 9.2.133–4.

43 See Hans Philipp, *Die historisch-geographischen Quellen in den etymologiae des Isidorus von Sevilla*, 2 vols (Berlin: Weidmannsche Buchhandlung, 1912–13), vol. 2, pp. 40–1.

44 Isidore, *Etymologiae*, 9.2.134–5.

45 Isidore, *Etymologiae*, 11.3.24–5.

46 Fabio Gasti, *L'antropologia di Isidoro: Le fonti del libro XI delle* Etimologie (Como: New Press, 1998), pp. 103–4. On

the antipodes as monstrous race see John Block Friedman, *The Monstrous Races in Medieval Art and Thought* (Cambridge: Harvard University Press, 1981; repr. Syracuse: Syracuse University Press, 2000), esp. pp. 11, 47–50.

47 Isidore, *Etymologiae*, 1.44.5.

48 Bede, *De temporum ratione*, in *Opera de temporibus*, ed. Charles W. Jones (Cambridge: Mediaeval Academy of America, 1943), p. 245. Bede gave an account of the theory of zonal division in his shorter works *De natura rerum* and *De temporibus*: *De natura rerum*, in *Bedae Venerabilis Opera Didascalica*, ed. C. W. Jones, Corpus Christianorum, Series Latina 123A (Turnhout: Brepols, 1975), esp. chapters 9, 46, 47, pp. 199–200, 228–31.

49 On the influence of Pliny and Isidore on Bede see W. D. McCready, 'Bede and the Isidorian Legacy', *Mediaeval Studies* 57 (1995) 41–73, esp. pp. 71–2, and Ribémont, *Les origines des encyclopédies médiévales*, pp. 259–71. Large slabs of Isidore's *Etymologiae*, including his references to the antipodes at 11.3.24–5, 14.5.17, and 9.2.133–4 were copied by Rabanus Maurus, *De universo*, *PL* 111, respectively cols 197, 352–3, 445: 7.7 ('De portentis'); 12.4 ('De regionibus'); 16.2 ('De gentium vocabulis').

50 *Die Briefe des heiligen Bonifatius und Lullus*, ed. Michael Tangl (Berlin: Weidmannsche Buchhandlung, 1916), pp. 178–9.

51 John Carey, 'Ireland and the Antipodes: The Heterodoxy of Virgil of Salzburg', *Speculum* 64 (1989), 1–10.

52 Carey, 'Ireland and the Antipodes', pp. 4–7. On the complex question of the interaction of native Irish traditions of supernatural subterranean habitation with learned Latin traditions see also Marina Smyth, *Understanding the Universe in Seventh-Century Ireland* (Woodbridge: Boydell Press, 1996), pp. 285–90.

53 As noted by Carey, 'Ireland and the Antipodes', p. 8.

54 See James F. Kenney, *The sources for the Early History of Ireland: Ecclesiastical* (New York: Columbia University Press, 1929), p. 524.

55 On the utterly anachronistic post-medieval identification of Virgil as a lone voice of proto-modern scientific thought crying in the wilderness of the Dark Ages, see especially Patrick Gautier Dalché, 'A propos des antipodes. Note sur un critère d'authenticité de la *Vie de Constantin* slavonne', *Analecta Bollandiana* 106 (1988), 113–19: pp. 114–17; also Valerie I. J. Flint, 'Monsters and the Antipodes in the Early Middle Ages and Enlightenment', *Viator* 15 (1984), 65–80, pp. 76–9.

56 Paris, BNF, MS lat. 12949, f. 30v.

57 The likely source of this remark is a passage inserted into Sallust's *Bellum Iugurthinum*: 'Maurique, vanum genus, ut alia Africae, contendebant antipodas ultra Aethiopiam cultu Persarum iustos et egregios agere'. The passage is listed by B. Maurenbrecher amongst the 'fragmenta dubia vel falsa' in his edition of Sallust's *Historiae*: *C. Sallusti Crispi Historiarum Reliquiae*, ed. Bertoldus Maurenbrecher (Leipzig: Teubner, 1891–3), pp. 207–8.

58 *Der Kommentar des Johannes Scottus zu den Opuscula Sacra des Boethius*, ed. E. K. Rand (Munich: Beck'sche Verlagsbuchhandlung, 1906), p. 72. For discussion of the authorship of these glosses see Marenbon, *From the circle of Alcuin*, pp. 119–20.

59 R.B.C. Huygens, 'Mittelalterliche Kommentare zum *O qui perpetua ...*', *Sacris Erudiri* 6 (1954), 373–427, pp. 389–90.

60 Ælfric, *De temporibus anni*, ed. Heinrich Henel, EETS o.s. 213 (London: Oxford University Press, 1942), p. 50.

61 Ælfric, *De temporibus anni*, p. 52. My emphasis.

62 See John Williams, 'Isidore, Orosius and the Beatus Map', *Imago Mundi* 49 (1997), 7–32. Edson provides a useful summary and analysis of the two branches, with reference to Williams' views: *Mapping Time and Space*, pp. 151–9. All the maps in Beatus manuscripts are reproduced in John Williams, *The Illustrated Beatus: A Corpus of the Illustrations of the Commentary on the Apocalypse*, 5 vols (London: Harvey Miller, 1994–2003).

63 Edson, *Mapping Time and Space*, p. 157.

64 Although Williams argues for Orosian, or at least non-Isidorean influence on early states of the map: 'Isidore, Orosius and the Beatus Map'.

65 Williams argues from this evidence that 'the "fourth part of the world" legend may have joined the Beatus map

tradition first as an emendation of the original map … in the tenth century': 'Isidore, Orosius and the Beatus Map', p. 18.

66 Burgo de Osma, Archivo de la Catedral, MS 1, ff. 34v–35; Isidore, *Etymologiae*, 11.3.23, where 'sciopodes' are listed immediately before antipodeans. Antipodeans are also conflated with 'sciopodes' in Lisbon, Arquivo Nacional da Torre do Tombo, Lorvão, f. 34 (bis v). On the confusion of 'antipodes' with 'sciopodes' and 'pedes latos' (broad feet) see Edson, *Mapping Time and Space*, p. 154.

67 von den Brincken, *Fines Terrae*, pp. 57–8 ('ein vierter Erdteil'); Arentzen, *Imago Mundi Cartographica*, pp. 110–11 ('Südkontinent'); Gautier Dalché, 'De la glose à la contemplation', pp. 751–2 ('un continent austral'); Woodward places Beatus maps within the category of 'quadripartite' world maps: 'Medieval *mappaemundi*', p. 357.

68 Williams, 'Isidore, Orosius and the Beatus Map', pp. 18–23; supported by Scafi, *Mapping Paradise*, pp. 110–11.

69 Williams, 'Isidore, Orosius and the Beatus Map', p. 20.

70 The maps in the Osma and Lorvão manuscripts. A similar point might be made about the eighth-century map in St Gall, Stiftsbibliothek, MS 237, which contains a strip of land termed 'terra inhabitabilis' (uninhabitable land) beyond the equator.

71 Williams, Edson, and Scafi all follow this translation: Williams, 'Isidore, Orosius and the Beatus Map', p. 18; Edson, *Mapping Time and Space*, p. 189 *n*28; Scafi, *Mapping Paradise*, p. 123 *n*142. See however *The* Etymologies *of Isidore of Seville*, trans. Stephen A. Barney, W. J. Lewis, J. A. Beach, Oliver Berghof (Cambridge: Cambridge University Press, 2006), p. 293: 'further inland toward the south'.

72 See Jacob, 'Il faut qu'une carte soit ouverte ou fermée', p. 37, on the 'progressive deconstruction' of Greek cartography in the Age of Discoveries: '[o]n passe d'une image globale, parfaite, circonscrivant le tout, à une image ouverte'. But this perfect, closed global image was perforated long before the sixteenth century.

73 Vatican City, BAV, MS Vat. lat. 6018, ff. 63v–64r. Richard Uhden argued that the map was a copy of the 'world map of Isidore of Seville': 'Die Weltkarte des Isidorus von Sevilla', *Mnemosyne*, 3rd ser. 3.1 (1935–6), 1–28. See Edson, *Mapping Time and Space*, pp. 61–2, for the revised view.

74 The inscription is hard to read and various suggestions have been made as to its wording. Gautier Dalché reads 'insola incognita ori solis iii[a] partes mundi', interpreting it as following the tendency of Beatus manuscripts to depict the southern temperate zone as an island: 'De la glose à la contemplation', p. 760. Other suggestions include: 'Insola incognita orisunt iiii partes mundi'; 'Insula incognita (h)omini (?); sunt IIII partes mundi' (von den Brincken, *Fines Terrae*, p. 51); 'insola incognita ori sunt (sed?) IIII partes mundi' (Uhden, 'Die Weltkarte des Isidorus von Sevilla', p. 6); 'insola incognita [ard]ori solis iiii pars mundi' (Williams, 'Isidore, Orosius, and the Beatus Map', p. 17).

75 von den Brincken, *Fines Terrae*, pp. 50–1, and Arentzen, *Imago Mundi Cartographica*, pp. 109–10 both see the influence of Isidore's *Etymologiae* 14.5.17; Uhden regarded the inscription as a reference to the four elements, therefore to Isidore's *De natura rerum* 11.1: 'Die Weltkarte des Isidorus von Sevilla', pp. 6–7. Williams questions the Isidorean connection, 'Isidore, Orosius, and the Beatus Map', p. 17.

76 Vatican City, BAV, Regin. lat. 123, ff. 143v–144r.

77 Patrick Gautier Dalché, 'Notes sur la "carte de Théodose II" et sur la "mappemonde de Théodulf d'Orléans"', *Geographia Antiqua* 3–4 (1994–5), pp. 91–106. The attribution to Theodulf was made by A. Vidier, 'La mappemonde de Théodulfe et la mappemonde de Ripoll (IXe–XIe siècle)', *Bulletin de géographie historique et descriptive* 3 (1911), 285–313. For recent discussion of the manuscript context of the map see Edson, *Mapping Time and Space*, pp. 80–6.

78 The text on the map is transcribed by Gautier Dalché, 'Notes sur la "carte de Théodose II"', pp. 97–9.

79 Gautier Dalché, 'Notes sur la "carte de Théodose II"', p. 99 *n*38. The principal source of the section in minuscule is Bede, *De natura rerum*, 9; the section in majuscule seems to be an adaptation of Isidore, *De natura rerum*, 10.2–4.

80 Gautier Dalché, 'Notes sur la "carte de Théodose II"', pp. 99, 102.

81 For the relevant poems of Theodulf see *Poetae latini aevi Carolini* vol. 1, ed. E. Dümmler (Berlin: Weidmann, 1890), pp. 544–8: Carmen 46 ('De septem liberalibus artibus in quadam pictura depictis'), ll. 77–8, and Carmen 47 ('Alia pictura, in qua erat imago terrae in modum orbis comprehensa'), ll. 41–54. Gautier Dalché's analysis of the verses

on the map permits him to identify them as consisting essentially of a single poem (the last 14 lines of Carmen 47, in fact a separate poem from the preceding 40 lines), plus two interpolated lines taken from the description of Geometry in Carmen 46, and to reject the hypothesis that their order in the Ripoll manuscript can be attributed to Theodulf: 'Notes sur la "carte de Théodose II"', pp. 100–2.

82 Edited by Gautier Dalché, 'Notes sur la "carte de Théodose II"', pp. 105–6; on the title see p. 95.

Chapter Five

Between Passage and Recessus: *the Antipodes 1100–1400*

At the beginning of the twelfth century the idea of passage to the antipodes appeared to Europeans as impossible as it had done for the previous millennium. By the end of the fourteenth century it not only appeared possible to many authors; some even boasted that they had achieved it. Two developments in the medieval matrix of learning and scholarly debate brought about this shift. The barriers to passage were weakened, first, by the introduction of Arabic texts to the Latin west, because Arab geographical tradition maintained that equatorial regions were habitable, even temperate. Second, the circulation of travel narratives, genuine and confected, encouraged the idea that passage to the south beyond the equator had been achieved. The erosion of barriers was gradual. It did not prompt the radical overhaul of the idea of the antipodes of the kind that would occur in the course of the fifteenth century and that would receive its defining moment at the beginning of the sixteenth, when passage to hitherto unknown regions on the other side of the earth was established beyond doubt. But the possibility of passage changed the meaning of the antipodes. Reachable, at least according to some theories and reports, they were no longer beyond the boundaries of sovereign power; nor were they beyond the word of God. In the works of the great thirteenth-century natural scientists Albertus Magnus and Roger Bacon, the antipodes started to function as a rebuke, rather than a check to political power: they marked the failure of states and their rulers to push at the barriers of knowledge. In fourteenth-century pseudo-travel narratives the antipodes were similarly invoked as an odd counterpart to crusade, proof of the paucity of Christian dominion and the need for its extension. The antipodes functioned in these texts not as the sign of boundaries, the marker of space where history ends and fable begins, but as a sign of uncertainty of categorization, and hence disorder, no longer opposite the known world, increasingly at its ends, and intermingled with its schemes of knowledge.

However, against the forces of expansion and completion of the world image that began to emerge on a variety of fronts – political, religious, intellectual – within Europe from the twelfth century, there flourished a countervailing impulse. The importance of the antipodes for most writers continued to lie in their conjectural and unverifiable nature, not only because of theological objections to passage and habitability, but because the idea of the antipodes as fundamentally beyond knowledge gave them a particular intellectual and literary potency. To retain the more symbolic and theoretical function of the antipodes was to preserve a space not simply of the geographically unknown, but of the unknowable – a space of depth.

Visions of passage

New directions to antipodal description were brought to Europe on the tide of learning that emanated from the Arab world in the twelfth century. The transmission of the *libri naturales* of Aristotle, among them *De caelo*, along with the works of Arab commentators and adaptors, altered perspectives on two interrelated questions: the habitability of regions at and beyond the equator; and the existence of landmasses equivalent to the ecumene in other parts of the world. On the first question, the work of the philosopher and physician Avicenna (Ibn Sīnā) became a key point of reference. Far from maintaining the uninhabitability of the torrid zone, Avicenna located the hypothetical city of Arīn on the equator, and even argued that equatorial regions were the most temperate on earth.[1] This claim gained credence from Claudius Ptolemy's *Almagest*, a work written in Alexandria c. 150 AD, translated from Greek into Syriac and Arabic in the late eighth and ninth centuries, and from Arabic to Latin between 1160 and 1175. While Ptolemy admitted ignorance of regions beneath the equator, he argued, as did Avicenna, that the climate there 'must be quite temperate' due to the swift motion of the sun at the equator and the mildness of the winters.[2] By extension it could be asserted that, if Avicenna was right that the equatorial region was the most temperate of all, it must be the site of the earthly paradise.[3] On the question of landmasses equivalent to the ecumene, the works of Aristotle (translated in the third quarter of the twelfth century, though of limited impact before 1200) offered two rather contradictory notions. The first of these was his reference in *De caelo* to inhabitants of the southern hemisphere, dwelling 'in the upper hemisphere and to the right, whereas we are in the lower and to the left' (2.2); the location of antipodal regions in the 'superior' (i.e. upper) part of the earth was rendered as 'nobilior' (more noble) in the Latin translation of the twelfth-century commentary of Averroes (Ibn Rušd), giving rise to the possibility that the earthly paradise was located even further south than the equator.[4] The second Aristotelian problem for conceptualization of the *antoecumene* was his theory of elements,[5] which led to a long and at times complicated debate about the relationship between earth and water. Since earth was the heaviest element it should, so it seemed, be completely enclosed within the lighter element of water; how to explain the obvious presence of expanses of earth above water had, by the fourteenth century, emerged as a difficult question for natural philosophers.

Contradiction of the system of zonal division came to the Latin west in the first quarter of the twelfth century, in the form of the *Dialogi* of Petrus Alfonsi, a converted Spanish Jew who constructed a series of dialogues between a Christian ('Petrus') and a Jew ('Moyses'). The dialogues, composed around 1110 and designed to explore the differences between Christianity, Judaism and Islam, and to demonstrate the superiority of Christianity, contain a digression in the first section. In arguing that God does not occupy a fixed position in the heavens, Petrus shows first that the positions of east and west are relative to each other, and then, at Moyses' request, outlines the organization of the known world on the basis of seven *climata*.[6] Arīn, the midpoint of the habitable earth according to Arab traditions (at the most temperate point of the equator, equidistant from both poles and the east and west), is situated in the first of the *climata*; six further *climata* extend up to the far north.[7] When Moyses points out that this scheme contradicts the division of the earth into five zones, he is given short shrift: the superbly temperate qualities of Arīn, produced by its position directly beneath the sun, dispel any possibility that it is uninhabitable; climatic division derives from the ancients, who divided habitable land into seven parts to match the number of planets.[8] Why is there 'uninhabitable' land to the south of Arīn, persists Moyses? Because, Petrus responds, the circuit of the sun is eccentric, causing it to come closest to the earth in the southern hemisphere, which was rendered as a result parched and infertile.[9] Satisfied with cosmography (two illustrative diagrams are provided for Moyses and the reader), the pair return to their theological dispute.

Petrus' denial of habitation beyond the equator was in general agreement with the views of Arab geographers, who tended to mark sub-equatorial regions on *climata* diagrams as desert, 'unknown', or who stated, as in one thirteenth-century instance, 'we possess no accounts of this hemisphere'.[10] However, in the Latin west the dismantling of the idea of torrid zone as a region of unendurable heat inevitably encouraged speculation about the extent of habitation within it: although influential twelfth-century texts such as Honorius Augustodunensis' *Imago Mundi* repeated the theory of zonal division and stated that human habitation was known only in the northern temperate zone ('solus solsticialis inhabitari a nobis noscitur'),[11] as early as 1141 it is possible to find the assertion in the works of Raymond of Marseilles that the torrid zone was inhabited well beyond the equator.[12] Similar rethinking of boundaries is evident in Hermann of Carinthia's *De essentiis* of 1143. Hermann was educated either in Paris or Chartres, but from at least 1138 he worked at the frontier of Arab and Latin cultures in Spain and southern France, where he was able to draw on sources hitherto unavailable in Europe, such as Ptolemy's *Almagest*, al-Battānī's *De scientia astrorum*, and Abū Ma'shar's *Introductorium magnum*, as well as more familiar texts like Boethius' *De consolatione philosophiae* and *De institutione arithmetica*, and Macrobius' *Commentarii*.[13] Despite a relatively restricted circulation, *De essentiis* contained ideas that, over the course of the next century and a half, became standard features of debate about the extent of terrestrial habitation. In the part of *De essentiis* dedicated to the generation of plants and animals, Hermann maintains a Cratesian division of the earth into four parts, while noting the lack of certain knowledge of all except 'our quarter'. He is particularly concerned with the measurement of the known world, and makes reference to its representation on a globe. Hermann notes that astronomical calculations have established that the extent of the known world from east to west (furthest India to furthest Libya) is 'almost a whole diameter of the earth's orb' (integra fere terreni orbis diametro interposita). As a consequence, he concludes, 'the people of each limit [i.e. the ends of India and Libya] are antipodal to each other' (utriusque termini populos antipodas ad invicem constituit).[14] Hermann's belief in the latitudinal extent of the ecumene enables him to use the term 'antipodes' to describe the relation between its eastern and western end. The northern boundary of the known world, however, remains just under 30 degrees beneath the north pole: a latitude marked by Scythia, the Rhipaean mountains, the Rubean woods and the marshes of Maeotis.[15] To the south, Hermann gives a contradictory account of the equatorial region. After asserting its uninhabitability due to excessive heat, he cites anonymous geographers ('girographi') who locate on the equator the islands of Arīn, Taprobana, and the six Fortunate Islands,[16] and adds the argument, attributed here to Ptolemy, that the climate at the equator is temperate, since the lateral movement of the sun over the region is swift, but close enough to prevent severe cold.

Passage to equatorial regions was not possible, according to Hermann, because the hot and dry nature of Sagittarius had created an infertile zone of around 50 days' journey in the area.[17] It was, instead, to the western part of the northern hemisphere that Hermann saw opportunities for overturning barriers:

> Ad subiectam vero plagam que infra nos inter circulum equinoctialem et polum borealem, nichil prorsus video quod transitum prohibeat ... nisi forte propter ignorantiam vie in tanti spatii impendio (cuius certam habemus dimensionem) quantum infra orizontem hemisperii nostri secernitur. ... In ea tamen parte non modica est opinio eam esse regionem quam Paradisum vocant
>
> As for the tract of land which is below us, between the equinoctial circle and the North Pole, I see nothing at all which prevents our crossing it ... unless perhaps we are prevented by our ignorance of the way in an area (whose dimensions we know definitely) of so great an expanse as is divided off below the horizon of our hemisphere. ... However, there is a substantial opinion that locates in that part the region which they call Paradise.[18]

The meeting of traditions Latin, Greek, and Arabic seems to have prompted Hermann to re-examine boundaries. He asserts that ignorance, rather than aquatic vastness, prevents travel across the expanse of the western ocean. And his suggestion that the earthly paradise might be found to the west of the known world similarly results from the juxtaposition of divergent traditions: one, more widespread, located the earthly paradise at the far east of the known world; the other placed it to the west of the ecumene. Men first came from the east, Hermann notes, but some historians suggest that 'in the West are the Islands of the Blessed ... so much more blessed than our lands that ... some have thought that they are an earthly paradise' (ab occidente vero Insule Fortunate ... tanto terris nostris beatiores sunt, ut, ... eas nonnulli Paradisum terrestrem arbitrati sunt).[19] The discourse of measurement and global mapping necessarily comes up against the limits of certain knowledge. But his access to literature not available to scholars in other parts of Europe allowed Hermann to adduce a greater range of possibilities with regard to *terrae incognitae* than they could, without ever resolving their contradictions into a definitive synthesis.

The political significance of the revision of theory undertaken by Hermann and others in the twelfth century only became apparent roughly one hundred years later. Until this point, it was possible to envisage access to antipodal regions in only two ways, ascent or descent: the celestial perspective granted to Scipio, or the subterranean passages leading to the antipodal underworld. But from the middle of the thirteenth century the possibility of travelling beyond the extent of the known world by breaching the torrid zone to the south began to be mooted with some seriousness. Such expeditions were not conceived without the simultaneous imagination of the role sovereign power might play in the discovery and appropriation of unknown land. In his *Liber de natura loci,* Albertus Magnus (c. 1200–1280) reviewed arguments for and against the possibility of inhabitation of the part of the world between the equator and the south pole (including, unusually, direct reference to Crates of Mallos).[20] Albertus not only affirmed the existence of habitable land there, but concluded that although it might be difficult, passage to it was not impossible. Contact with people existing beyond the equator in southern climes was impeded by a vast and sandy desert, but it would be amazing ('mirum videtur') if there were no points of access to the southern part of the world.[21] The 'lower' hemisphere, moreover, was divided in the same manner as the upper one, had regions uninhabitable because of cold and heat, and habitable regions divided by climate ('distingui per climata'), just like the 'upper' one.[22] Crucially, Albertus argued for the variability of climate within the so-called torrid zone, allowing for the possibility that some parts of it might lend themselves to inhabitation and passage.[23] Albertus noted on more than one occasion the failure not only of historiography, astronomy, or philosophy but also of kings to penetrate the barriers to the southern part of the earth. This inability of 'the messengers and armies of kings' (nuntii regum et exercitus) to reach such regions he attributed to the aridity of the vast and wide desert to the south, or to a group of quasi-magnetic mountains to the south of the equator that exert an irresistible attraction on human flesh. But Albertus did not rule out the possibility of royal commerce – and conquest – in the future.[24]

The sense of the potential for state- and church-sponsored exploration of unknown regions evident in Albertus' work is also one of the defining characteristics of the section of Roger Bacon's *Opus Maius* (1267) devoted to the extent of habitation on the earth. In this work, the compilation of which was encouraged by Pope Clement IV, Roger (c. 1214–c. 1292) argues strongly that an awareness of geography is necessary for the purposes of exegesis and conversion, and in order to protect the Christian faith against the onslaught of the forces of Antichrist and his allies. According to Roger, knowledge of place is of the highest utility ('locorum mundi cognitionis maxima utilitas est'): 'man can know the constitutions, natures, and qualities of all things of the world which derive from the property of place' (possit homo scire complexiones omnium rerum mundi et naturas et proprietates quas a virtute loci contrahunt).[25] But if knowledge of place constitutes a philosophical, theological, and political ideal, ignorance of place is the measure of the imperfections of philosophy and polity 'amongst the Latins' (apud Latinos). Not only do large parts of the ecumene remain unknown,[26] there is reason to believe that the unknown parts of the world (the southern hemisphere, and the underside of the northern hemisphere) are habitable. The thrust of Bacon's argument is that the ecumene extends beyond a quarter of the globe: the size of the ocean between the westernmost point of Spain and the easternmost point of India is 'small' (Aristotle), 'navigable in a very few days with a favouring wind' (Seneca);[27] moreover, land extends beyond the seven *climata* ('et ante climata et post', i.e. to the north towards the pole, and to the south beneath the equator).[28] Contrary to certain authorities,[29] who claimed that only one-sixth of the earth was inhabited because the rest of the sphere was covered by water, or who confined the ecumene to a quarter of the sphere, Bacon cited in support of his theories the 'fourth book of Esdras', which records that only one-seventh of the earth is covered with waters.[30] Natural philosophical reasoning dictated that if the known world extends to 66 degrees north of the equator (the latitude of the Scottish islands and Norway), a similar area of the two quarters of the world beyond the equator should be habitable.[31] Mathematical calculations suggested that the sun passes closer to the southern hemisphere, but far from indicating the impossibility of habitation, this raised the possibility that there might be more land there than in the northern hemisphere.[32] Bacon went on to note the statement made by Aristotle in *De caelo et mundo* (2.14) and confirmed by Averroes that the region beyond the Tropic of Capricorn was 'the loftier and better part of the earth' (superior pars in mundo et nobilior),[33] and therefore contained a great deal of habitable land. It was also possibly the location of paradise. Should the evidence of pagan philosophy not suffice, Bacon adduced the authority of the *Hexaemeron* of Ambrose, in which dwellers south of the equator are said to cast shadows to the north at certain times of the year.[34] Bacon admitted that this part of the world 'is not, as far as we know, described by any author; the people of the region are nowhere given a name; and nowhere are we told that they have visited us or we them' (tamen non invenimus apud aliquem auctorem terram illam describi, nec homines illorum locorum vocari, nec quod ad nos venerunt, nec nostri ad eos).[35] All the same, he attempted to give the unknown regions of the earth some form of visual representation.

The geographical section of the *Opus Maius* contained three 'figurae': a diagram of the northern hemisphere; a representation of the disposition of sea to land; and a world map, now lost.[36] The second of these images, described as a 'figura aquae', shows a strip of ocean running between two larger gatherings of water at the northern and southern poles, and dividing the eastern and western ends of the earth ('principium Indiae' and 'principium Hispaniae').[37] India and Hispania are understood here to extend beyond the equator: Bacon refers to evidence that Spain and Africa once formed a continuous landmass, and, following Pliny, that India extends further south than the Tropic of Capricorn.[38] From the evidence of the verbal description in the *Opus Maius* Bacon's world map was based on the seven *climata*: it included 'famous cities' positioned according to their longitude and latitude, the extent of each *clima* in miles and degrees, and the length of the longest day in each *clima*; the map also appears to have shown land beyond ('ante') the southernmost *clima*, and perhaps also beyond the northernmost.[39]

Bacon wrote from the position of a natural philosophy that was strongly allied to the interests of royal and ecclesiastical power. Unhappy with the traditional range of sources, he included in the *Opus Maius* the most up-to-date information concerning the extent of the habitable world (in particular, the reports of Franciscans, such as William of Rubruck, sent on missions of conversion to the east). He added this recent material to a base of knowledge constructed around a careful (but highly selective) reading of classical Greek (Aristotle), Roman (Seneca, Pliny the Elder) and late antique authors (Ptolemy, Orosius, Isidore), scripture (including apocrypha) and church fathers (Jerome, Ambrose, Basil), and later commentators, astronomical and geographical authorities, many Arab (Averroes, Alfragani), some diverse and unexpected (Aethicus Ister, the 'Toledo' tables, the prophecies of Merlin). The geographical section of the *Opus Maius* is, like many comparable works of this period, a synthesis – but it is one that leaves loose ends. Above all, there is an underlying fantasy of the contact that experience of unknown regions might bring. And in this context it is significant that Bacon, like Albertus, affirms the possibility of achieving such contact: he cites 'Ptolemy' (the 'Liber de dispositione sphaerae') to show that exploration beyond the *climata* to the equator has taken place, thanks to the backing of the kings of Egypt, and would still be possible, if the pursuit of such knowledge were supported by present-day princes. Alexander and Aristotle are frequently mentioned by Bacon as a model of the interrelationship of philosophy and expansionist polity; they are given, at one point, a muted echo in Nero-Seneca;[40] by implication, Roger awaits his Macedonian.

As its use of William of Rubruck's narrative shows, Bacon's work was itself the beneficiary of travel by Europeans to regions previously known only at several removes. Travel narratives enhanced the possibility of the eye-witness account, something that constituted a new development in the antipodal tradition, where vision had normally been divinely bestowed, or generated by theoretical speculation. Experience was, however, a slippery phenomenon in a genre in which authorial personae ranged from the elusive to the anonymous. As a consequence, perhaps, other more secure forms of authority (named, ancient, approved) were not discarded. The change that the emergence of personal accounts of travel made to the representation of the *antoecumene* was that the notion of travel to the antipodes was no longer the preserve of satire. As the edges of the known world were pushed outwards, and impenetrable barriers of heat, cold, and ocean seemed increasingly permeable, the antipodes began to figure as the furthest extent of knowledge, rather than the epitome of ignorance. For example, in the *Divisament dou monde*, the Venetian merchant Marco Polo's late thirteenth-century account of his travels and his record of the political, commercial, and ethnological disposition of the east, the island of lesser Java is noted for a thing that, Marco feels, will seem marvellous to every man: 'this island is so far towards the south that one does not see any part of the Pole Star' (une couse que bien senblera a cascun merveillose couse ... ceste ysle est tant a midi, que la stoille de tramontaine ne apert ne pou ne grant).[41] It is probable that this passage does not represent Marco's actual experience, being instead a piece of armchair travelling based on reports Marco had heard or read while at the court of Kubilai Khan.[42] But the significance of the story is its assertion of contemporary travel experience beyond the equator, and the possibility of its use as evidence that passage to the southern hemisphere had been achieved.

What Polo (or his scribe/editor, Rustichello) thought would be a marvel to his readers, by 1330 was a matter of straightforward if vigorous assertion. The author of the anonymous *Directorium ad Faciendum Passagium Transmarinum*, an energetically intolerant tract that calls on Latin Christians to liberate the Holy Land, recorded his travels beyond the equator while preaching the faith. Having given reasons to support his claim to have reached the equator, he asserts that he progressed further south, until he saw the Antarctic pole at an elevation of about 24 degrees. Others, merchants and 'men worthy of faith', had gone so far south as to see an elevation of 54 degrees.[43] The relevance of this for a resumption of the crusade to Jerusalem? It proves that land is inhabited beyond the *climata*, to the north and the south; that Asia is greater than commonly thought; that 'it is not foolish or false to mark out the antipodes' (quod non est friuolum neque falsum Antipodes assignare); and finally that true Christians (excluding, that is, the Greek Church) comprise not even one-twentieth of the world's population.[44] The same curious conjunction between the holy land and places literally on the other side of the world, the sense that the path to Jerusalem might somehow lie through the antipodes, was to recur in the thought of Christopher Columbus. In a fourteenth-century context the assertion of cross-equatorial passage is surprising and radical, but no longer, it seems, within the category of *mirabilia* as it was for Marco Polo; in the *Directorium* it is part of an expanding, and explicitly Christian, world vision.

Further verisimilitude was added to the idea of passage to the antipodes around the middle of the fourteenth century by the immensely popular travel narrative of John Mandeville. Mandeville's antipodal digression begins where Marco Polo's journey reaches its furthest point. In the isle of Lamary, where nudity, communism, and cannibalism are the key features of social life, Mandeville reports that men see the Antarctic star rather than the Transmontane, the lodestar of the north. This fact demonstrates the sphericity of the earth, and the possibility of navigation 'alle aboute the world and abouen and benethen'.[45] At this point Mandeville asserts personal experience, inserting a series of observations of the height of the stars that he made with an astrolabe while on his travels in parts north (Brabant, Germany, Bohemia) and south (Libya, other isles and lands beyond that country). These observations attest the symmetrical nature of the northern and southern lodestars:

> Et se ie eusse trouue nauie et compaignie pour aler plus auant, ie cuide estre certain que nous eussions veu toute la rondesse du firmament tout entour
>
> And yif I hadde had companye and schippyinge for to go more beyonde, I trowe wel in certeyn that wee scholde haue seen alle the roundness of the firmament alle aboute.[46]

One might see this confidence in the possibility of circumnavigation as a bold debunking of the theory of zonal division, since regions of unbearable heat or cold are tacitly swept aside in Mandeville's vision of the globe. Equally, it is consistent with the strand of scientific theory that questioned the validity of the concept of uninhabitable zones. What gives the passage immediacy is the recasting of learned speculation in the genre of travel narrative: Mandeville's dramatic, counter-empirical, strategy is to rewrite theory as experience. A standard explanation of the antipodes ('ceuls qui demeurent dessouz nous sommes pie contre pie'/ 'thei that dwellyn vnder vs ben feet ayenst feet') turns into the assertion that

> la terre de Prestre Iehan, empereur dinde, est dessouz nous. … Et ont la le iour quant nous auons la nuit, et aussi au contraire il ont la nuit quant nous auons le iour
>
> the londes of Prestre Iohn, emperour of Ynde, ben vnder vs … and han there the day whan wee haue the nyght, and also high to the contrarie thei han the nyght whan wee han the day.[47]

'We' here refers specifically to the English and Scottish, who are held to be at the westernmost end of the earth, opposite its easternmost extent, India, and equidistant from Jerusalem, described not only as the centre of the earth, but also its highest point. Mandeville's design of the antipodes – far east and far west are diametrically opposed and the notion of an antipodal continent in the northern hemisphere 'beneath' the ecumene is dispensed with – echoes suggestions that had been in currency since at least Hermann of Carinthia in the twelfth century.[48] To illustrate this proposition Mandeville inserts not, this time, experience, but anecdote: a story, heard when he was a young man, of a traveller who ventured to India and continued beyond until he encountered an island where 'son langaige' (his owne langage) was spoken. The anecdote is interpreted by Mandeville as the record of circumnavigation ending in return to a defamiliarized homeland (implicitly the British isles), rather than an encounter with another, parallel, world.[49] It leads him onto a reaffirmation, perhaps based on Macrobius, John of Sacrobosco, or Gossouin of Metz, of the impossibility of falling from the underside of the earth.

Mandeville's antipodes are simultaneously everywhere and nowhere. The image conjured up is indeed that of a terraqueous globe: this is not the globe of Crates of Mallos, divided into four landmasses, each cut off from the other; nor is it the globe of Aristotelian theorists who posited a greatly reduced ratio of land to sea, and the consequent impossibility of antipodal landmasses.[50] It is a vision of the world that draws upon at least four sources (Macrobius' *Commentarii*, John of Sacrobosco's *De sphera*, Brunetto Latini's *Tresor*, and the *Directorium ad faciendum passagium transmarinum*), but shapes them idiosyncratically and purposefully.[51] The novelty of the image of Mandeville's world is evident from the attempt of a redactor to redraw it. In the Vulgate Latin version of Mandeville's work, composed around the turn of the fifteenth century, Jerusalem has been moved from the centre of the entire world (where, in the vernacular versions, Mandeville placed it) to the centre of the northern habitable region, while references to circumnavigation and antipodeans are omitted.[52] This attempt to normalize Mandeville's geography reveals what makes it, against the odds, innovative: not only is a vision conjured up of the globe from pole to pole, and from far west to far east, but this vision urges the possibility of travel across its length and breadth. What is figured, then, is an intellectual and practical push to the ends of the earth. This is a dramatization of scientific speculation using the idea of the traveller, and the stubborn assertion of travel narrative (I saw, I measured). It presages many of the great themes of fifteenth-century geographical debate: the extent of the known world; the habitability of torrid and frigid zones; the subsumption of the antipodes in the known world; the complex negotiation between experience and authority in fashioning a world image. It reforms, criss-crosses, and encircles the globe. But the centre and the summit of Mandeville's world image remains Jerusalem.[53]

It is possible to see further development of the complicated relationship between travel narrative and cosmology – in which each fed off the other, but with travel accounts increasingly destabilizing ordered schemes – in a brief reference to the antipodes in another pseudo travel narrative *El Libro del conoscimiento de todos los reinos* (The Book of Knowledge of All Kingdoms). The book, recently dated to the last quarter of the fourteenth century, combines geo-political description with a marked interest in heraldry. Its anonymous author asserts that 'wise men' (sabios) call the Indian sea the South Sea (el Mar Meridional), and that:

> E deste mar fasta el Polo Antartico es una grand tierra que es la deçima parte de la faz de la tierra, e quando el sol es en Tropico de Capricornio pasa el sol sobre las cabeças de los pobladores, a los quales llaman los sabios antipodas. Et son gentes negras quemadas de la grand calentura del sol, pero que es tierra en que son muchas aguas que salen del Polo Antartico. Et llaman los sabios a esta tierra Trapovana, et confina con la ysla de Java …

> from this sea to the Antarctic pole there is a great expanse of land that is one tenth of the face of the earth, and when the sun is in the tropic of Capricorn it passes over the heads of the inhabitants, which the wise men call the antipodeans. And they are black people, burnt from the great heat of the sun, but it is a land in which there are many waters that come from the Antarctic pole. And the wise men call this land Trapovana, and it borders on the island of Java ...[54]

No longer the product of abstract reasoning, no longer unreachable, the antipodes function as the extension of the ecumene as far as the south pole: the reference to them seems casual, emerging from an account of travel in the Indian sea (specifically the kingdom of 'Viguy'), and blurred with the description of the extremities of the known world (Java, Taprobana).[55] Its rather confused synthesis draws on numerous thirteenth- and fourteenth-century debates about the habitability of torrid regions, and the possibility that the sun may pass closer to the southern hemisphere than to the northern – hence the notion of extreme heat at the southern tropic. At the same time, around two centuries of speculation about the possibly temperate nature of antipodal regions, including suggestions that paradise and the source of the Nile could be found in the southern hemisphere, lead to the assertion that antipodal land contains 'many waters' (i.e. the four rivers emanating from paradise).[56] A further conflation, this time ethnographic, appears to connect the antipodes with southern Africa, whose inhabitants were regularly described as black due to the heat of the sun. However, the name of the land is clearly a corruption of Taprobana, and its context in a description of southern Asia, alleged proximity to Java and general features (forty-five regions in the two islands, in which pepper and other spices are harvested, griffins and crocodiles) work to enfold the antipodes within the ecumene, and within the boundaries of ecumenical division. The passage concludes with a blazon of the arms of the king of Viguy: 'and the king has as his insignia a silver flag with a gold pale' (E el rey dende ha por señales un pendon de plata con un baston de oro).[57] King of the antipodeans? Only for a moment, the confused moment in which cosmological debate meets narrative description of space, and the order of the 'book of knowledge' temporarily breaks down. Insistence upon access to the other side of the earth – as theorized by natural scientists such as Albertus Magnus and Roger Bacon, and as apparently realized by the authors of travel narratives such as *The Book of John Mandeville* and the *Libro del conoscimiento* – unsettled the clear distinction between known and unknown parts of the world, but they did not, for the time being, provide a clear reformulation of the question of the antipodes.

The antipodes as *recessus*

Narratives of travel beyond the torrid zone did not prompt a sudden and dramatic redrawing of the world image because, in order to be given validity, such assertions of experience had to be assimilated to a theoretical framework. Polo's Java made its first appearance on maps in the fourteenth century,[58] but in general disparate reports of land and habitation beyond the equator provided insufficient authority to overturn existing models, or even to demand with any urgency their reformulation. More significantly, the discourse of expansion and assimilation to the ecumene remained outside of, if not necessarily contrary to, the mainstream understanding of the antipodes as a space that could only be apprehended by intellect and imagination. In the remainder of this chapter, I will discuss the philosophical tradition that developed around the question of the antipodes, beginning in the twelfth century with the Neoplatonist William of Conches' understanding of spaces beyond the known world as susceptible to 'philosophical reading'. That tradition reached its peak in thirteenth-century scholastic debates over the antipodes and related issues, such as the possibility of habitation at the equator, or the ratio of earth to land. Accompanying the deployment of the antipodes as emblem of a mode of philosophical thought and debate was the development of the meta-cartographical trends already evident before the twelfth century. The space of the antipodes on the world image became in some instances the place for lengthy written statements of theory, not as the result of any kind of *horror vacui*, but rather because of a desire to explain and – one might go so far as to suggest – celebrate the capacity of theory to construct and attest places beyond the known world. Yet there was another way in which the lack of physical passage to the antipodes gave the concept particular resonances. An important part of the role of the antipodes as *recessus* – a term that could describe a depth, or remote place, but also a retreat, or withdrawal from the world – was the development of the antipodes as space between earth and heaven, the possible location of inferno, purgatory, or the earthly paradise. In the work of Dante and Petrarch, I will argue, the antipodes were used as a third term, a means of mediating between life on earth and the life to come, vice and salvation, ambition and virtue. The theme of antipodes as *recessus* inevitably complicated, perhaps even retarded, the discourse of exploration and expansion that had begun to sound increasingly loud from the thirteenth century.

Renewed interest in the question of the antipodes in learned circles seems to have stemmed initially from a revival of interest in Macrobius, Martianus Capella, and Neoplatonic thought generally, which began in the second half of the eleventh century and gathered pace in the first half of the twelfth.[59] Previous revivals had shown that the antipodal element in Neoplatonic texts contradicted aspects of the Christian tradition, and representation of the antipodes carried inherent risks that demanded modification and mediation. In this regard the emergence of dialectic as a mode of intellectual expression in which heterodox ideas could be articulated and debated, if not approved, legitimized the presentation of antipodal theory *as theory* – not as belief. The work of William of Conches, traditionally associated with the 'school of Chartres',[60] epitomises philosophical interest in Macrobius and Martianus, and contains notable adaptations of, and responses to, classical antipodal theory. The use of dialectic was complemented by the development and proliferation of the encyclopedia as a means of organizing knowledge. Even before the emergence of Chartrian humanism the capacity of a lively interest in late antique cosmology to generate new and striking images of the world was evident in the work of the encyclopedist Lambert of Saint-Omer. For both William and Lambert, the map of Macrobius remained the basis for representation of the entire world, with *antoecumene* alongside ecumene. Rather than copying Macrobius, however, both sought, in different ways, to probe and supplement his map, and in particular to examine the notion of a southern temperate zone.

The *Liber Floridus*, a deeply idiosyncratic and richly illustrated encyclopedia, was compiled by Lambert, a canon of the church of Our Lady in Saint-Omer, principally between 1112 and 1115, although it was subject to revision up to 1121. Lambert's interests ranged widely, and his book, regarded by an earlier generation of scholarship as the product of a disordered mind, has been convincingly shown to be the product of a coherent scheme, albeit one that operated on the basis of a fundamentally associative logic, and that was subject to almost continual alteration and supplementation.[61] The *Liber Floridus* survives in an autograph manuscript, as well as in nine copies dating from the twelfth to the fifteenth centuries, an unusually large number for a work of this kind.[62] It contains not only material concerned with geography, time-reckoning, cosmography and astronomy, but also botany, and history both local (focused on Flanders and Saint-Omer) and universal, the whole 'pervaded by an emphatic sense of allegory and eschatology'.[63] Among its many geographical and cosmological illustrations (executed, it is thought, by Lambert himself), the book contains a series of zonal representations of the earth. These representations are, for the most part, explicitly derived from the text of Macrobius' *Commentarii*,[64] but they show a considerable degree of innovation. In them Lambert depicts a southern temperate zone marked either with the words 'unknown to men of our race', or 'unknown to the sons of Adam' (incognita hominibus nostri generis/ filiis Ade incognita); a strategy consistent with the comments of Augustine. However, in a map which is missing from the autograph manuscript of the *Liber Floridus*, but which is contained in seven later copies of the work (including the earliest and most faithful of all the copies: Wolfenbüttel, Herzog August Bibliothek, MS Gudeanus lat. 1),[65] a more expansive series of comments is placed in the empty space of the unknown part of the world (Plate 1). The map is designed to combine what might be described as geometric and ecumenical traditions of representing the earth, an intention signalled by its dual title of 'Spera geometrica' and 'Hormista regnorum mundi'.[66] Oriented to the east, it consists of two hemispheres divided by a representation of the equatorial torrid zone, which coincides with the centre of the codex. On the left, a fairly detailed ecumene is surrounded by a series of islands, ranging from the semi-legendary 'Thyle' (Thule) in the north-west to Taprobana in the south-east. Asia occupies the top half of the known world, extending from 'India ultima' in the east to Palestine, Pamphylia and Phrygia at its westernmost extent, while beneath it the Mediterranean divides Europe from Africa, the two connected by the circular island of Sicily. The representation of the ends of the known world on this map is also fairly characteristic of ecumenical cartographic tradition. In far southern Africa a strip of land has been drawn to show the 'Terra Ethiopum' and the 'Deserta Ethiopie'; the strip extends further east to southern Asia, where a 'place of dragons, and serpents and cruel beasts' (locus draconum et serpentium et bestiarum crudelium) has been marked. At the far east of the image the earthly paradise is represented as a peninsula, from which its four rivers (Tigris, Euphrates, Nile and Ganges) flow into the world.

The map bears witness to a number of topics of particular interest to Lambert: not only the earthly paradise, subject of several entries in the *Liber Floridus*,[67] but also the natural world (tigers, griffons, parrots, lions and the phoenix are marked in Asia as well as the serpents), history, and the figure of Alexander (the '32 kingdoms' enclosed by him are marked in the far north-west of Asia). The right-hand side of the map is devoted to the unknown part of the world: it is filled with a slab of land, equal in size to that of the known world on the left, but unaccompanied by islands, and with a wavy, rather than a smooth coast-line. In opposition to the northern half's busy divisions and locations stands a monolithic chunk of text:

> Plaga australis temperata. sed filiis ade incognita, nichil pertinens ad nostrum genus. Mare namque mediterraneum, quod ab ortu solis ad occidentem defluit et orbem terre diuidit humanus oculus non uidit; quoniam solis ardore semper illustratum qui desuper per lacteum currit circulum, accessus repellit hominum, nec ulla ratione ad hanc zonam permittit transitum. Hanc inhabitare phylosophi antipodes autumant, quos a nobis diuersitate temporum diuisos asserunt, nam cum estate torremur, illi frigore congelantur. Nobis uero septentrionalia sydera cernere permissum est et illis penitus denegatum. Nulla alia astra sunt que illorum obtutibus denegentur, et que simul cum illis oriuntur simul ueniunt in occasum, et dies noctesque sub una longitudine patiuntur. Solsticii autem celeritas et sol per brinam [sic: brumam] properando reuertens, bis hyemem per illos inducit.[68]

> Southern region, temperate but unknown to the sons of Adam, extending in no way to our race. For the human eye does not see the sea that lies between the lands, which flows from the east to the west and divides the world; since it [the sea] is always bright with the heat of the sun, which from above runs through the Milky Way, it repels the approach of men, nor does it permit transit to this zone by any means. Philosophers assert the antipodeans inhabit this zone, whom they say are divided from us by the diversity of seasons. For when we are burning with heat, they are congealing with cold. To us it is moreover permitted to make out the northern stars, and to them it is completely denied. There are no other stars which are denied to their gaze, and those stars that for them rise at the same time, set at the same time, and they experience days and nights of equal length. However, the frequency of the solstice [differs], since the sun, returning to reach the winter solstice, induces winter twice for them.

This somewhat garbled account conflates Martianus Capella's passage on the experience of inhabitants of the antipodes (6. 605–6) with his passage on inhabitants of equatorial regions (6. 607), who were thought to experience days and nights of equal length, and two summers and two winters per year.[69] Nevertheless, its function and the sentiments it expresses are clear. The passage outlines a conceptual bridge between the two landmasses. It begins with 'genus nostrum', moves to the topos of inaccessibility and repulsion due to cosmological forces and the equatorial sea that separates the lands of the temperate zones, and finally, licensed by the invocation of 'philosophi', it theorises habitation. Imagination of the inhabitants of the austral region encompasses their experiences of the seasons and the stars, their climatic and cosmological perspective. By its conflation of the text of Martianus, the inscription in the 'plaga australis' also manages to attribute both identity and difference, symmetry and a-symmetry to the southern zone: their experiences of cold and heat are the reverse of ours (identity), but they cannot see the northern stars and experience two winters (difference). The result of Lambert's confusion, or misunderstanding of his late antique sources? Probably. But the idea of the southern land seems to demand a summary of possibilities, acts of theoretical balancing; in its evocative space, doubt and assertion accompany admissions of ignorance and curiosity.

A third dimension to this map reveals how flexible representation of unknown land could be. For, in the northern hemisphere, reference is made to another region beyond the ecumene. The largest and most western island, directly opposite the representation of paradise in the far east, is marked with the legend 'Hic antipodes nostri habitant sed noctem diversam diesque contrarios perferunt et estatem …' (here our antipodeans live, but they experience different night and contrary days, and summer). The island signifies the underside of the northern hemisphere (hence 'our' antipodeans), land separated not by the torrid zone, but by the expanse of Ocean, and the frigid Arctic region. The text (again derived from Martianus Capella) seems incomplete: it has been suggested that the island is meant to be imagined as continuing on the underside of the map.[70] One could alternatively posit that this incompleteness denotes a different mode of representation: symbolic and speculative, rather than strictly topographic, a mode that here includes the representation of paradise as well as the antipodes. The multiplication of *anto-ecumenical* representation on this map results, at any rate, in an infiltration of the ecumene; at the margins of the known world is located the presence of the unknown, or more precisely that which textual authority (biblical in the case of paradise, philosophical in the case of the antipodes) assumes and asserts, islands that represent the possibility of place rather than its actuality.

The antipodes on the world map of Lambert reveal a complex balancing act between different sources of authority. If his world image derives from classical cosmography, as transmitted by authors such as Macrobius, Martianus, and Isidore, its antipodal content acknowledges Augustine's exclusion of the space from Christian history. That position, in which the idea of the antipodes must be located between divergent streams of authority, is also visible in the 'philosophical reading' of the antipodes undertaken in the work of William of Conches. William's precise connection with Chartres is uncertain, but his close ties with the grammarian Bernard of Chartres, and his deep and systematic engagement with natural philosophy are well established.[71] William produced glosses on both the *Timaeus* (in Calcidius' Latin translation) and Macrobius' *Commentarii*, evidence of his response to these texts not simply as a commentator, but as a teacher.[72] Both these works were important sources for William's major works of philosophical instruction, the *Philosophia* (written c. 1125–1130) and the *Dragmaticon* (c. 1144–1149). The *Dragmaticon* is similar, even identical, to the *Philosophia* in many passages, but it refashions the material in the earlier text as a dialogue between a 'Dux' and a 'Philosophus', and contains significant revisions in some areas.[73]

William discusses the question of the antipodes and their inhabitants most explicitly in the fourth book of the *Philosophia*, and book six of the *Dragmaticon*. He notes the existence of two temperate parts of the world, one on the northern side of the torrid zone, the other beyond it. At this point he raises the question of habitation, noting that 'we believe one [temperate zone] alone to be inhabited by men, not all' (unam tamen tantum ab hominibus inhabitari credimus, nec totam).[74] William immediately goes on to state that philosophers speak of the inhabitants of regions on both sides of the torrid zone – not because they are there, but because it is possible that they are there: 'non quia ibi sint, sed quia ibi esse possunt'.[75] Consequently, he will speak of those antipodal inhabitants, 'which we do not believe exist, for the purpose of understanding philosophical reading' (de illis, quos non credimus esse, propter intellectum lectionis philosophicae dicamus).[76] As Irene Caiazzo has noted, these remarks slightly tone down William's more positive statement in his glosses on Macrobius' *Commentarii* that 'Macrobius does not affirm that there are men there, but he says they should be believed to exist because of the similar climate' (Macrobius non affirmat ibi esse homines, sed dicit esse credendum propter consimilem temperiem).[77] On its surface, William's reversion from belief ('credendum esse') to possibility of existence ('possunt esse') appears to represent a compromise between the influence of Augustine, and the Neoplatonic cosmological theory that William was concerned to convey to his readers. The result of this simultaneous disavowal of antipodal inhabitation as literally beyond belief, and the insistence on the antipodes as an integral part of philosophical thought, is that the antipodes become emblematic of the *lectio philosophica*. The antipodes are, in William's formulation, a space of intellection: coterminous with a philosophical tradition, they can be apprehended only through reading.[78] By taking this position William authorizes the discussion of antipodal inhabitants: such discussion cannot be omitted if the process by which the *lectio* is comprehended is to take place successfully. It is not that reason overcomes or sidelines belief here, but rather that two different significations are given to the habitation of known and unknown parts of the world – the former is subject to belief and reason; the latter to reason alone.[79] The antipodal 'they' are hypothesized, disbelieved, yet intellectually crucial – a philosophical figment. How, then, are they and their putative land to be represented, and what is their relation to the ecumenical 'we'?

In the *Philosophia* William was concerned to elaborate degrees of difference: days and nights are of equal duration across the northern hemisphere, even if they occur at different times; the 'antoeci' in the southern hemisphere experience day and night at the same time as the inhabitants of the known world, but seasons at opposite times; the antipodeans of the antoeci (those in the 'lower part' of the southern hemisphere) are different in every regard – 'we and the antipodeans of the antoeci have neither summer nor winter at the same time as each other, neither day nor night' (nos vero et antipodes antoecorum neque simul aestatem neque hiemem, neque diem, neque noctem habemus).[80] This exposition of antipodal theory leads onto a brief account of the known world, and its division between Asia, Africa, and Europe, illustrated by a map which in most manuscripts shows an image of the world divided into zones, with an encircling ocean, a central torrid zone, and two frigid zones (see fig. 14). The northern hemisphere is divided into the three continents, with major seas and rivers marked, while the southern is empty of topography, usually marked simply 'temperata', 'temperata zona', or 'temperata australis'. Unlike Lambert of Saint-Omer's 'Spera geometrica', William's world map does not mark antipodal inhabitation, reserving discussion of the possibility to the accompanying written text.

The *Dragmaticon* largely reproduces the *Philosophia*'s account of the relation of the known world to antipodal spaces and inhabitants, but it adds an extended proof of the rotundity of the earth, against 'certain bestial people, believing more in perception than reason, who maintain that the earth is flat' (Quidam bestiales, plus sensui quam rationi credentes, dixerunt terram esse planam).[81] This section develops and illustrates a crucial aspect of William's discussion of the differences and similarities of the experience of peoples in different places. Two striking diagrams are used by William to prove the sphericity of the earth. The first of these, the 'circulus solis', shows the circuit of the sun passing directly over a city of the east ('civitas orientalis') and a city of the west ('civitas occidentalis').[82] On certain manuscripts these cities are given 'eastern'

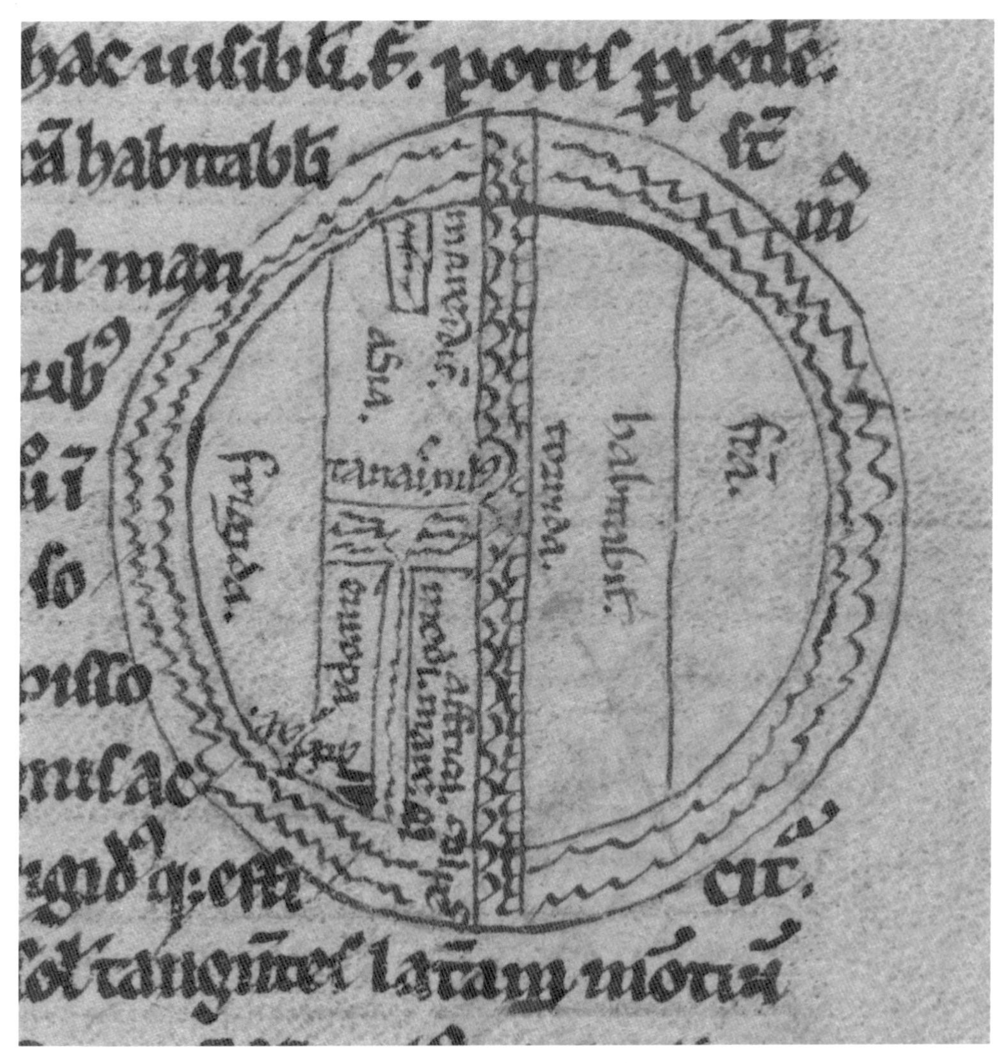

Fig. 14. London, BL, MS Arundel 377, f. 133r. World map in William of Conches, *Dragmaticon*. East at the top. The map is divided into five zones. The two frigid zones (both marked 'frigida') are located in the far north and far south. The southern temperate zone is marked simply 'habitabilis', next to an uninhabitable central zone ('torrida'). The temperate northern zone is filled with a representation of the ecumene, featuring division into three parts ('Asia', 'Europa', 'Affrica'), the Mediterranean, Tanais and Nile rivers, the Indian sea ('mare indicum') with the western extent of the known world marked by 'calpes' in Africa, and 'athlas' in Europe. Wavy lines mark the equatorial and encircling oceans. Similar zonal maps appear in William's *Philosophia*.

and 'western' features: BL Arundel 377, for instance, represents the eastern city as a tower with six spikes, and the western as a square with a church spire emerging (fig. 15). However, the majority of manuscripts emphasize the identity of the cities rather than any cultural or religious differences. In making such points William manages to dismantle the opposition between east and west, since it becomes easier to see east and west as points on a circuit, rather than on the linear plane of the ecumene. But the distinction between north and south remains: a second

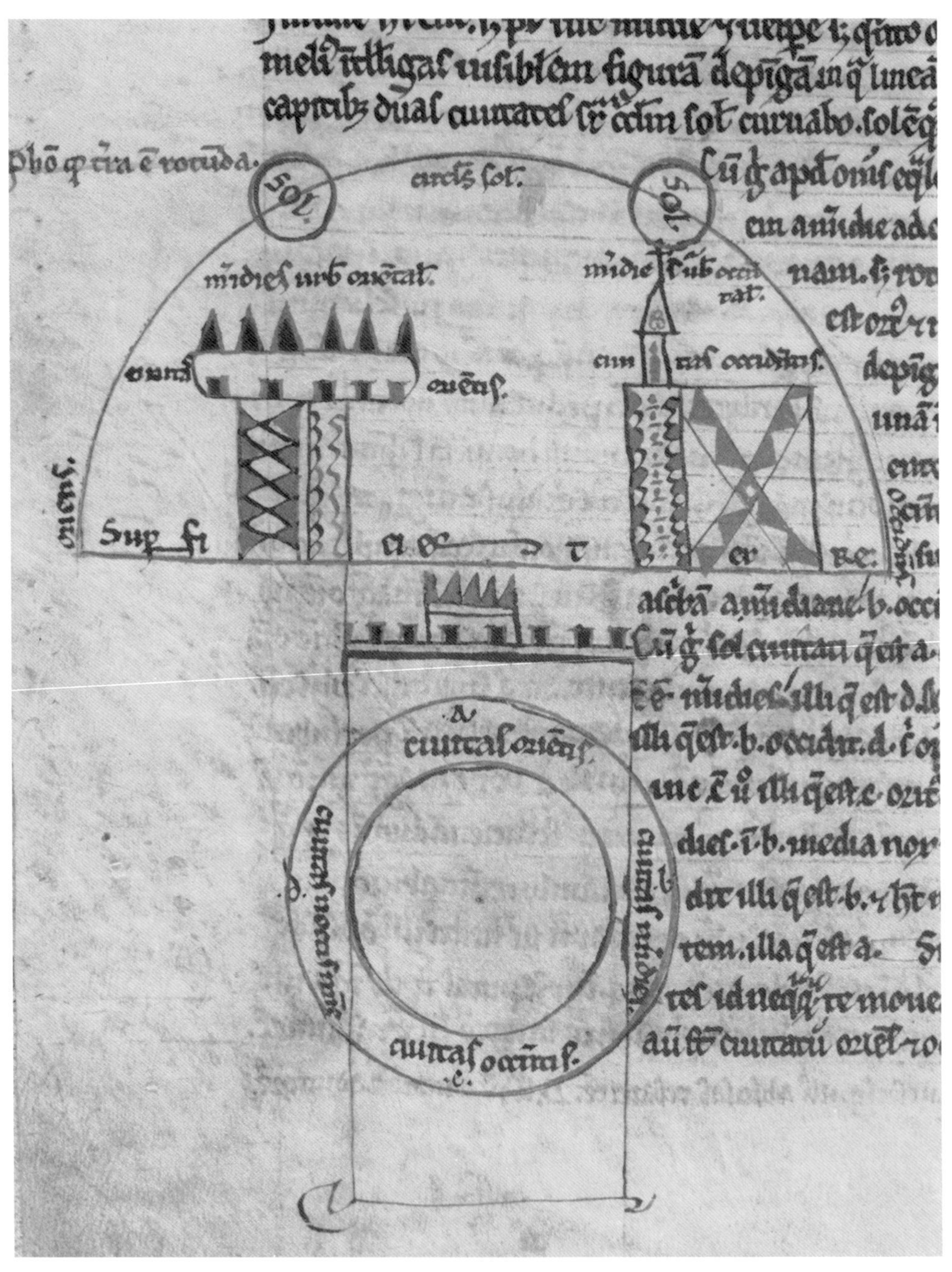

Fig. 15. London, BL, MS Arundel 377, f. 131v. East-west and four cities diagrams in a late twelfth- or early thirteenth-century manuscript of William of Conches, *Dragmaticon.* In the upper image two cities are marked east and west; the arc of the semicircle represents the passage of sun ('circulus solis') above them. Representation of the two cities differs: the eastern is a tower with six spikes, the western a square with a church spire emerging. Beneath, a circle marked a-b-c-d shows the city of the east ('ciuitas orientis') at the top, the city of the west ('ciuitas occidentis') at bottom, and the city of midnight and the city of midday in the north and south, indicating the different experience of time.

illustration shows four putative cities within the known world – of the north, south, east, and west (fig. 15).[83] These cities mark the points that the sun has reached at different times of its circuit: the fact that people in the cities observe the rising and setting of the sun (and midday and midnight) at different times proves that the earth is round, since these differences would not exist on a flat earth. The diagrams are designed to show different experience of the sun and time within the known world, but their principle is an antipodal one – that a corollary of the rotundity of the earth is the existence of people who stand in different regions, and whose existence can be imagined in terms of shared and opposite experiences. Indeed, in his subsequent discussion of the northern-hemisphere antipodes William refers back to his 'figure of four cities, which we drew above' (figura quatuor ciuitatum, quam superius depinximus) to illustrate his point.[84] True to his desire to enhance a *lectio philosophica* more than a *lectio geographica*, William does not use actual cities to illustrate his points about the sun's movement across the earth: he resists, in other words, the application of theory. So while experience is invoked, it is entirely hypothetical: no mention is made of travellers' reports, either contemporary or contained in classical texts such as Pliny; there is no real interest in the notion of testing such theory against experience; and, finally, there is no desire to push back the frontiers of knowledge through exploration. In William's work the antipodes represent the depth of philosophical thought, and their presence is necessary for the full expression of that thought, even if the existence of antipodeans cannot be believed. To exclude them from the *lectio philosophica* would be to sunder an integral part of philosophical tradition, the image of the world as transmitted and explicated by sources such as Macrobius and Martianus Capella.

William's inclusion of the antipodes within his 'lectio philosophica' is quite likely to have been an influence on the authors of two late twelfth-century Latin poems, both of whom specifically term the space of the antipodes a 'recessus'. The association of the antipodes with the death of Alexander the Great, familiar from Lucan, received an exuberant reworking in the *Alexandreis*, an epic poem of considerable popularity and influence, usually dated to 1178–82,[85] composed by the notary to the Archbishop of Reims and former schoolmaster, Walter of Châtillon. In book nine of the *Alexandreis*, Alexander announces his intention to travel beyond the known world to conquer the antipodes: 'I hasten to penetrate the shores of the antipodes and to see another Nature' (Antipodum penetrare sinus aliamque uidere/ Naturam accelero).[86] In book 10, Nature, furious at Alexander's planned transgression, alarms Leviathan (Satan) into thinking that the king will perform the prophesied leading forth of the dead from the underworld: 'he will not leave that Chaos intact, and he will seek to behold the depths of the antipodes and the sun of another Nature' (istud/ Non sinet intactum Chaos Antipodumque recessus/ Alteriusque uolet nature cernere solem).[87] From this outrage to natural order emerges a plot to poison Alexander. In his penultimate speech, following the submission of the known world, the king confirms Nature's fears by urging his followers to turn their ambitions towards the antipodes:

> Nunc quia nil mundo peragendum restat in isto,
> Ne tamen assuetus armorum langueat usus,
> Eia, queramus alio sub sole iacentes

> Antipodum populos ne gloria nostra relinquat
> Vel uirtus quid inexpertum quo crescere possit
> Vel quo perpetui mereatur carminis odas.
> Me duce nulla meis tellus erit inuia. uincit
> Cuncta labor. nichil est inuestigabile forti.[88]

> Now because nothing remains to be accomplished in this world, and lest the familiar use of arms languish, quick! Let us seek peoples of the antipodes lying beneath another sun, lest our glory and virtue leave untried that from which they can grow and merit songs of perpetual renown. While I am leader no land will be impenetrable to my men. Hard work conquers all. Nothing is inscrutable to the brave.

The phrase 'Antipodumque recessus' was echoed in the *Architrenius* (Arch-Weeper), a poem by Walter's contemporary Jean de Hauville. There Cupidity, mother of Avarice, damned to produce an endless thirst unquenchable by material good, is said to 'teach the skilful hand to steal; her eyes see into the depth of the Antipodes, both the night's days and the sinister shadows, the secret places of the heart's desire' (rapinis/ Artifices factura manus, visura recessus/ Antipodum noctisque dies umbrasque sinistras/ Ardentis secreta sinus).[89] The meaning of de Hauville's phrase is unclear: is the 'recessus Antipodum' simply a figure for inversion, for seeing so far into the night that day in the opposite hemisphere is perceived? Or are the antipodes being used here to invoke a supreme darkness? In the *Alexandreis* the desire to penetrate recess becomes the sign of excess, the sign of an anti-Ciceronian boundary breaking. That which Alexander envisages is precisely what the elder Scipio declares impossible and undesirable: *uirtus* by means of the expansion of *gloria*, 'songs of perpetual renown', passage to the impassable beyond the columns of Hercules, the burning interior of Libya, beyond India and the Ganges as far as people living 'under another sun'. Yet the folly of Alexander in this version seems not so much his act of overreach as his abdication of power. Precisely as the point of the 'Somnium Scipionis' is the retention of power by means of the restriction of ambition for glory to the limits of the Roman world, so the Alexander legend suggests that the abandonment of the ecumene for another people, another sun, and not least for 'odes', is not to expand but to renounce power. Indeed the lingering implications of subterranean locations held by the word 'recessus', as well as Nature's reference to bringing forth the dead, indicate that, along with the known world, Alexander's desire is to leave life behind. For in Walter's poem Alexander also acts as an avatar of Christ. The desire to reach all humanity, to spread a name to the ends of the earth and to release the dead from hell is the sign of proleptic divinity. That shadow of a greater glory hovers over book ten of the *Alexandreis*, including its antipodal references, tingeing Alexander's challenge to nature with ironic humour, and making its monstrosity comprehensible, indeed admirable, as well as laughable.

Walter's deployment of the antipodes as *recessus* echoes not only the debates of twelfth-century philosophy, but also a series of contemporary narratives in which the antipodes appear literally as a recess: a region beneath the earth that allowed passage to lost worlds – and to legendary rulers. Popular accounts of subterranean regions and kingdoms, where other worlds and heroic figures from the past might be found, were mediated through the works of learned culture, being either included in the histories of erudite clergy, well aware of the scientific traditions that located antipodal regions in other parts of a spherical world, or incorporated into romance.[90] Gerald of Wales, Walter Map and Gervase of Tilbury all told stories of people – usually unlettered, such as swineherds or servants – who stumble upon an underworld kingdom, notable either for its association with Arthur,[91] the alternation of its climate with the upper world,[92] or its inhabitation by homunculi.[93] It seems likely that these stories of subterranean kingdoms were the product of an 'imaginative synthesis' through which inherited beliefs and legends (in these cases Welsh and Irish) were fused with cosmological theory (evidenced by references to scientific concepts such as the antipodes, and hemispheres).[94] They were then given currency as narratives or anecdotes told at several removes from the teller. The antipodes become, here, a kind of learned joke, recorded and perpetuated precisely as the 'fabula' derided by Augustine, Isidore, and Bede.[95] The antipodal motif was a favourite of Arthurian romance: Chrétien de Troyes recorded the dwarf Bilis, 'king of the Antipodes' (rois d'Antipodés) amongst Arthur's baronage in *Erec et Enide*,[96] while in Étienne de Rouen's poem *Normannicus Draco* (1169–70) Arthur himself, following his departure to Avalon, receives the lordship of the antipodeans (antipodum ... jus), where he 'rules the lower hemisphere ... the other part of the world' (emisperium regit inferius ... Altera pars mundi).[97] From this domain he challenges Henry II of England to desist from his military interventions in Brittany.[98]

In these stories the antipodes provide a means of formulating and de-centring sovereign power. The fantasizing of Arthur's territorial expanse may be seen within the context of an 'imitatio Alexandri' tradition, where the extension of imperial power is associated with the exploration of geographic mysteries: the source of the Nile, the ends of the earth, the other side of the earth.[99] This is evident particularly in *Normannicus Draco*, where Arthur's rule over the antipodeans is said to be a feat that 'neither the zeal of Alexander nor Caesar could attain' (nec Alexandri potuit, nec Cesaris ardor).[100] But these subterranean kingdoms seem to constitute less the imitation of empire than its shadow and, at times, its lost conscience.

A space of the intellect

The deployment of the *antoecumene* in various twelfth-century contexts – natural philosophy, romance, epic, history – confirms the function of hypothesized *terrae incognitae* as a necessary element of the whole, but one that inevitably disrupts wholeness. Whether the product of a *lectio philosophica*, a part of the world image asserted by philosophers, a popular legend, or an element of classical literary tradition that continued to invite adaptation, the antipodes derived potency from their position beyond the thresholds of knowledge, yet within the world. They could not be known, but attempts to show and discuss the world recurred to their possibility because to think spherically was to think of the other side of the world, and to consider its habitation. For this reason scholarly discussion of geographical and cosmological matters continued to include *antoecumenical* spaces. The works of William of Conches, in particular, helped the antipodes become a standard topic for debate and argument in the thirteenth-century scholastic tradition. Such debate was often highly formulaic, but it ensured that the multiplicity of theories about *antoecumenical* space and its habitation was preserved without resolution. In some cases – most notably the *Liber introductorius* of Michael Scot and a miscellaneous compilation of the St Albans monk John of Wallingford – scholastic discussion was augmented by world images, in which spaces beyond the known world acted, as they had done for Lambert of Saint-Omer, as sites for expressions of theory rather than the endpoint of passage. Indeed by around 1300 the antipodes had emerged, in the work of Ramon Lull, as a sign of intellect – representative of phenomena that could be apprehended only by rational thought, as opposed to observation and experience. One could attribute the reluctance to insist on the possibility of penetration into *antoecumenical* spaces to a medieval *mentalité*, predisposed to closure and repetition rather than criticism of existing schemes of knowledge. It is more edifying, however, to cease to think in terms of oppositions between progress and stasis, open and closed, exploration and tradition. Discussion of antipodal regions prior to the fifteenth century was intellectually adventurous, at times experimental, and by no means static, even – perhaps especially – when it reaffirmed the unknowable nature of the antipodes. As I will argue in the following pages, throughout the thirteenth and fourteenth centuries new modes of thought – both scholastic and humanist – constructed new meanings for the antipodes, repositioning their significance within a Christian discourse increasingly preoccupied with the relationship between the known world, inferno and the earthly paradise.

The great popularity and influence of John of Sacrobosco's *De sphera* – an early thirteenth-century university textbook still in use in the seventeenth century –[101] indicates one reason why discussion of places beyond the known world continued in spite of their potentially heterodox nature: the topic became entrenched as a topic of scholastic disputation. The mechanics of disputation demanded that arguments be put in favour of habitation at and beyond the equator, and in favour of antipodeans, even if these propositions were to be rejected. In fact, Sacrobosco himself was relatively reticent on these matters: to prove that the earth is round he points to the evidence of a hypothetical man travelling south, and seeing previously hidden stars as he goes; he explains that only one of the 'world poles' can be visible from the northern hemisphere; he cites the different experience of sun and shade at 'situs noster' and at Ethiopia.[102] However, several commentaries on Sacrobosco's work sought to examine *antoecumenical* questions in some detail. In one of the earliest commentaries, attributed with no great certainty to Michael Scot, the question is posed 'whether the zone that is between the tropic of Capricorn and the Antarctic circle is temperate'.[103] On the affirmative side, the commentator uses the arguments that since nature does nothing in vain, land would not be created to be uninhabitable; on the other hand, the impossibility of passage between the northern and southern temperate zones due to the torrid zone prevents affirmation. The commentator goes on to ask whether there are mortal or immortal men in the southern temperate zone: if immortal they do not consist of the four elements, and are not descended from Adam; but only a heretic would suggest they are mortal. Since 'God was born and died for us in our habitable region', the same would have had to have happened for them in their region, a falsehood, since Christ did not become flesh twice and die twice.[104] Yet arguments could run in the other direction: some forty years later Robertus Anglicus was able to come down on the side of habitability at and beyond the equator by citing Avicenna and Isidore, as well as his own experience as a native of England, located, according to the ancient division of the *climata*, beyond the northernmost habitable boundary.[105] Robertus was similarly confident about the habitability of the earth's surface between the equator and the southern pole, although with the important qualification 'unless water should impede' (nisi aqua impediat), an implicit acknowledgment of Augustine's comments; he added the speculation that a smaller portion of land would be habitable than in the ecumene, due to the excessive heat in the area underneath the tropic of Capricorn.[106]

The *Liber introductorius* attributed to Michael Scot confirms the indeterminate nature of thirteenth-century thought on the antipodes, but it also reveals a richness not immediately evident in the formulaic terms of debate preserved in the commentaries on *De sphera*. The discussion of places beyond the ecumene in the *Liber* – the most detailed after William of Conches – floats several different possibilities for what might be contained in the *antoecumene*, including the tentative location of inferno and the earthly paradise in the torrid zone. While Michael's summary of explanations for *antoecumenical* space represents the depth of scholastic debate and the development of the *lectio philosophica*, it also represents the attempt of an individual to penetrate the recess – to see into the depth of the antipodes – and in that regard it begins to clarify the context in which Dante could imagine his Inferno, Purgatorio, and Paradiso to occupy a precise (and antipodal) place within and on the earth.

The *Liber introductorius* is divided into four distinctions: the first, devoted to astronomical matters, includes a discussion of the zones of the earth and the heavens; the second and third distinctions are concerned with astrology, including its practical use by astrologers. The fourth distinction, on the human soul, survives only as a fragment.[107] Assessment of the text is complicated by its survival in two versions, one considerably longer than the other, the chronology of which is uncertain.[108] Manuscript evidence suggests a moderate level of dissemination of the full text, but a much wider dissemination of extracts from it.[109] Does the *Liber*, at least in its longer version, date from Michael Scot's time at the court of Emperor Frederick II from 1227 to 1235, where he appears to have been employed as an astrologer?[110] Or is it, as recent research has suggested, not the work of a single author at a particular time, but instead the product of a 'fashion' – a mode of compiling astrological treatises, in vogue in northern Italy in the thirteenth century, by interpolation and verbose elaboration upon a core of material ultimately derived from Arabic sources?[111] Firm conclusions on these matters remain difficult, but a picture is gradually emerging of the *Liber introductorius* as representative more of the topics, theories, and debates within learned circles over the course of the thirteenth and early fourteenth centuries than scientific learning at the court of Frederick c. 1230. On the other hand, the notion of a composite text is problematized by the *Liber*'s strident use of the first-person pronoun: while not inconsistent with the thesis that a core text compiled by Michael was subject to numerous interpolations, such strong use of the authorial voice indicates that, at times, the *Liber* expresses the views of an individual, rather than a scholarly tradition – or more precisely the views of an individual in response to a scholarly tradition.

The discussion of the terrestrial and celestial zones in the first distinction of the *Liber introductorius* follows descriptions of the terrestrial and celestial spheres, with a particular focus on the zodiac and eclipses, and the seven *climata* of the northern hemisphere.[112] The exposition of the division between the zones, and the position of the equator and tropics is standard enough, including quotation from both Virgil's *Georgics* and Ovid's *Metamorphoses* to establish the equation between celestial and terrestrial zones.[113] Where the *Liber* becomes unusual is in its use of a more sophisticated version of the standard diagram to illustrate the zones, and in its chapter 'De notitia divisionis çone habitabilis nostre' (On knowledge of the division of our habitable zone) that follows the diagram, notable for its reference to several contradictory and heterodox theories. The diagram is almost wholly theoretical (fig. 16). Devoid of toponyms apart from Asia, Europa, and Africa, it shows the equator and tropics, the paths of the sun and moon, and clearly indicates zonal division, with each zone subject to unusually extensive description. The northern frigid zone, never receiving the rays of the planets due to its distance and declination, is not inhabited, 'nor can it be usefully cultivated'. The northern temperate zone is by contrast 'usefully cultivated and trodden upon by men' (cultiuata utiliter et ab hominibus calcata); ocean surrounds and is dispersed throughout it. The torrid zone, described as burnt up by the rays of the sun, is also marked 'locus inferni', in part a representation of the etymological connection between 'inferno' and 'infra', i.e. the zone within, or between the other zones.[114] The southern temperate zone is marked with two inscriptions:

> Temperata pars mundi uersus austrum et habitabilis licet ignoretur a nobis qui sint habitantes eam.
>
> Temperate and habitable part of the world towards the south, although we do not know who its inhabitants are.
>
> Gerit hec çona omnem similitudinem nostre çone que habitatur.
>
> This zone exhibits every resemblance to our zone which is inhabited.

The description of the southern frigid zone matches that of the northern, except that it omits reference to its length and declination and instead adds 'light is obscure there as in early evening' (ideo ibi est lux obscura ut in primo sero).

Two elements of the diagram are immediately striking: the preparedness to acknowledge symmetry of habitation in the southern temperate zone; and the location of the 'locus inferni' in the torrid zone. The diagram shows a tripartite division not only of the northern temperate zone but also of the southern, implying the 'every resemblance' (omnem similitudinem) of the caption: if we have three *partes* and a surrounding ocean, so do they. The context for this diagram, as the accompanying text makes clear, is certainly the post-Sacrobosco discussion of antipodes: there is no explicit mention of Augustine, although it may be that 'calcata' in the diagram's description of the northern hemisphere is a subtle reference to *De civitate dei* 16.9 ('aduersa pedibus nostris calcare vestigia'). A third significant feature of the diagram is brought out in the 'expositio' that follows it. Here it is emphasized that the zones must be understood to extend 'above and below', that is, right around the surface of the sphere, not simply over one half of it. Numerous diagrams throughout the *Liber* carefully represent the 'pars orbis sub terra' as well as the 'pars orbis super terram'.[115] This is one of several roughly contemporary examples of the manipulation of the zonal diagram in order to represent the full extent of the underside

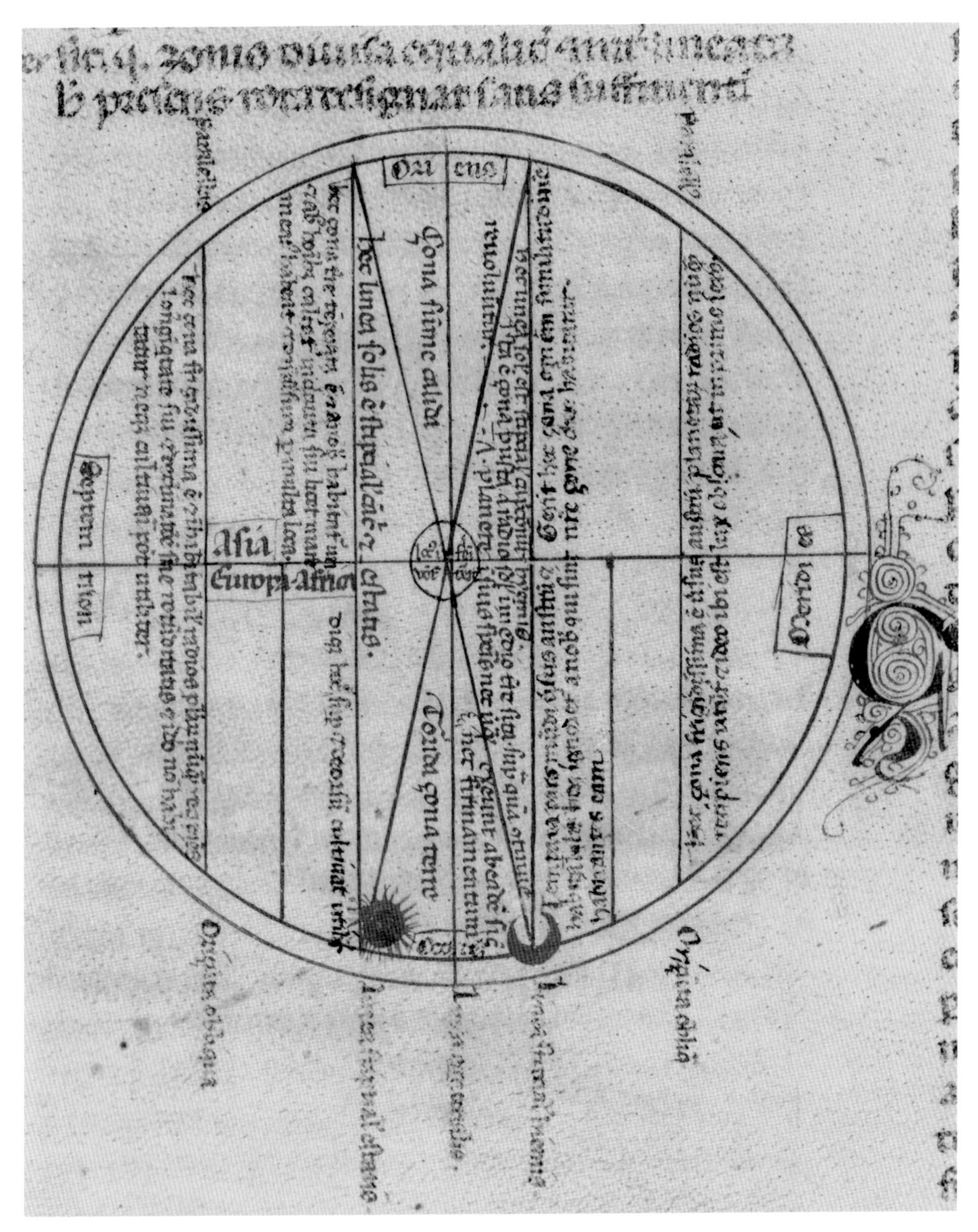

Fig. 16. Munich, Bayerische Staatsbibliothek, Clm 10268, f. 45r. World map from a manuscript (c. 1300) of Michael Scot, *Liber introductorius*. East at the top. This image shows the equator and tropics, the paths of the sun and moon, and clearly indicates zonal division, with each zone subject to unusually extensive description. The northern frigid zone, never receiving the rays of the planets due to its distance and declination, is not inhabited, 'nor can it be usefully cultivated'. The northern temperate zone, divided between Asia, Europe, and Africa, is by contrast 'usefully cultivated and trodden upon by men' (cultiuata utiliter et ab hominibus calcata). The torrid zone is described as burned up by the rays of the sun, and also marked 'locus inferni'. The inscriptions in the southern zone note its inaccessibility but also its symmetry with the known world. Its division into three parts mimics that of the northern temperate zone.

of the earth, or the 'western' hemisphere, from pole to pole. It is the product of cosmological thinking, resulting from the need to represent the revolution of the zodiac around the entirety of the globe on a plane surface. But it also leads to a reconceptualization of the known world: 'we know that there are inhabitants both above and below and they hold their feet around ours, such as the Indians, Egyptians, and so forth' (scimus quod quidam sunt incole tam de subtus quam desuper et tenent suos pedes circa nostros ut indi, egyptii et cetera).[116] This probable elaboration of William of Conches' discussion of east and west in the northern hemisphere reformulates the notion of the antipodes along the same lines mooted by Hermann of Carinthia, so that they are no longer unreachable peoples, but far-off dwellers within the same ecumene.[117]

What of hell? In the discussion of the zones contained in the 'expositio' and the chapter 'De notitia divisionis çone habitabilis nostre', Michael advances four theories with regard to the southern temperate zone, the torrid zone, and the frigid zones.

1. Demons and other spirits separated from their bodies inhabit the frigid zones: 'et eas inhabitant demones et multe anime a corporibus separate'.[118]
2. At death, souls leave human bodies, pass through fire and water, and enter a new body created in the southern temperate zone where they remain for 1000 years: 'fuerint quidam dicentes quod quam cito anima hominis separatur de corpore in ista zona ipsa transit per ignem et aquam et intrat noviter in corpus novum quod creatur in altera zona permanens etate completa mille annorum'.[119]
3. The frigid zones contain the pains of hell that are of extreme cold, while the torrid zone contains the pains of hell that are of extreme heat: 'zone extreme continent qualitates penarum inferni que sunt frigus quod descendit in centrum. et media zona continet penas caloris inferni quod est animabus ignis inconsumptilis et invisibilis ut patet in terminis'.[120]
4. The torrid zone is the land of the living, the promised land: 'çona vero perhusta est terra viventium corporaliter sive terra vere promissionis'.[121] Terrestrial paradise is located in torrid zone, a theory represented in a subsequent diagram, in which 'paradisus deliciarum' is marked at the centre of the torrid zone (f. 46ra); there the air is more temperate, the zone is higher than the others due to the shape of the globe ('globositas'); in the torrid zone is also found the fountain of life, dividing into four rivers, as well as the promised land of India, where the body of St Thomas is found, and where St Brendan is reputed to have journeyed with his companions.[122]

Michael, in airing these theories, does not necessarily agree with them. Indeed, he outlines the theory of the southern temperate zone as a repository of souls prior to judgement day only to deny it vehemently. He personally ('ego personaliter oppinor') believes the southern temperate zone to be cultivated and fully inhabited, since God would not create anything in vain. By whom it is inhabited cannot be known, only conjectured: 'And I, wanting to know this, never yet was able to learn except by supposition' (Et ego hoc volens cognoscere nunquam adhuc nisi oppinando potui experiri).[123] Nevertheless, Michael makes the interesting suggestion that the antipodes cannot be inhabited by one race of men alone because they receive different winds: 'but from ours we know that that [southern zone] in its entire circuit is cultivated and inhabited not by one race of people but by diverse races, who depending on their dwelling place have east, south, and west winds both advantageous and disadvantageous in the span of their year' (de nostro vero scimus quod ipsa in toto circuitu sui collitur et habitatur non ab uno genere gentium sed a diversis que secundum earum permanentiam habent ortum solis meridiem et occasum ventos etiam sibi utiles et inutiles etatibus sui anni).[124] The principle of symmetry generates a new conviction within antipodal theory: just as the diversity of climate in the known world produces diverse peoples, so the antipodes must be multiracial. Here an element of bold speculation is apparent, cutting through the accumulation of possibilities.

The *Liber* is marked both by the strength of the desire to know and the concomitant recourse to the first-person singular in the face of an absence of scriptural or other authority. This 'I' emerges from the thwarted intercourse between we and they to record experience by opining, and the experience of opining. The position of the commentator is contradictory: wanting to know, he is compelled to approach knowledge by conjecture, knowing that conjecture cannot be proved. That action of shaking off the collective, the 'we' that knows and is ignorant (and that knows its ignorance), leaves nothing in its place but the individuated writer, possessor of no knowledge 'unless I by opining ...'.

The diagrams of the *Liber introductorius* represent two broad directions that characterize the representation of the antipodes on the thirteenth-century *imago mundi*. Images of the antipodes could be deployed within the context of didacticism: the lengthy, popular, and copiously illustrated cosmographical treatise of Gossouin of Metz, *L'image du monde*, composed in the 1240s, included a diagram showing the passage of two men setting out from the same point on the globe, walking in opposite directions, and coming face to face on the opposite side of the earth. While this is a purely theoretical exposition of the rotundity of the earth ('comme une mouche iroit entour une pomme reonde'),[125] and promises no encounter with antipodal peoples, it nevertheless uses visual means to imagine the relationship between ecumenical and antipodal space (fig. 17). The other function of the antipodes on the world image at this time continued to be that of recess, as a space of internal cosmographical meditation. In the world map drawn by John of Wallingford not long before his death in 1258 the southern part of the world is a site for the type of theorization contained in Michael Scot's *Liber introductorius*. John located the image immediately before his excerpts from the chronicles of various historians, in particular his own work, and that of his contemporary at the monastery of St Albans, Matthew Paris (fig. 18).[126] The diagram is a climatic, rather than a zonal map of the world; it shows the known world divided into the seven *climata*, each of which is carefully marked within the diagram and named directly beneath it: from south to north, the '*clima* of the Indians', Ethiopians or Moors, Egyptians, the '*clima* of the Jerusalemites', the *clima* of the Greeks, Romans, Franks. An eighth *clima* has been added to accommodate England, Ireland, and 'northern parts'. The scheme is ultimately derived from Arabic sources, although adapted to Christian interests: while the centre of the earth remains the city of 'Aren' (in the '*clima* indorum'), the centre of the northern hemisphere is Jerusalem.[127] Floating outside the map, above the designation of east ('oriens'), masquerading it almost seems as the heading of the page, is the single word 'paradisus'.

As a consequence of the *climata* format the map does not show a torrid zone, but rather a

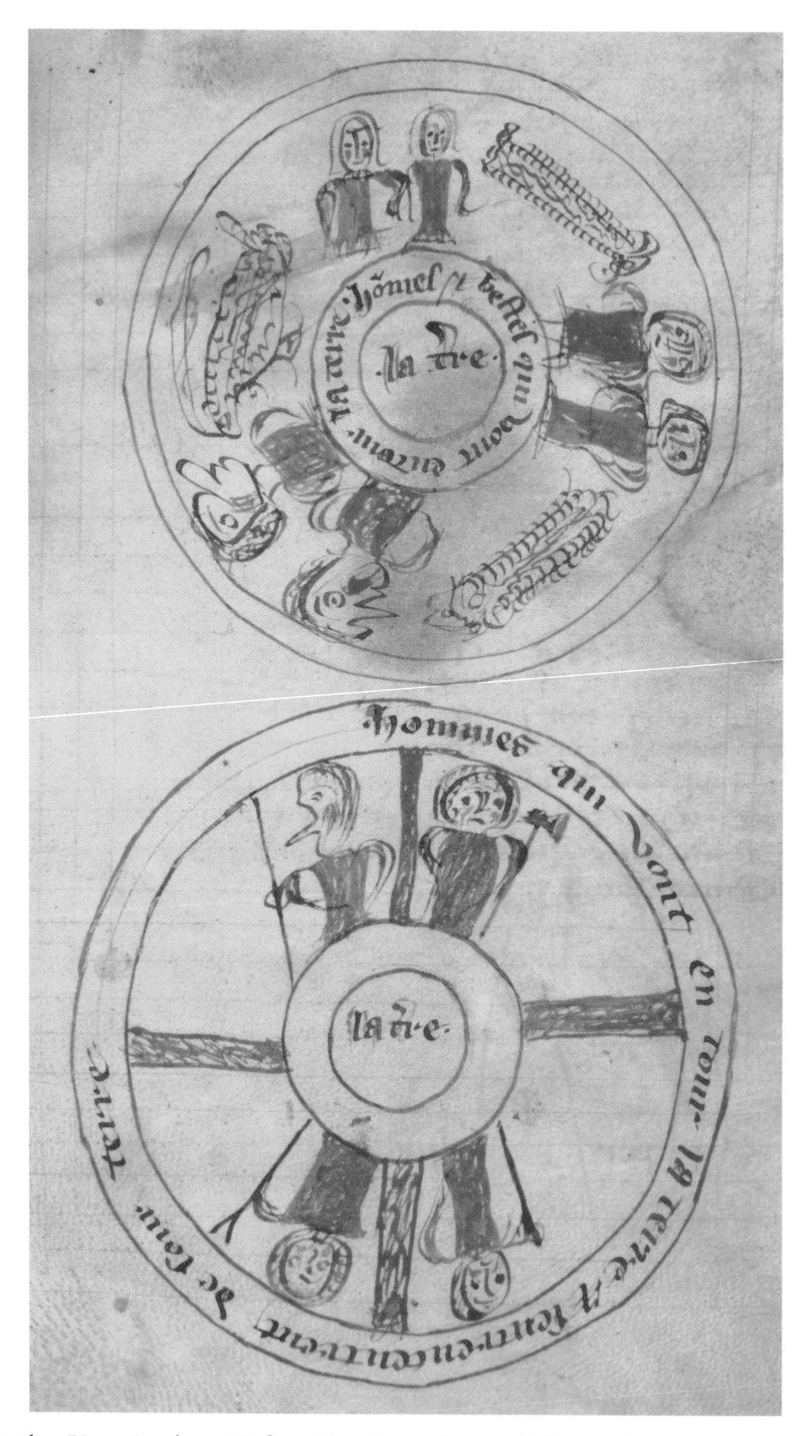

Fig. 17. London, BL, MS Royal 20.A.III, f. 54r. These images from a mid-fourteenth-century manuscript illustrate the thirteenth-century poem of Gossouin of Metz, *L'image du monde.* Above, the passage of men and beasts around the earth is depicted ('Hommes & bestes qui vont entour la terre'); below, the principle of the antipodes is demonstrated by four figures standing on opposite sides of the earth.

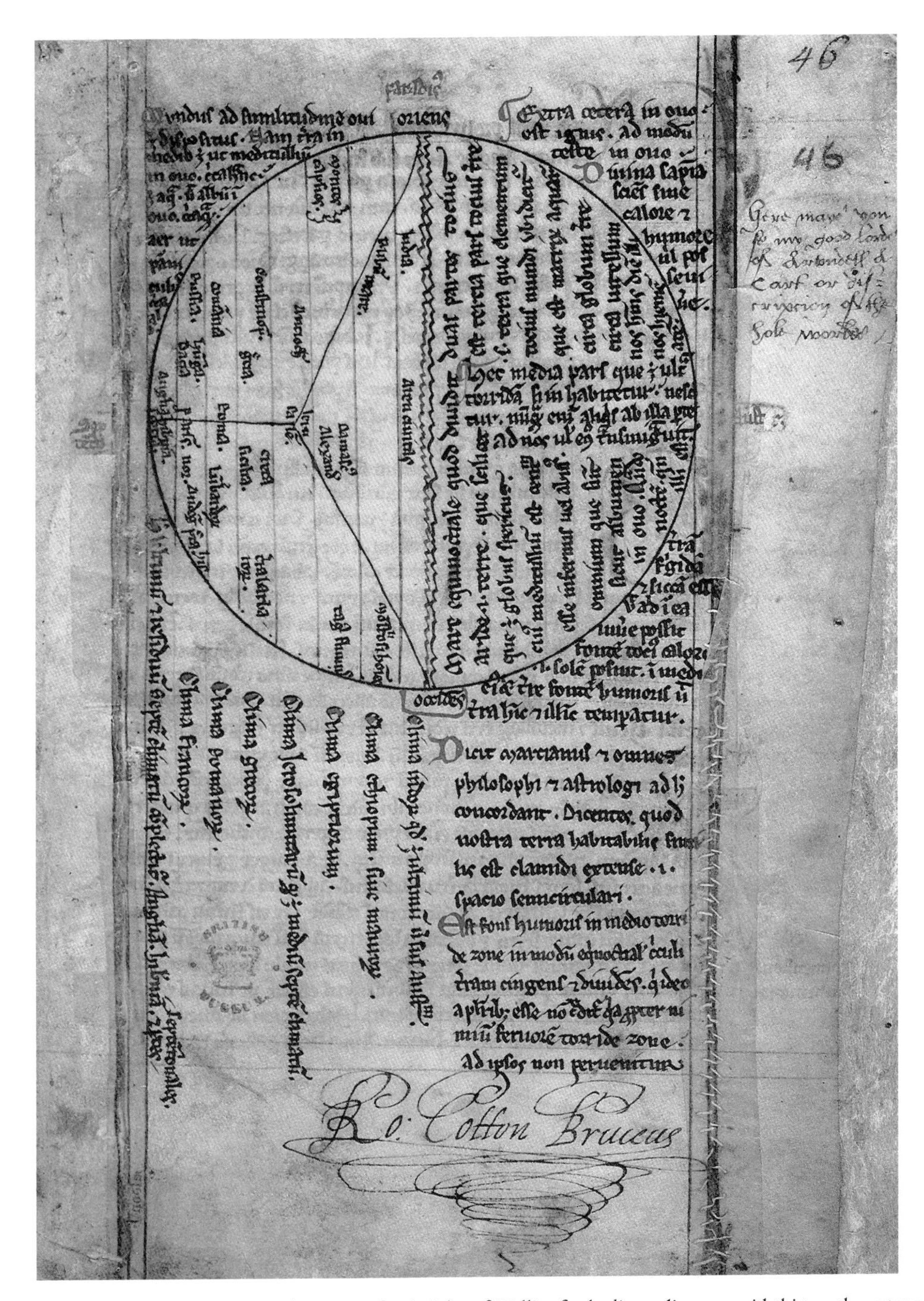

Fig. 18. London, BL, MS Cotton Julius D.VII, f. 46r. John of Wallingford, *climata* diagram, mid-thirteenth century. East at the top. The northern hemisphere is divided into horizontal climatic zones, beginning at the '*clima* of the Indians', then followed by the *climata* of the Ethiopians, Egyptians, Jerusalemites, Greeks, Romans, and Franks. The standard seven *climata* are supplemented by an eighth – the 'last and remainder of the seven *climata*', which contains Anglia, Hibernia, and 'northern parts'. 'Aren civitas' appears at the centre of the earth, and Jerusalem at the centre of the northern hemisphere. In the southern hemisphere, a central, horizontal band of text notes that it is unknown whether the part of the earth beyond the torrid zone is inhabited, since no-one has ever travelled there, and no-one from there has ever travelled here. The vertical text identifies the equinoctial sea as a 'third part of the whole earth', and suggests that an 'inferno or abyss' is located at the earth's centre; that the earth's centre is the source of all waters that surround the earth, like the white around the yoke in an egg; and that 'when we have day it is night for them; when summer for us, winter for them'.

ribbon of equatorial ocean separating southern from northern hemispheres. The southern half of the world, in contrast to the schematically divided north, is completely filled with two units of text, one horizontal, the other vertical. The vertical text, confident in tone, conflates a number of ideas: it identifies the equinoctial sea as a 'third part of the whole earth', which divides the two (southern and northern) landmasses, all of which together constitute a 'spherical globe'; it suggests that an 'inferno or abyss' is located at the earth's centre; that the earth's centre is the source of all waters that surround the earth like the white around the yoke in an egg; and finally it notes in the same way as Lambert of Saint-Omer that 'we' have day when an unspecified 'they' have night, and that our winter is their summer.[128] The shorter, horizontal text is by contrast a statement of ignorance: 'It is not known if this middle part which is beyond the torrid part is inhabited, for no-one ever crossed from that part to us, nor from us to them' (Hec media pars que est ultra torridam si in habitetur nescitur. numquam enim aliquis ab illa parte ad nos uel econtra transmigrauit). As with the *Liber Floridus*' 'Spera geometrica', the combination of reasoning, speculation, and frank ignorance that fills the southern part of this map is no more than a continuation of the text that surrounds three quarters of the map. This text in fact forms a continuous series of remarks, running from the top left of the image to the bottom right. It begins by reciting the notion of the world as egg, moves on to explain the role of divine wisdom in establishing habitable parts of the earth, notes that philosophers and astrologers agree with Martianus Capella that the earth is shaped like a cloak ('similis est clamidi', although the reference is apparently to Macrobius), and posits the existence of the source of the ocean at the centre of the torrid zone, while noting that many object to this proposition. As a whole, the page assembles, and to a certain extent attempts to assimilate, a variety of theoretical propositions concerning the ratio of earth to land, the shape and extent of the known world, the validity of the concept of the 'torrid zone' and the nature of equatorial regions, as well as the existence of inhabitants in the southern hemisphere. The map itself seems almost overwhelmed by the weight of theory that surrounds and penetrates it. It is as if, in the absence of communication, passage or migration, and therefore of knowledge, that outer, explanatory, text has seeped into the geographic space of the unknown land. Devoid of topography, the southern hemisphere becomes a recess in the sense that it is a retreat from geography to cosmology as the only mode of apprehension. Yet its function is noticeably different to that of the terrestrial paradise. The *antoecumene* is not a floating signifier, hovering between inclusion in the ecumene and location outside but within view of the known world; instead it is space without demarcation. Without subdivision, without cities, rivers, *climata*, without inhabitation, what appears to remain is a dialogue beyond the surface: a discussion of the fundamentals of sun, stars, sea and earth, and their interaction.

Since the idea of the antipodes represented the (unachievable) fulfillment of geography – the description of the entire earth – it came by the end of the thirteenth century to be deployed not simply as a subject for debate, but as theory's sign. The *Declaratio Raimundi* is a dialogue written by the Mallorcan polymath Ramon Lull in Paris in 1298, in response to the Bishop of Paris' condemnation in 1277 of 219 Averroist theses of philosophers at the University of Paris on the grounds that they were against theology. The antipodes function in the dialogue as a means of illustrating the distinction between human 'imaginatio' (imagination, but perhaps better translated as impression) and 'intellectus'. The *Declaratio* takes the form of a scholastic disputation between Raimundus and Socrates, whom he encounters in a wood on the outskirts of Paris. The ostensible purpose of the dialogue, in which Socrates advances the 219 propositions condemned by the Bishop in order for Raimundus to dispute them, is to demonstrate the erroneous nature of the opinions of certain 'philosophers and their followers' damned by the venerable Bishop.[129] However, its objectives are not, for the most part, polemical; rather, the *Declaratio* works (contrary to the 1277 condemnation) to revive the notion of a concordance between theology and philosophy.[130]

Prior to the commencement of the dialogue, Raimundus sets out three 'positions', to which he recurs during the debate. The third of these positions concerns 'transcendent points', namely the proposition 'that the intellect by its properties transcends to the comprehension of the truths of matters that inferior powers are not able to grasp' (quod intellectus per suam uirtutem transcendat ad intelligendum rerum ueritates, quas potentiae, quae sub ipso sunt, attingere non possunt).[131] To illustrate this statement Raimundus cites the idea of the antipodes, a concept that can be grasped by the intellect, but not by the 'imaginatio':

Sicut imaginatio hominis, quae imaginatur, quod antipodes inferius cadant; et sicut gustus infirmus, qui dulcem saporem pomi, quod infirmus comedit, attingere non potest. Sed humanus intellectus sensitiuam et imaginatiuam transcendit, quando intelligit, quod hoc, quod uidetur imaginationi de antipodibus, uidelicet quod inferius cadant, esset supra ascendere. Item, quando intelligit hoc, quod sensitiuae infirmae uidetur esse amarum, in pomo dulci esse dulce.[132]

Just like the imagination of a person, which envisages that the antipodeans fall down below, and just like the diseased taste of a sick person, which cannot experience the sweet flavour of an apple. But the human intellect transcends the felt and the envisaged, when it comprehends that instead of appearing to fall downwards, the antipodes should seem to the imagination to rise upwards. The same, when it comprehends that what in a sweet apple seems to be bitter to the infirm taste is in fact sweet.

Raimundus' pairing of the antipodes with the apple appears to have been carefully deliberated: the one a theory involving some conceptual difficulty; the other a physical object, easily comprehended. The sick person is aligned with the doubter of the antipodes, both unable to understand a truth that defies appearances and experience. The intellect overrides the sense – the impression that those on the other side of the world must fall, that the apple is bitter. Once directed by intellect, however, true imagination can be attained: it can comprehend the sweetness of the apple, and the perspective of the antipodeans. (In his *Tractatus novus de astronomia*, Lull suggests what the antipodal perspective might be, asserting that the imagination placed in the region of the antipodeans 'would imagine that people in Paris were falling towards the heavens': imaginaretur, quod homines Parisiis caderent uersus caelum.)[133] Their intellectual function established, the antipodes are used by Raimundus throughout the *Declaratio* to refute arguments of Socrates that are held to be the product of 'imaginatio' rather than 'intellectus': that since God is in everything he knows nothing other than himself; that the generation of humans is a circular process; that before the creation of the world there was a vacuum.[134]

As a sign of the intellect, the antipodes defy assumptions, false reasoning; they mark the division between sophistical or 'infirm' philosophy, and one used to reveal theological truths. Why select the antipodes to serve this purpose? Undoubtedly in part because they would have been a familiar topic to Lull and his audience, as attested by several references to antipodal theory in other of his works composed between 1297 and 1305, particularly the *Tractatus nouus*.[135] Since it was a commonplace to assert that the antipodes could be known by reason but not by experience, and that down for us was up for them, the concept may have seemed satisfyingly emblematic. Does their deployment also mark an attempt by Lull to keep certain scientific theories within the bounds of orthodoxy? It is surely significant that, apparently unconcerned by its contradiction of scripture, Lull emphasizes the validity of antipodal theory while writing as a champion of the bishop against pagan and heterodox philosophizing. At the very least this shows that the antipodes could be appropriated to different contexts of theoretical debate. They could, that is, emerge from the space of the cosmological treatise in order to animate various kinds of learned discussion, especially in the harness of an orthodox rationality. And, as Lull's use of the idea suggests, what made the antipodes disturbing for those who periodically opposed them also made them attractive to scholastic thought: they revealed the power of the intellect to turn the world upside-down.

Between heaven and hell: the antipodes as a third term

For much of the thirteenth century, debate about the habitation of antipodal regions ran alongside another debate about unknown land: the question of the location of the earthly paradise. Between these two debates there existed numerous points of intersection. Like the antipodes, paradise was conceived of as a place that existed within the world, but one that could not be reached by mortals – since, following scriptural authority, it was held to be surrounded by flame (Genesis 3.24) – and hence for which no experience could be adduced. As the preceding discussion has mentioned, the earthly paradise could be represented on maps, at times in ways revealing of its connections with antipodal space. Unlike the antipodes, however, it was a space for which unquestionable textual authority existed. The traditional location of paradise was at or beyond the easternmost extent of the known world, following Genesis (2.8) and Isidore's *Etymologiae* (14.3.2: 'Paradisus est locus in orientis partibus constitutus'), but the weakening of the hypothesis of the torrid zone in the twelfth and thirteenth centuries led to speculation that it may be located on the equatorial line, or even further south.[136] Around the beginning of the fourteenth century that speculation was given limpid expression in the form of a *quaestio*: 'whether the earthly paradise may be the site of those we call antipodeans' (utrum paradisus terrestris sit locus illorum quos uocamus antipodes). The anonymous author constructed syllogisms on either side of the equation (e.g. all habitable places that are beyond the quarter that we inhabit are antipodal; paradise is not in our quarter; therefore paradise is in the antipodes) before adducing the works of Macrobius, Sacrobosco, Seneca (the *Quaestiones naturales*), and finally Augustine to settle the matter against the proposition.[137] As Patrick Gautier Dalché has argued, what is exceptional is not the conclusion, but the explicit nature of the formulation and consideration of the question.[138] Far from a departure, this seems to represent the culmination of scholastic approaches to the intersection of natural philosophy with spiritual and theological concerns. Nor, as Michael Scot's *Liber introductorius* makes clear, were debates about habitation at the equator and in the southern hemisphere limited to the question of the earthly paradise: in the 1270s, a disputation amongst friars preachers in Rome included the question 'whether inferno may be in the centre or near the centre of the earth' (an infernus sit in centro uel iuxta centrum terre).[139] Paradise, inferno, purgatory ... and the antipodes: as the limits of the known world expanded, and more significantly as new theoretical models allowed for a plurality of answers to such fundamental questions about the extent of humanity on the earth, spaces that had always been thought, often represented, but perforce never known became tantalizingly near at hand, just beyond the reach of exploration, but at the same time mobile, their locations and the directions towards them suddenly fluid.

The connections between Dante's novel treatment of the antipodes and scholastic discussions of natural phenomena are incontrovertible. In the *Convivio*, Dante explicitly acknowledged his debt to scholastic natural philosophy (Albertus Magnus' *De natura loci*) and classical poetry (Lucan's *De bello civili*) when outlining antipodal theory to explain the relationship between the north and south poles.[140] Dante's intriguing innovation was to imagine a northern polar city, named Maria, twinned with Lucia, a city located on the spot on which a stone would fall when dropped from the south pole. The two cities are described in terms of their distances from each other, from Rome, and from the equatorial *clima* of the Garamantes (a classically-derived African people), tellingly reiterating the axis between Rome and Africa found in Lucan – and in Cicero's 'Somnium Scipionis'. To Dante is also attributed *De situ et forma aque et terre* (or the *Quaestio de aqua et terre*), a response to Aristotelian debate about the relation between the elements of earth and water, and the extent of land not covered by water. According to Aristotle's theory of elements as set out in *De caelo*, the sphere of water should exceed in size and depth the sphere of earth, so that all earth should be covered by water and hence be uninhabitable – something manifestly not the case. *De situ et forma aque et terre*, written at Verona in 1320, purports to be an account of a disputation on this problem held at Mantua.[141] It begins by setting out and then dismantling the case that water is naturally higher than land, and that the exposure of habitable land must be explained by the fact that land and water do not possess the same centre, so that water is eccentric to land. The author concludes that water is not higher than land, and that they share the same centre. Using diagrams, he argues for the possibility of a mixture of elements, and rejects the notion of elevated land beneath the equator.[142] The significance of this work for *Inferno* lies in a question raised towards the end of *De situ*: 'Why was the elevation of the hemisphere on this [i.e. northern] side rather than the other?' (Quare potius elevatio emisperialis fuit ab ista parte quam ab alia?)[143] An answer is provided in *Inferno* 34, where Virgil explains to the poet that, when Lucifer fell, the earth, in terror, shifted elevation from the southern to the northern hemisphere:

> Qui è da man, quando di là è sera;
> e questi, che ne fé scala col pelo,
> fitto è ancora sí come prim'era.
> Da questa parte cadde giú dal cielo;

e la terra, che pria di qua si sporse,
per paura di lui fé del mar velo,
e venne a l'emisperio nostro; e forse
per fuggir lui lasciò qui loco vòto
quella ch'appar di qua, e sú ricorse.

Here it is morning when it is evening there, and this one [i.e. Lucifer] who made a ladder for us with his hair is still fixed as he was before. On this side he fell down from Heaven; and the earth, which before stood out here, for fear of him made a veil of the sea and came to our hemisphere; and perhaps in order to escape from him that which appears on the side [i.e. the mountain of purgatory] left here the empty space and rushed upwards.[144]

Previously in *Inferno,* Dante had hinted at the absence of human inhabitation beyond the ecumene, when Ulysses, explaining his intent to sail beyond the columns of Hercules, taunted his companions that they surely did not wish to deny themselves the experience 'di retro al sol, del mondo sanza gente' (of what lies behind the sun, of a world without people).[145] Both texts broach the possibility of elevated land in the other hemisphere: in *De situ,* the restriction of elevated land to the northern hemisphere is explained as a mystery beyond human comprehension; in *Inferno,* it is attributed to the ultimate celestial put-down. What is significant, though, is the imagination of a different pre-Luciferan world: this goes beyond the familiar idea that there are worlds beyond the known. The idea that prior to Lucifer's fall exposed land existed in the southern hemisphere but not the northern posits the priority of an unknown, antipodal world to the ecumene. More radical still, according to this scheme, is the thought that, after Lucifer, the known world is in fact the unknown reformed, turned inside-out.[146]

The antipodal geography alluded to in *Inferno* 26, when on the other side of the world Ulysses glimpses 'a mountain, dark in the distance' (una montagna, bruna/ per la distanza) (133–4), is elaborated in *Purgatorio.* There it emerges that the mountain seen by Ulysses is purgatory, that it is located directly opposite to Jerusalem, and that at its summit is found the earthly paradise. Here too Dante plays with the idea of the antiquity of this other part of the world, since the poet's entry to it is represented as a rediscovery:

I' mi volsi a man destra, e puosi mente
a l'altro polo, e vidi quattro stelle
non viste mai fuor ch'a la prima gente.
Goder pareva 'l ciel di lor fiammelle:
oh settentrïonal vedovo sito,
poi che privato se' di mirar quelle!

I turned to the right and considered the other pole, and I saw four stars never seen except by the first people. The sky seemed to rejoice in their flames: Oh northern site, widowed because deprived of gazing on those![147]

Commentators confidently glossed the four stars as the four cardinal virtues (Prudence, Justice, Temperance, and Fortitude), and Dante's son and commentator, Pietro, explained that the poet intended to figure the opposition between the two poles in terms of good and evil:

> fingit hic se primo egressum de tali figurato Inferno revolvisse cum visu ad alium polum anctarticum, ubi dicitur esse alia stella tramontana, oppositum directo huic nostro polo artico, in quo vult ostendere et denotare quod homo, ita volens de malo ad bonum transire, debet primitus suum amorem volvere ad oppositum mali, quod est ipsum bonum et virtuosum esse[148]

> here he imagines himself, having departed from Inferno, turning his sight to the other, Antarctic pole, where other polar stars are said to be, directly opposite to this our Arctic pole. By which he wants to show and take note that a man, wishing in such a way to pass across from evil to good, should first turn his love to the opposite of evil, that is to be himself good and virtuous.

The belief that the location of Purgatory should be as far as possible from 'our habitation', and cut off from it by the impediments of mountains and seas, is consistent with Augustine's view that the terrestrial paradise was 'most remote' from human knowledge and inhabitation.[149] Hence the emphasis on opposition, although there is also obviously a parallel at work, between the poles ('Maria' and 'Lucia'), and between Jerusalem and Purgatory/Paradise ('hanno un solo orizzòn/ e diversi emisperi' 4. 70–1). As with the imagined fall of Lucifer, antipodal space is the site of historical origins, witnessed by the first people but by no others. Dante's references to the antipodes in *Inferno* and *Purgatorio* make it clear that, for him, normal inhabitation of the southern hemisphere was poetically impossible: part of the world, the antipodes also mediated between earth and heaven, origins and end. They were the site of pre-lapsarian humanity and of the path to salvation, where memory of pain and loss is renewed, and where the turn can be made from life on earth to the life to come.

Dante's mode of rethinking antipodal space as beyond yet between – a zone of transgression, exploration and refuge (Ulysses), or a means of mediating terrestrial and celestial life (Purgatorio, Parasdiso) – was an influence, although never explicitly acknowledged, in the multiple references to the idea in the writing of Francis Petrarch. Petrarch invoked the antipodes in apparently incidental fashion in six of his letters;[150] he also included descriptions of zonal theory in his epic poem *Africa*, and in the *Secretum*, a meditation on love and the pursuit of glory couched in the form of a dialogue between Augustine and one 'Franciscus'. As recent criticism has noted, for Petrarch the antipodes represented the fault-line of a 'divided knowledge', between paganism and Christianity, antiquity and modernity.[151] While he enjoyed placing such divisions in a state of confrontation, Petrarch had no wish to overcome or dissolve them, and for this reason the antipodes remain comprehensible in his work only as a term existing outside of established schemes of knowledge. Petrarch's awareness of classical and medieval debate about the antipodes is attested by a marginal comment in his manuscript of Pliny's *Naturalis historia* (Paris, BNF, MS lat. 6802). Opposite Pliny's comment that there is a great debate between the lettered and the uneducated on the question of whether men stand with their feet against each other ('undique homines conversisque inter se pedibus stare') (2.161), Petrarch wrote: 'About these things there is no dispute in the works of Cicero and many others; Augustine argues against robustly' (De his apud Ciceronem multosque alios nulla questio est; contra ualide disputat Augustinus).[152] The antipodes emerge, then, as a clear point of division not between the well-read and the ignorant but, in a provocative transformation of Pliny's dichotomy, between two of Petrarch's most important literary and moral models. The use of scholastic terminology here (questio; contra ... disputat) suggests that, like Dante, Petrarch encountered the question of the antipodes within the context of university debate of the thirteenth and fourteenth centuries. However, in his writings Petrarch reveals his consciousness of the antipodes primarily as part of a classical literary inheritance; as such, his deployment of the antipodal motif is oblique: it functions as a nodal point between *Romanitas* and Christianity, and between empire and poetic glory and their virtuous renunciation. His aim was not to resolve the antipodes, but to exploit the possibilities and significances of their disputed, ambivalent status.

Petrarch's deployment of zonal theory in *Africa* and the *Secretum* hovers between emulation of Cicero's and Lucan's use of antipodal spaces as a rebuke to acquisitive glory and Virgilian incitement to cross barriers, or at least to figure provocatively their transgression. The first two books of *Africa* replay the dream of Scipio, with the significant alterations that in this version the dreamer is the elder Scipio (Publius Cornelius Scipio 'Africanus'), guided by his father (Gnaius Cornelius Scipio), and that the last two sections of the 'Somnium Scipionis' (on the immortality of the soul and the structure of the heavens) are omitted. The function of the access to global vision remains the same: Scipio is made to see the small extent of the Roman empire, the large amount of land over which it does not rule, the frigid and torrid zones over which it cannot rule, and the other temperate zone, separated by heat and ocean – 'but to you barred' (sed vobis est invia). These factors, combined with the diversity of human languages and customs, and the wide extent of peoples, prohibit world domination: 'Whose glory commends him to Taprobana at the same time as it resounds off the Irish shore?' (Quem sua Toprobani commendet gloria et idem/Litus ad Hibernum resonet).[153] In the *Secretum*, Augustine echoes the elder Scipio in the 'Somnium Scipionis' by instructing Franciscus on the zonal division of the earth, and emphasizing the tiny extent of the known world. He insists on the absence of any inhabitation in the southern temperate zone, citing in the process not only his own work (*De civitate dei*), but even some lines from *Africa* book 2.[154] All this is meant to demonstrate to Franciscus (here aligned with the younger Scipio) the futility of his hopes for worldly fame. Yet the rebuke is characteristically ambivalent: Augustine instructs Franciscus to 'abandon Africa' (meaning both Petrarch's poem of that name and his investment in Roman imperial glory), but the advice acts also as an advertisement and a memorial.

In his letters, Petrarch recurred to the idea of the antipodes, often in connection with the ends of the known world (particularly Taprobana), so as to probe the conjunction between intellectual and political ambition, but also as a means of expressing his own complicated attitudes towards *patria*. In one sense the antipodes were, for Petrarch, a sign of imperial excess. In a letter to Urban V (*Seniles* 11.17), he quotes Lucan's digression on 'mad' Alexander in book 10 of *De bello civili* as part of a series of exempla on the theme of the intervention of death in the deeds of great men: 'Alexander the Macedonian, having overrun Asia and conquered all that part of the world, was threatening Carthage on this side, and on the other, as Lucan says, "prepared to bring his fleet into the ocean", seeking – for all I know – Taprobana or the Antipodes' (Alexander Macedo, perarata Asia, totaque illa terrarum parte perdomita, hic Carthaginensibus minabatur, illinc, ut Lucanus ait: *Oceano classes inferre parabat.* Thaprobanem, an Antipodas petiturus nescio).[155] On the other hand, the antipodes could figure as a place of refuge.

A letter to Neri Morando of Forlì imagines not the passage of writing from the antipodes to the known world, in the manner of Tiberianus' letter from the antipodeans, but instead the passage of vice in the other direction. The letter begins with the image of the world in terminal decline, with each age worse than the one before. Following conventional imagery of life on earth as death, and a prison, Petrarch identifies vice as an enemy greater than war: the vices are immortal and multiply as the human grows older.[156] The problem for Petrarch and Neri is that as virtuous people they must live amongst not only those who do evil, but those who positively exult in it:

> Quo enim patet fuga, quo ibimus, quo non nos sceleratorum acies ac signa precesserint, ubi non pessimis moribus imperium partum atque firmatum sit? Trans occeanum navigandum erat, nisi quia credibile est vitia nostra iampridem ad antipodas descendisse. Ad celum potius evolandum, nisi nostris ad terram ponderibus premeremur[157]

> So where does escape lie? Where will we go? Where do the army and standards of the wicked not lead the way, where is empire not established and held firm with the very worst customs? It would be necessary to cross the ocean, except that it is believable that our vices have long since descended to the antipodes. Rather we would have to fly away to the heavens, if we were not pressed by our weights to the earth.

Again, consistent with Augustine, passage to the antipodes is incredible, patently fantastic – or is it? Throughout the letter the overwhelming quality of vice is constrasted with physical violence, which can be prevented or at least defended against by walls, fortresses, and shelters. The martial imagery (army, standards) used to describe the vices seems intended to heighten this contrast: there is no stopping this army, except by divine assistance. Society has reached such depths that no barriers can restrain its depravities, perhaps not even the barrier between our world and the antipodes, a barrier that is simultaneously one between truth and fiction. This description of vice is echoed at the conclusion of the letter by a passage devoted to the omnipotence of gold: like vice, gold is unstoppable, a force that corrupts institutions and friendships, overturns social hierarchies (makes rulers of slaves and slaves of rulers) and even nature itself (dries up rivers, traverses lands, agitates seas, levels mountains). This diatribe concludes what is perhaps the letter's ultimate contrast, between Christian and pagan sociey. Despite its opening image of steady decline, and its vision of a continuity of evil from the Roman empire to the present day, the letter is punctuated with references to the redeeming qualities of the Christian God, qualities not possessed by chastity-breaking Jupiter, he of the golden shower. Hence the shadowy, unspoken, presence of Augustine in the letter. Petrarch is not simply using the antipodes as a means of self-reflection, in the manner of classical satirical literature; instead, he deploys the antipodes as a space of innocence in contrast to the corrupting forces of the known world. Petrarch and Neri, would-be fugitives across the ocean, are exempt from the criticism. Their innocence from vice aligns them with the antipodeans, under siege from northern-hemisphere wickedness. The fantasy of passage was traditionally imperialist, because what was envisaged as passing to the antipodes was the glory and conquests of the known world. But in this fleeting formulation Petrarch raises a darker possibility: the conquest of the antipodes by an empire of sin. What the letter seems implicitly to lament is the failure of evangelical Christianity, the failure of virtue, to overcome and indeed replace this vicious empire.

Following Augustine, any antipodeans cannot be reached by the word of God, and therefore cannot be human; but following the logic of satire, the antipodeans cannot help but be reached by human depravity.

The letter to Neri invokes a theme consistent throughout Petrarch's writing, and one that is connected with his antipodal allusions: alienation from *patria*.[158] Petrarch comments approvingly on his homeland's accommodation to the Holy Roman Emperor: 'I read with more happiness both that obedience was not denied to the Roman emperor, and that if there is sorrow for neglected liberty anywhere in the entire world at this moment, it is in my fatherland' (letius legi, et romano principi obedientiam non negari et siqua iam toto orbe neglecte libertatis cura est, eam in patria mea esse).[159] This approval evidently comes from outside the native land, and that kind of absence emerges as the connecting thread that runs through Petrarch's consolatory letter to Giovanni Colonna on the difficulties of life (*Familiares* 6.3.57). At 'the source of the Sorgue' (ad fontem Sorgie), Petrarch writes to Colonna, tethered by gout to his homeland in Tivoli. The writer is, in other words, not only outside Italy, site of his own *patria*, but explicitly outside Avignon – outside, he puns, both court and cares (curas curie). This self-location is depicted very clearly at the conclusion of the letter, in the aquatic journey Petrarch imagines 'father' Colonna making from the river Aniene near Tivoli to the Sorgue, from Italy to France.[160] Against this dynamic of estranged filiation, the antipodes represent the abandonment of lineage and history. They are the destination and perhaps by implication the product of the 'mind that does not know how to stand still':

> recense, pater, ab adolescentia cursus tuos et animum stare nescium; videbis, ut frenum equo indomito, sic tibi necessariam podagram. ... Isses extra nostre habitabilis zone terminos, transisses occeanum, adisses antipodas; nullum tibi discurrendi finem, qua polles in reliquis, ratio fecisset.[161]

> review, father, your journeys from adolescence and your mind that does not know how to stand still; you will see that gout is necessary to you, as a rein to a wild horse. ... You would have gone beyond the boundaries of our habitable zone, you would have crossed ocean, you would have reached the antipodes; your reason, which prevails in other matters, could have brought no halt to your wandering.

Gout, reason's ally, keeps Colonna where he is. The extended fantasy of the untethered Colonna sees him not only on hypothetical journeys to the antipodes, but also in a variety of places and all-action poses: swimming in the Nile, the Indian Sea, and the Don river, crossing the Rhipaean mountains or the Hercynian pass, traversing a world rather than a European geography, 'vagrant always, a nomad on the earth' (vagus semper et profugus super terram).[162]

The mind that knows how to stand still, praised here, the mind of reason and *patria*, seems at first glance the antithesis of the intellect praised in Petrarch's famous letter to Philippe de Vitri (*Familiares* 9.13), in which the chaplain to the cardinal Gui de Boulogne is castigated for having complained about his master's Italian 'exile' from Paris. Drawing heavily on Boethius' *De consolatione Philosophie* (Book 1, Metrum 2), Petrarch's attack on Philippe in the letter gains purchase from the youth-age contrast. Philippe, Petrarch perceives, has grown old in mind as well as body: his young self eagerly sought knowledge of India, Taprobana, the unknown extremities of the eastern Ocean, and (in the other direction) Ultima Thule, since Ireland and the Orkneys seemed too close. More than that, Philippe's learning should have drawn him on to the study of the stars, including those of the southern hemisphere:

> Quid autem miri si angusta animo literatissimi hominis terra erat, in hunc assidue celi verticem qui supra nos gelido temone convertitur, inque illum alterum quem siqui sunt antipodes, australi clarum regione suspiciunt, in obliquum denique solis callem inque fixas et errantes stellas infatigabili studio conscendentis?[163]

> It would not be surprising if the earth seemed narrow to the mind of this most learned man, embarking continually with indefatigable study towards the pole of heaven which above us is turned around by the icy beam, and towards that other pole, bright in the southern region, to which antipodeans, if they exist, look up, and finally towards the oblique path of the sun and the fixed and the wandering stars.

The use of the adjective 'obliquus' here seems to echo the description of the zodiac in Virgil's *Georgics*; the perspective is the inverse of that of the 'Somnium Scipionis' – here the gaze is from the earth to the heavens, and the medium of access is learning rather than the dream vision. The reference to study and the stars opens up a discourse about *patria*: the learned man, Petrarch maintains, does not lament his exile, because he finds his native land wherever he may be. Further, to lament exile is weak, unmanly, as can be seen by an array of examples of heroic figures who left their *patria* (in order to return in triumph) – Scipio, Alexander, Pompey, Caesar, Neoptolemus, Ulysses, Aeneas. It is the unlearned man, the peasant, who properly remains on his land ('in proprio rure consistere'), since his business is to know and work the land; the noble spirit desires to observe the lands and customs of many men. *Patria* is not, ultimately, renounced: Philippe's devotion to Paris is mocked by constrasting French and Roman history, and the letter ends with a virtual *laus Italiae* in the form of a description of Gui's passage through northern Italy to Rome and back to France. Mental passage to and through *patria* is, in other words, more than enough. This seems consistent with the Ciceronian model: the gaze beyond *patria* will be brought back to it, and from there to the heavens. But if the antipodes figure the gaze beyond native land as they did in Cicero's 'Somnium Scipionis' (and many of the other political deployments of the motif), this is nevertheless a gaze that itself emanates from outside of, or even beyond, *patria*: the gaze of the lettered, without home, whose home is everywhere. The antipodes are the sign, in other words, of intellectual exploration, of the voyaging of the eager – but perhaps too eager – mind.

If from a scientific point of view there is nothing new in Petrarch's writing of the antipodes, from a literary perspective the realignment of antipodal tradition is significant. It is not simply that classical material is Christianised, but that the antipodes appear as a point of continuity rather than marker of division between classical pagan and Christian texts, a flawed continuity exploited by Petrarch without ever entirely resolving its troubling qualities. Augustine's renunciation of the antipodeans in *De civitate dei* is made into a renunciation of glory in the *Secretum*; the denial of inhabitation outside of the ecumene becomes an opposition to public, epic poetry, and glorifying histories. As with Dante, part of the appeal of the antipodes seems to have been their unverifiable nature, allowing them to be construed as fiction, or as the product of a speculative cosmographical tradition (as in the rendition of zonal theory in *Africa*) – but in any case as wholly conceptual. And, unlike the heavens, the antipodes existed outside of theology. That somewhat precarious, unlicensed but not unthinkable, position gave the antipodes a poetic force. A perhaps-land of 'siqui sunt' and the subjunctive mood, the antipodes offered a place to go instead of heaven, and more particularly a means of mediating yet also complicating and undoing the relationship of the human to the divine.

In different ways both Dante and Petrarch participated in a discourse of passage, imagining journeys to the other side of the earth. They did so, however, in such a way as to preserve the fantastical, mythological, or theological implications of passage: the antipodes remain the destination of Alexander or Ulysses, or a privileged vision mediated by classical and divine authority. As a third term the antipodes remain essentially unassimilable. As I have suggested, that position offered valuable intellectual as well as poetic possibilities. The antipodes could continue to represent the politically unobtainable; they could therefore represent the abdication of power. At the same time, as the sign of the possible but uncertain, the antipodes invited speculation from those who wished to specify a location for infernal and paradisiacal regions within the world but beyond the ecumene. Finally, their incorporation within a philosophical tradition gave the antipodes a particular status: that of a *quaestio*, unresolved, comprehensible only through the exercise of rational thought. If one was prepared to accept the reality of passage, however, each of these possibilities could be read differently. From the perspective of *accessus*, the antipodes represented the politically unobtained, the failure – but not the impossibility – to direct the energies of the European state outward. They represented the next frontier of evangelization. They were a question, not for scholastic disputation, but for travellers and missionaries. These different functions and meanings of the antipodes were evenly balanced at the end of the fourteenth century. But in the fifteenth century the intellectual and political bases of the representation of unknown lands changed fundamentally.

Notes

1 *A Treatise on The Canon of Medicine of Avicenna Incorporating a Translation of the First Book*, trans. O. Cameron Gruner (London: Luzac, 1930), Book 1, thesis 3, no. 34 (p. 61).

2 *Ptolemy's Almagest*, trans. G. J. Toomer (London: Duckworth, 1984), 2.6 (p. 83).

3 For thirteenth-century assertions of the location of paradise at the equator or further south see Scafi, *Mapping Paradise*, pp. 174–82; Patrick Gautier Dalché, 'Le paradis aux antipodes? Une *Distinctio divisionis terre et paradisi delitiarum* (XIVe siècle)', in *'Liber largitorius': Études d'histoire médiévale offertes à Pierre Toubert par ses élèves*, ed. Dominique Barthélemy and Jean-Marie Martin (Geneva: Libraire Droz, 2003), pp. 615–37, esp. pp. 632–34. The location of paradise at or beyond the equatorial region was rarely mooted prior to 1200. However, it was a topic of discussion as early as the second decade of the thirteenth century: Gervase of Tilbury noted that the location of paradise beyond the torrid zone, inaccessible due to the heat of the sun, was 'a view that can be defended' (id defendi potest): Gervase of Tilbury, *Otia imperialia*, ed. and trans. S. E. Banks and J. W. Binns (Oxford: Clarendon Press, 2002), 1.10, pp. 66/67.

4 *Averrois Cordubensis commentum magnum super libro De celo et mundo Aristotelis*, ed. Francis J. Carmody, Rüdiger Arnzen and Gerhard Endress, 2 vols (Leuven: Peeters, 2003), 2.296.

5 See for example *De caelo*, 2.4 [287a–b], or the pseudo-Aristotelian *De mundo: Translationes Bartholomaei et Nicholai*, ed. William L. Lorimer, rev. L. Minio-Paluello (Bruges: De Brouwer, 1965), 392a30–393a10.

6 Petrus Alfonsi, *Dialogi*, *PL* 157, col. 545.

7 *Dialogi*, col. 547. On the Arab derivation of Arīn from the ancient Indian Ujjain see Gerald R. Tibbetts, 'The Beginnings of a Cartographic Tradition', in *The History of Cartography*, vol. 2.1: *Cartography in the Traditional Islamic and South Asian Societies*, ed. J. B. Harley and David Woodward (Chicago: University of Chicago Press, 1992), pp. 90–107: p. 103.

8 *Dialogi*, col. 547.

9 *Dialogi*, col. 548.

10 See, for example, Ahmet T. Karamustafa, 'Cosmographical Diagrams', pp. 71–89: 78–9, and Gerald R. Tibbetts, 'Later Cartographic Developments', p. 146, both in *The History of Cartography*, vol. 2.1. Compare André Miquel, *La géographie humaine du monde musulman jusqu'au milieu du 11e siècle*, 3 vols (Paris: Mouton, 1967–80), vol. 2 (1975), pp. 21–4; Konrad Miller, *Mappae Arabicae*, 6 vols (Stuttgart: Miller, 1927), vol. 5.

11 Honorius Augustodunensis, *Imago Mundi*, ed. V. I. J. Flint, *Archives d'histoire doctrinale et littéraire du Moyen Âge* 49 (1983), 7–153: p. 51. Flint's edition is of the text of 1139.

12 Patrick Gautier Dalché, 'Le renouvellement de la perception et de la representation de l'espace au XIIe siècle', in *Renovación intelectual del Occidente Europeo (siglo XII)* (Pamplona: Gobierno de Navarra, 1998), pp. 169–217: pp. 103–4.

13 See Hermann of Carinthia, *De essentiis*, ed. and trans. Charles Burnett (Leiden: Brill, 1982), pp. 4–5 for biography and pp. 16–43 for discussion of Hermann's sources. Burnett suggests that Hermann had access to the *Almagest* prior to its translation c. 1175, either in Arabic or in another, unidentified translation, on which Hermann himself was possibly working. Burnett also discusses the likelihood of Hermann's indirect access to Aristotelian works. On Hermann's career, significance, and other works see also Burnett, 'Hermann of Carinthia', in *A History of Twelfth-Century Western Philosophy*, ed. Peter Dronke (Cambridge: Cambridge University Press, 1988), pp. 386–404, and idem, 'Arabic into Latin in Twelfth Century Spain: the Works of Hermann of Carinthia', *Mittellateinisches Jahrbuch* 13 (1978), 100–34.

14 Hermann, *De essentiis*, trans. Burnett, Book 2, 77rF (pp. 216/217). For further comments on Hermann's conception of longitude see Gautier Dalché, 'Le renouvellement de la perception et de la representation de l'espace au XIIe siècle', p. 202.

15 Hermann, *De essentiis*, p. 214.

16 Hermann, *De essentiis*, p. 214. The orthography of 'girographi' is suggestive of the term's unfamiliarity to the Latin tradition.

17 Hermann, *De essentiis*, pp. 220–2. Burnett comments on the markedly different attitude to *astrologi* shown by William of Conches: *De essentiis*, pp. 24–5.

18 Hermann, *De essentiis*, trans. Burnett, p. 222/223.

19 Hermann, *De essentiis*, trans. Burnett, pp. 222/223–224/225. Hermann here draws on Isidore's *Etymologiae* (14.3.2; and 14.6.8); Isidore noted the association of the *Fortunatae Insulae* with the earthly paradise, but argued that the two should not be confused: see Scafi, *Mapping Paradise*, pp. 47–8.

20 Albertus draws on the discussion of the two Ethiopias of Homer in the works of Crates ('Karites') to establish the possibility of sub-equatorial habitation. Albertus was aware of Crates thanks to Gerard of Cremona's twelfth-century translation of Geminus' *Introduction to the Phemomena*, which formed the basis for an introduction to the *Almagest*: Albertus Magnus, *De natura loci*, ed. Paul Hossfield, *Opera Omnia* v.ii (Aschendorf: Monasterii Westfalorum, 1980), 1.7: p. 13.

21 *De natura loci*, 1.7: p. 14.

22 *De natura loci*, 1.12, p. 21.

23 See *De natura loci*, 1.6–7, pp. 9–10, 14. Albertus' ideas on these matters were subsequently repeated and developed in the commentary on John of Sacrobosco's *De sphera* completed in 1337 by the Dominican Hugh of Castello: see Nathalie Bouloux, *Culture et savoirs géographiques en Italie au XIVe siècle* (Turnhout: Brepols, 2002), pp. 24–5.

24 *De natura loci*, 1.7, p. 14. The belief in magnetic mountains or a magnetic sea in equatorial regions appears to have become reasonably widespread in Europe from the twelfth century. It appears, for example, in Robertus Anglicus' thirteenth-century commentary on Sacrobosco's *De sphera* and in Dominicus de Clavasio's fourteenth-century *Quaestiones super Aristotelis De caelo et mundo.*

25 Roger Bacon, *The 'Opus Maius' of Roger Bacon*, ed. John Henry Bridges, 2 vols (Oxford: Clarendon Press, 1897), vol. 1, pp. 300–1.

26 Bacon, *Opus Maius*, vol. 1, p. 293.

27 Bacon, *Opus Maius*, vol. 1, p. 290.

28 Bacon, *Opus Maius*, vol. 1, pp. 296–7.

29 Bacon here cites the *De dispositione sphaerae* attributed to Ptolemy; the work is in fact the introduction to the *Almagest* based on Gerard of Cremona's translation of Geminus' *Introduction to the Phemomena*: David Woodward and Herbert M. Howe, 'Roger Bacon on Geography and Cartography', in *Roger Bacon and the Sciences: Commemorative Essays*, ed. Jeremiah Hackett (Leiden: Brill, 1997), pp. 199–222, p. 207 *n*25.

30 Bacon, *Opus Maius*, vol. 1, p. 291. Bacon refers to IV Ezra 6.42: 'On the third day thou didst command the waters to be gathered together in the seventh part of the earth; six parts thou didst dry up and preserve, in order that (issuing) from them there might serve before thee those who both plough and sow': *The Apocrypha and Pseudepigrapha of the Old Testament in English*, ed. R. H. Charles, 2 vols (Oxford: Clarendon Press, 1913), vol. 2, p. 578.

31 Bacon, *Opus Maius*, vol. 1, p. 293.

32 Bacon, *Opus Maius*, vol. 1, pp. 293–4.

33 Bacon, *Opus Maius*, vol. 1, pp. 294, 307.

34 Bacon, *Opus Maius*, vol. 1, pp. 307–8. See Saint Ambrose, *Opera Omnia di Sant'Ambrogio*, vol. 1*: Hexameron (i sei giorni della creazione)*, ed. C. Schenkl, trans. Gabriele Banterle (Milan: Biblioteca Ambrosiana, 1979), Dies IV, ser. VI, chapter 5.23, pp. 216–18.

35 Bacon, *Opus Maius*, vol. 1, p. 307.

36 Bacon, *Opus Maius*, vol. 1, pp. 292–7.

37 Bacon, *Opus Maius*, vol. 1, p. 294. See Woodward and Howe, 'Bacon on Geography and Cartography', p. 207.

38 Bacon, *Opus Maius*, vol. 1, pp. 292, 306, 309–10.

39 Bacon, *Opus Maius*, vol. 1, pp. 306–7. See Woodward and Howe, 'Bacon on Geography and Cartography', pp. 215–17; there is no evidence to suggest that the map included a graticule of parallels and meridians.

40 Bacon, *Opus Maius*, vol. 1, p. 291.

41 Marco Polo, *Le divisament dou monde*, ed. Gabriella Ronchi (Milan: Mondadori, 1982), chapter 166 (p. 543).

42 John Larner, *Marco Polo and the Discovery of the World* (New Haven: Yale University Press, 1999), p. 95.

43 *Directorium ad faciendum passagium transmarinum*, ed. C. R. Beazley, *American Historical Review* 12 (1907), 810–57; 13 (1907), 66–115: pp. 821–2: 'videbam polum antarcticum circa viginti quatuor gradus eleuatum. Ab isto loco ulterius non processi. Mercatores vero et homines fide digni passim ultra versus meridiem procedebant usque ad loca ubi asserebant polum antarcticum quinquaginta quatuor gradus eleuari'.

44 *Directorium ad faciendum passagium transmarinum*, p. 822.

45 *Mandeville's Travels*, ed. M. C. Seymour (Oxford: Clarendon Press, 1967), p. 132.

46 The French text is the Continental Version of Mandeville's *Book*, edited in *Mandeville's Travels: Texts and Translations*, ed. Malcolm Letts, 2 vols (London: Hakluyt Society, 1953), vol. 2, p. 332. The Middle English translation is taken from *Mandeville's Travels*, ed. M. C. Seymour (Oxford: Clarendon Press, 1967), p. 133. On Mandeville's calculations, and symmetry see Iain Higgins, *Writing East: The 'Travels' of Sir John Mandeville* (Philadelphia: University of Pennsylvania Press, 1997), pp. 135–9.

47 *Mandeville's Travels*, ed. Letts, vol. 2, pp. 332–3; *Mandeville's Travels*, ed. Seymour, p. 134.

48 See Brunetto Latini, *Li Livres dou Tresor*, ed. Francis J. Carmody (Berkeley: University of California Press, 1949), p. 98.

49 *Mandeville's Travels*, ed. Letts, vol. 2, p. 333; *Mandeville's Travels*, ed. Seymour, p. 135.

50 See Christiane Deluz, *Le livre de Jehan de Mandeville: Une 'géographie' au XIVe siècle* (Louvain: Institut d'Études Médiévales de l'Université Catholique de Louvain, 1988), p. 183.

51 Higgins, *Writing East*, p. 133; compare Deluz, *Le livre*, p. 189.

52 Higgins, *Writing East*, pp. 141–2.

53 A point emphasized by Higgins, *Writing East*, pp. 132–42.

54 *El Libro del conoscimiento de todos los reinos (The Book of Knowledge of All Kingdoms)*, ed. and trans. Nancy F. Marino (Tempe: Arizona Center for Medieval and Renaissance Studies, 1999), section 87, pp. 72/3–74/5. A reference to Avignon suggests that the text may not have been composed (or at least completed) before late 1378; it must have been in circulation by 1402 when the authors of *Le Canarien* appear to make reference to it: see Marino's discussion, pp. xxxii–xxxviii. The text has usually been dated to the third quarter of the fourteenth century.

55 *Libro del conoscimiento*, pp. xliv–xlviii.

56 The book also reports the view that the earthly paradise was located in the high sierras of the Antarctic pole, from which flowed the Nile: *Libro del conoscimiento*, pp. 56/7, 60/1.

57 *Libro del conoscimiento*, pp. 74/5.

58 The Catalan Atlas, produced in Mallorca in the late 1370s or early 1380s, is the earliest extant map that bears the unequivocal influence of Polo's travels: Woodward, 'Medieval *mappaemundi*', p. 315. One manuscript of Polo's book was supplemented by a diagram that shows the ecumene (with the Mediterranean but no other features and no toponyms marked) and *antoecumene*, divided by equatorial and outer Oceans. The diagram illustrates a brief rendition of theories about the distribution of land and water and the zonal and quadripartite division of the earth: Stockholm, Kungliga Biblioteket, Cod. Holm. M.304, f. 200v.

59 See Pierre Duhem, *Le système du monde: Histoire des doctrines cosmologiques de Platon à Copernic*, 10 vols (Paris: Libraire scientifique A. Hermann, 1915), vol. 3, pp. 62–119; Edouard Jeauneau, *'Lectio Philosophorum': Recherches sur l'Ecole de Chartres* (Amsterdam: Hakkert, 1973), pp. 267–308; Albrecht Hüttig, *Macrobius im Mittelalter: Ein Beitrag zur Rezeptionsgeschichte der Commentarii in Somnium Scipionis* (Frankfurt am Main: Lang, 1990), pp. 75–144. For a reassessment of the meaning of 'Platonism' in the twelfth century see John Marenbon, 'Twelfth-century

Platonism: Old Paths and New Directions', in his *Aristotelian Logic, Platonism, and the Context of Early Medieval Philosophy in the West* (Aldershot: Ashgate, 2000), XV (pp. 1–21).

60 For a summary of the debate about the existence and intellectual originality of the 'school of Chartres', see William of Conches, *Glosae super Boetium*, ed. L. Nauta (Turnhout: Brepols, 1999), pp. xvii–xxi. While R. W. Southern had some success in questioning the extent and influence of the 'school of Chartres', his attempts in *Medieval Humanism and Other Studies* (Oxford: Blackwell, 1970) to characterize the thought of its luminaries – including William – as 'old' and at 'the end of the road' required significant modification.

61 Albert Derolez, *The Autograph Manuscript of the* Liber Floridus*: A Key to the Encyclopedia of Lambert of Saint-Omer*, Corpus Christianorum, Autographa Medii Aevi 4 (Turnhout: Brepols, 1998), pp. 181–3. Following a detailed codicological study of the autograph manuscript, Derolez identified no less than thirteen phases of Lambert's work on the *Liber Floridus*: *Lambertus qui librum fecit: een codicologische studie van de Liber Floridus-autograaf (Gent, Universiteitsbibliotheek, handschrift 92)* (Brussels: Verhandelingen van de Koninklijke Academie voor Wetenschappen, 1978) (summarized in English on pp. 469–79 as 'The Genesis of the *Liber Floridus* of Lambert of Saint-Omer'). For the earlier view of Lambert see Léopold Delisle, *Notice sur les manuscrits du 'Liber Floridus' de Lambert, chanoine de Saint-Omer* (Paris: Imprimerie nationale, Klincksieck, 1906) and Yves Lefèvre, 'Le Liber Floridus et la littérature encyclopédique au moyen âge', in *Liber Floridus Colloquium: Papers read at the international meeting held in the University Library Ghent on 3–5 September 1967*, ed. Albert Derolez (Ghent: E. Story-Scientia, 1973), pp. 1–9.

62 See Derolez, *Autograph Manuscript of the* Liber Floridus, pp. 185–9.

63 Derolez, *Autograph Manuscript of the* Liber Floridus, p. 183.

64 For the context of these images in the autograph manuscript see Derolez, *Autograph Manuscript of the* Liber Floridus, pp. 57, 100, 157, 158.

65 Derolez, *Autograph Manuscript of the* Liber Floridus, p. 186; Delisle, *Notice sur les manuscrits du 'Liber Floridus' de Lambert*, p. 92 (no. 173); Danielle Lecoq, 'La Mappemonde du Liber Floridus ou La Vision du Monde de Lambert de Saint-Omer', *Imago Mundi* 39 (1987), 9–49: p. 11.

66 See Lecoq, 'La Mappemonde du Liber Floridus', p. 16; Arentzen, *Imago Mundi Cartographica*, pp. 88–94. The term 'hormista' derives from one of the medieval titles of Orosius' *Seven Books of Histories against the Pagans.*

67 The autograph manuscript contains a full-page picture of paradise, as well as notes on its four rivers: Derolez, *Autograph Manuscript of the* Liber Floridus, p. 75. See also Scafi, *Mapping Paradise*, p. 144.

68 Wolfenbüttel, Herzog August Bibliothek, MS Gudeanus lat. 1, f. 70r.

69 The reference to the 'sons of Adam' in the opening sentence of this passage seems to acknowledge Augustine, but 'nichil pertinens ad nostrum genus' draws on Cicero's 'nihil ad vestrum genus' (*De re publica* 6.21). The section from 'mare' to 'transitum' may be a dilution of portions of Macrobius' commentary, e.g. his discussions of the term 'meridies' (2.5.19), the Ocean (2.9.4–6), the Milky Way (1.15.7), and the sun (2.8.3–4). For the correspondence with the text of Martianus see Richard Uhden, 'Die Weltkarte des Martianus Capella', p. 103.

70 von den Brincken, *Fines Terrae*, p. 76. For the argument that the island represents the southern, rather than northern, half of the western hemisphere see Edson, *Mapping Time and Space*, p. 110; Arentzen, *Imago Mundi Cartographica*, pp. 92–3.

71 Joan Cadden, 'Science and Rhetoric in the Middle Ages: The Natural Philosophy of William of Conches', *Journal of the History of Ideas* 56 (1995), 1–24; Tullio Gregory, *Anima Mundi: La filosofia di Guglielmo di Conches e la scuola di Chartres* (Florence: Sansoni, 1956); see also Charles Burnett, *The Introduction of Arabic Learning into England* (London: British Library, 1997), pp. 31–60. John of Salisbury famously described William as the best grammarian after Bernard: *Metalogicon*, ed. J. B. Hall and K. S. B. Keats-Rohan, Corpus Christianorum, Continuatio Mediaevalis 98 (Turnhout: Brepols, 1991), 1.5.

72 See Jeauneau, 'Gloses de Guillaume de Conches sur Macrobe: Note sur les manuscrits' and 'Macrobe source du platonisme chartrain', in *'Lectio Philosophorum'*, pp. 267–300. On the intellectual context and impact of William's glosses on Macrobius see Irene Caiazzo, *Lectures médiévales de Macrobe: Les* Glosæ Colonienses super Macrobium (Paris: Vrin, 2002), pp. 46–111. For a discussion of the canon and chronology of William's oeuvre, which also included glosses on Boethius' *Consolatio* and Priscian's *Institutiones* see Nauta's assessment in *Glosae super Boetium*,

pp. xxii–xxv.

73 William of Conches, *Dragmaticon Philosophiae*, ed. I. Ronca (Turnhout: Brepols, 1997), p. xix. On the title 'Dragmaticon' (a form of dialogue) see Ronca's comments, pp. xiv–xvi. The fullest discussion of the differences between the *Philosophia* and the *Dragmaticon* remains Dorothy Elford, 'Developments in the natural philosophy of William of Conches: a study of his *Dragmaticon* and a consideration of its relationship to the *Philosophia*' (University of Cambridge, unpublished doctoral dissertation, 1983), esp. pp. v–xviii. See also Elford's 'William of Conches', in *A History of Twelfth-Century Western Philosophy*, ed. Peter Dronke (Cambridge: Cambridge University Press, 1988), pp. 308–27, esp. p. 315.

74 William of Conches, *Philosophia*, ed. and trans. Gregor Maurach (Pretoria: University of South Africa, 1980), 4.1.5.

75 *Philosophia*, 4.1.5.

76 *Philosophia*, 4.1.5; compare *Dragmaticon*, 6.3.6: 'de illis, quos esse non credimus, propter intellectum philosphicae lectionis ac si ibi sint disseramus.'

77 Caiazzo, *Lectures médiévales de Macrobe*, p. 267 *n*2. For later twelfth-century denunciation of the idea of antipodal habitation, possibly directed against William, see Godfrey of St Victor, *Microcosmus*, ed. Philippe Delhaye, 2 vols (Lille: Facultés Catholiques, 1951), vol. 1, 1.52, p. 70, which contains a digression undertaken 'for the purpose of extirpating the dangerous error of certain of our natural philosophers'. By contrast, in a possible reference to William of Conches, Godfrey's contemporary Alexander Neckham states that, 'philosophically speaking', we are as much under the feet of the antipodeans as they are under ours, before repeating Augustine's dismissal: Alexander Neckham, *De naturis rerum*, ed. Thomas Wright, Rolls Series 34 (London: Longman, 1863), pp. 159–60.

78 On the significance of the term 'lectio' see Jeauneau, 'La lecture des auteurs classiques à l'Ecole de Chartres durant la première moitié du XIIe siècle', in *'Lectio Philosophorum'*, p. 301: 'Il s'agit de "lire" – c'est-à-dire de comprendre et de faire comprendre – les auteurs classiques'.

79 For a related discussion see Italo Ronca, 'Reason and Faith in the *Dragmaticon*: The Problematic Relation between *philosophica ratio* and *diuina pagina*', in *Knowledge and the sciences in medieval philosophy*, ed. Simo Knuuttila *et al.*, 3 vols (Helsinki: Luther-Agricola Society, 1990), vol. 2, pp. 331–41. Ronca stresses William's separation of the domains of reason and sacred authority.

80 *Philosophia*, 4.2.10; compare *Dragmaticon* 6.4.7: 'Nos vero et antipodes antoecorum neque tempore anni neque die neque nocte communicamus'.

81 *Dragmaticon*, 6.2.2. For a discussion of the anti-flat earth passage see Elford, 'Developments', pp. 175–9, who notes William's use of Calcidius, Macrobius, Martianus Capella, and Bede in the passage, as well as evidence of his knowledge of Seneca's *Quaestiones naturales*.

82 *Dragmaticon*, 6.2.3.

83 *Dragmaticon*, 6.2.4.

84 *Dragmaticon*, 6.4.5.

85 For discussion and summary of the dating and influence of the poem (it survives in over two hundred manuscripts) see Walter of Châtillon, *Alexandreis*, ed. Marvin L. Colker (Padua: Antenore, 1978), pp. v–xxxviii; and *The* Alexandreis *of Walter of Châtillon: A Twelfth-Century Epic*, trans. David Townsend (Philadelphia: University of Pennsylvania Press, 1996), pp. xiv–xv.

86 Walter of Châtillon, *Alexandreis*, ed. Colker, 9.569–70.

87 Walter of Châtillon, *Alexandreis*, ed. Colker, 10.98–100. On the importance of Lucan's account of the death of Alexander for Walter's depiction of Nature see Maura K. Lafferty, 'Nature and an unnatural man: Lucan's influence on Walter of Châtillon's concept of nature', *Classica et mediaevalia* 46 (1995), 285–300, and Lafferty's expanded discussion of this topic in *Walter of Châtillon's* Alexandreis*: Epic and the Problem of Historical Understanding* (Turnhout: Brepols, 1998), pp. 141–69.

88 Walter of Châtillon, *Alexandreis*, 10.312–319.

89 Johannes de Hauvilla (Jean de Hauville), *Architrenius*, ed. and trans. Winthrop Wetherbee (Cambridge: Cambridge

University Press, 1994), 5.246–49; Wetherbee's translation, p. 129, slightly modified. The possible echo of Lucan, 'Scythicosque recessus' (*De bello civili*, 8.216) is noted by Paul Gerhard Schmidt in his 1974 edition. De Hauville (c. 1150–c. 1208), a master at the cathedral school of Rouen, dedicated the *Architrenius* to Walter of Coutances in 1184.

90 See further the discussions in Moretti, *Gli antipodi*, pp. 88–98; Patrick Gautier Dalché, 'Entre le folklore et la science: la légende des antipodes chez Giraud de Cambrie et Gervais de Tilbury', in *La leyenda: antropología, historia, literatura* (Madrid: Universidad Complutense, 1989), pp. 103–14, esp. 107–8; R. S. Loomis, 'King Arthur and the Antipodes', *Modern Philology* 38 (1941), 289–304.

91 Gervase, *Otia imperialia*, 2.12, pp. 334–7, where Arthur is found within Mount Etna by a bishop's servant.

92 Gervase, *Otia imperialia*, 3.45, pp. 642–4, cites Robert of Kenilworth as the source of this story, in which a Derbyshire swineherd follows a sow into a cave, retrieves it from the lord of the land ('preposito terre illius'), and returns to 'our hemisphere'. Elsewhere in the *Otia* Gervase follows Isidore's location of the antipodes at the ends of the known world ('in finibus'), between the Red Sea and Ocean, but crucially omits Isidore's qualifier, 'fabulose': 2.4, p. 216.

93 Gerald of Wales, *Itinerarium Kambriae*, ed. James F. Dimock, Rolls Series 21 (London: Longman, 1868), 1.8, pp. 75–7, where a priest, Eliodorus, admits to Bishop David (1148–76) that as a boy he had been led by 'homunculi duo, staturae quasi pygmaeae' into a subterranean kingdom, inhabited by 'men of the smallest stature' who pass frequently into the upper hemisphere, the mendacity of which they are nevertheless staunch critics. Compare Walter Map, *De nugis curialium*, ed. and trans. M. R. James, rev. C. N. L. Brooke and R. A. B. Mynors (Oxford: Clarendon Press, 1983), Dist. 1, ch. 11, pp. 26–31, where King Herla enters an underworld kingdom to fulfill a pact with the king of the Pygmies, and finds that three subterranean days there equal over two–hundred years in the upper world.

94 The phrase is used by John Carey, 'Ireland and the Antipodes', p. 10.

95 Gautier Dalché, 'Entre le folklore et la science', p. 113.

96 Chrétien de Troyes, *Erec and Enide*, ed. and trans. Carleton W. Carroll (New York: Garland, 1987), ll. 1955–73.

97 *Le Dragon Normand et autres poèmes d'Étienne de Rouen*, ed. Henri Omont (Rouen: Société de l'histoire de Normandie, 1884), 2. 1165–8 (pp. 112–13).

98 The kings exchange letters, but a physical confrontation is averted by the death of Matilda (2. 955–1309). Loomis concurs with J. S. P. Tatlock, 'Geoffrey and King Arthur in *Normannicus Draco*', *Modern Philology* 31 (1933), 1–18, 113–25, that Étienne's deployment of the Arthurian legend is intended to flatter Henry, and to mock belief in Arthur's promised return: 'Arthur and the Antipodes', pp. 289–90.

99 Moretti, *Gli antipodi*, p. 95.

100 *Normannicus Draco*, 2. 1169 (p. 113). Tatlock outlines the extent of allusions to Alexander in the poem and argues that Arthur is mockingly figured as Darius, with Henry II in the role of Alexander: 'King Arthur in *Normannicus Draco*', pp. 9–11.

101 On Sacrobosco and the tradition of commentaries on his work see Lynn Thorndike, *The Sphere of Sacrobosco and Its Commentators* (Chicago: Chicago University Press, 1949), pp. 1–75; Olaf Pedersen, 'In Quest of Sacrobosco', *Journal for the History of Astronomy* 16 (1985), 175–221; Owen Gingerich, 'Sacrobosco as a Textbook', *Journal for the History of Astronomy* 19 (1988), 269–73.

102 Thorndike, *Sphere of Sacrobosco*, pp. 82, 87, 107. All further references to commentaries on Sacrobosco are to this edition.

103 Thorndike, *Sphere of Sacrobosco*, pp. 321–2. The question of the authorship of the commentary is discussed by Thorndike, *Sphere of Sacrobosco*, p. 23.

104 Thorndike, *Sphere of Sacrobosco*, p. 321

105 Thorndike, *Sphere of Sacrobosco*, pp. 188–93.

106 Thorndike, *Sphere of Sacrobosco*, p. 193. This point echoes the theory advanced by Robert Grosseteste in *De sphaera*:

Thorndike, *Sphere of Sacrobosco*, pp. 11–12. See Robert Grosseteste, *De sphaera*, in *Die philosophischen Werke des Robert Grosseteste, Bischofs von Lincoln*, ed. Ludwig Baur (Münster i. W.: Aschendorffsche Verlagsbuchhandlung, 1912), pp. 23–4. Robert's views appear to have been known by the Franciscan John Pecham, who considered and rejected the possibility of inhabitation beyond the known quarter of the world in a treatise produced c. 1277–9, shortly before he became Archbishop of Canterbury: Bruce MacLaren, 'A Critical Edition and Translation, with Commentary, of John Pecham's *Tractatus de sphera*' (University of Wisconsin-Madison, unpublished doctoral dissertation, 1978), pp. 17, 29–30, 194–201. Cecco d'Ascoli's commentary on Sacrobosco, also printed in Thorndike's collection, shows the debate continuing into the fourteenth century, with little change in the terms of argument: Ptolemy and Avicenna are still cited to support the case for habitability at and below the equator – but it is important to note that Virgil, Ovid, and Plato remain on the roster of authorities (cited in support of the uninhabitable nature of the torrid zone).

107 Glenn Michael Edwards, 'The *Liber introductorius* of Michael Scot' (University of Southern California, unpublished doctoral dissertation, 1978), pp. xxxiv–xxxix. Edwards argues that the *Liber introductorius* was seen by Scot as the first book of a trilogy, followed by the *Liber particularis* and the *Liber physiognomie* (pp. xxii–xxiii).

108 Glenn M. Edwards, 'The Two Redactions of Michael Scot's "Liber introductorius"', *Traditio* 41 (1985), 329–40, argues that the long version was written by Scot c. 1230 and is anterior to the shorter one, which he regards as a digest. These findings are not accepted by Charles Burnett, '*Partim de suo et partim de alieno*: Bartholomew of Parma, the Astrological Texts in MS Bernkastel-Kues, Hospitalsbibliothek 209, and Michael Scot', in *Seventh Centenary of the Teaching of Astronomy in Bologna 1297–1997*, ed. Pierluigi Battistini, Fabrizio Bònoli, Alessandro Braccesi, Dino Buzzetti (Bologna: Clueb, 2001), pp. 37–76, who does not regard the *Liber* in its present forms as the work of a single author; Silke Ackermann has also suggested that the relation between the four manuscripts of the *Liber introductorius* (or the *Liber quatuor distinctionum*) and its fragments is more complicated than Edwards thought: 'Bartholomew of Parma, Michael Scot and the set of new constellations in Bartholomew's *Breviloquium de fructu tocius astronomie*', in *Seventh Centenary of the Teaching of Astronomy in Bologna*, pp. 77–98.

109 The *Liber* survives in four manuscripts (two of each redaction); fragments have been noted in around thirty manuscripts: Edwards, 'The Two Redactions', pp. 331–2.

110 This is the conclusion of Edwards, 'The Two Redactions', pp. 338–9. On Scot and the court of Frederick see Lynn Thorndike, *Michael Scot* (London: Nelson, 1965); Folker Reichert, 'Geographie und Weltbild am Hofe Friedrichs II', *Deutsches Archiv für Erforschung des Mittelalters* 51 (1995), 433–91; Charles Burnett, 'Michael Scot and the transmission of scientific culture from Toledo to Bologna via the court of Frederick II Hohenstaufen', *Micrologus* 2 (1994), 101–26. Charles Homer Haskins' account of science at the court of Frederick remains an entertaining if dated overview: *Studies in the History of Mediaeval Science* (Cambridge: Harvard University Press, 1924), pp. 242–71.

111 Burnett, '*Partim de suo et partim de alieno*', pp. 37–76. See also Burnett's 'Michael Scot and the transmission of scientific culture', pp. 101–26. The precise role of Bartholomew of Parma in the transmission, and perhaps alteration of the *Liber introductorius* of Michael Scot remains unresolved.

112 There is no edition of the *Liber introductorius*, although the chapter 'De notitia divisionis zone habitabilis nostre' is edited by Reichert, 'Geographie', pp. 484–8. All references are to the long redaction of the *Liber* contained in Munich, Bayerische Staatsbibliothek, Clm 10268. Where the text is illegible or uncertain I have supplied readings from Munich, Bayerische Staatsbibliothek, Clm 10663, a copy of Clm 10268 made c. 1700.

113 Munich, Clm 10268, f. 44vb.

114 Reichert, 'Geographie', p. 455.

115 E.g. Munich, Clm 10268, ff. 49ra, 133va.

116 Munich, Clm 10268, f. 45rb.

117 The influence of William's *Philosophia* and *Dragmaticon* on the *Liber* has been noted by Piero Morpurgo, 'Fonti di Michele Scoto', *Accademia nazionale dei Lincei* ser. 8, 38 (1983), 59–71: p. 67.

118 Munich, Clm 10268, f. 45rb.

119 Munich, Clm 10268, f. 45va. The sources of such notions were not restricted to Christian thought, as Servius' commentary on Virgil's *Georgics* 1.243 shows: *In Vergilii bucolica et georgica commentarii*, p. 187.

120 Munich, Clm 10268, f. 45va.

121 Munich, Clm 10268, f. 45va.

122 Munich, Clm 10268, f. 45va–vb. On Michael's discussion and location of hell and paradise see Reichert, 'Geographie', pp. 454–9. Reichert notes that the connection between India and St Brendan's island is their position at the far east and west of the ecumene.

123 Munich, Clm 10268, f. 45va.

124 Munich, Clm 10268, f. 45va.

125 *L'Image du monde de Maitre Gossouin*, ed. O.H. Prior (Lausanne and Paris: Librairie Payot, 1913), p. 93.

126 For a list of contents of the manuscript see *A Catalogue of the Manuscripts in the Cottonian Library deposited in the British Museum* (London, 1802), pp. 15–16. Ff. 50v–53 formerly contained one version of Matthew Paris' map of England and Scotland: Richard Vaughan, *Matthew Paris* (Cambridge: Cambridge University Press, 1958), pp. 241, 243.

127 For further nomenclature see Anna-Dorothee von den Brincken, 'Die Klimatenkarte in der Chronik des Johann von Wallingford – ein Werk des Matthaeus Parisiensis?', *Westfalen* 51 (1973), 47–56: pp. 48–51; von den Brincken, *Fines terrae*, p. 111; and Chekin, *Northern Eurasia in Medieval Cartography*, pp. 202–3. Von den Brincken argues that the map was initially intended to accompany Matthew Paris' *Chronica majora*, and should be ascribed to him.

128 London, BL, MS Cotton Julius D.VII, f. 46r: 'Mare equinoctiale quod diuidit duas partes tocius aride id est terre. que scilicet est tercia pars tocius terre que est globus spericus. sed terra que elementum cuius meditullium est centrum tocius mundi vbi dicitur esse infernus uel abissus. que est matrix aquarum omnium que sunt circa globum terre sicut albumen circa uitellum in ouo. Quando nos habemus diem illi noctem. quando nos hiemem illi estatem' (the equinoctial sea which divides two parts of the whole, that is of dry land, and so is a third part of the whole earth which is a spherical globe. But land is also an element, whose middle is the centre of the whole world where inferno or abyss is said to be. And it is source of all the waters which surround the globe like white around the yoke in an egg. When we have day they have night. When we have winter, they have summer). The egg-cosmos analogy was a relatively common feature of medieval cosmology: see Peter Dronke, *Fabula: Explorations into the uses of myth in medieval Platonism* (Leiden: Brill, 1974), pp. 79–99; Rudolf Simek, *Altnordische Kosmographie: Studien und Quellen zu Weltbild und Weltbeschreibung in Norwegen und Island vom 12. bis zum 14. Jahrhundert* (Berlin: de Gruyter, 1990), pp. 72–81.

129 *Declaratio Raimundi*, ed. Theodor Pindl-Büchel, in *Raimundi Lulli Opera Latina 76–81*, Corpus Christianorum Continuatio Mediaevalis 79 (Turnhout: Brepols, 1989), p. 253.

130 As Pindl-Büchel argues: *Declaratio Raimundi*, pp. 231–41.

131 *Declaratio Raimundi*, p. 260.

132 *Declaratio Raimundi*, p. 260. The examples of the antipodes and the apple had appeared a few years previously, in Lull's *Arbor Scientiae*, again illustrating human imaginative power: *Arbor Scientiae*, ed. Pere Villalba Varneda, 3 vols, *Raimundi Lulli Opera Latina* 65, Corpus Christianorum Continuatio Mediaevalis 180 (Turnhout: Brepols, 2000), vol. 2, p. 459; vol. 3, p. 1294.

133 *Tractatus novus de astronomia*, ed. Michela Pereira, in *Raimundi Lulli Opera Latina 76–81*, 1.2.3 (p. 143).

134 *Declaratio Raimundi*, pp. 267, 270, 275, 318, 392, 398.

135 *Tractatus novus*, 1.2.3, 1.2.5, 5.2.1 (pp. 142–3, 166, 214); *Liber de lumine*, dist 2 and 3, ed. Jordi Gayà Estelrich, in *Raimundi Lulli Opera Latina 106–113*, Corpus Christianorum Continuatio Mediaevalis 113 (Turnhout: Brepols, 1995), pp. 33, 48; *Liber de regionibus sanitatis et informitatis*, dist. 1 and 3, and *De ascensu et descensu intellectus*, dist. 7, both in *Raimundi Lulli Opera Latina 120–122*, ed. Aloisius Madre, Corpus Christianorum Continuatio Mediaevalis 35 (Turnhout: Brepols, 1981), pp. 75, 105, 125. See also Lull's work of 1308, the *Liber de centvm signis Dei*, ed. Aloisius Madre, in *Raimundi Lulli Opera Latina* 131–3, Corpus Christianorum Continuatio Mediaevalis 114 (Turnhout: Brepols, 1998), 5.3.

136 See Scafi, *Mapping Paradise*, esp. pp. 160–90.

137 The unedited text, the 'Distinctio divisionis terre et paradisi delitiarum', is summarized by Gautier Dalché, 'Le

paradis aux antipodes?', pp. 620–5. It continues by considering a second, related *quaestio*: 'whether the place (of paradise) may be suitable for human habitation' (utrum sit locus conueniens habitacioni humane).

138 'Le paradis aux antipodes?', pp. 636–7.

139 Gautier Dalché, 'Le paradis aux antipodes?', p. 631. Albertus Magnus thought the answer was no, but admitted the lack of authorities to support his conviction.

140 Dante Alighieri, *Convivio*, ed. Franca Brambilla Ageno, 2 vols (Florence: Casa Editrice Le Lettere, 1995), 3.5.9–22.

141 Ragazzani summarizes arguments about the authenticity of the texts, and concludes that the question is now 'unattackably' resolved in favour of Dantean attribution: Dante Alighieri, *Questio de aqua et terra*, ed. Severino Ragazzini and Luigi Pescasio (Mantua: Editoriale Padus, 1978), pp. 37–43. Although his arguments against the attribution of *De situ* to Dante have not found lasting favour, the intellectual background to the work is helpfully set out by Bruno Nardi, *La caduta di Lucifero e l'autenticità della 'Quaestio de aqua et terra'* (Turin: Editrice Internazionale, 1959). See more recently Bouloux, *Culture et savoirs géographiques*, pp. 27–31.

142 The leading Aristotelian philosopher of the fourteenth century, John Buridan, subsequently examined the same question in his *Quaestiones* on Aristotle's *De caelo*, a work probably compiled in the early 1330s, and perhaps as early as 1328–30, as part of his teaching of the text at Paris: *Ioannis Buridani Expositio et Quæstiones in Aristotelis De cælo*, ed. Benoît Patar (Leuven: Peeters, 1996), pp. 115–16. Buridan's original solution was to argue that earth's centre of gravity is not identical with its centre of magnitude, because the portion of the earth that was exposed to the air became lighter than the portion that remained under the sea. This a-symmetricality of the earth meant that some part of it would continue to be exposed, but the implication of the theory is that all land beyond the ecumene must be covered by water: *Expositio et Quæstiones in Aristotelis De cælo*, p. 416. Buridan's ideas were taken up by a number of his followers, some more competent than others: see particularly Nicolas Oresme, *Le Livre du ciel et du monde*, ed. Albert D. Menut and Alexander J. Denomy, trans. Menut (Madison: University of Wisconsin Press, 1968), Book 2, chapter 7, ff. 87a–89a, pp. 346–52; on Domenicus de Clavasio's *Quaestiones super Aristotelis De caelo et mundo* see Patrick Gautier Dalché, 'L'influence de Jean Buridan: l'habitabilité de la terre selon Dominicus de Clavasio', in *Comprendre et maîtriser la nature au Moyen Âge: Mélanges d'histoire des sciences offerts à Guy Beaujouan* (Geneva: Droz/Paris: Champion, 1994), pp. 101–13.

143 *De situ et forma aque et terre*, ed. Giorgio Padoan (Florence: Le Monnier, 1968), p. 76.

144 Dante Alighieri, *La Commedia secondo l'antica vulgata*, ed. Giorgio Petrocchi, 4 vols (Milan: Mondadori, 1966–7), vol. 2, *Inferno*, 34.118–26; translation from Dante Alighieri, *The Divine Comedy, vol. 1: Inferno*, trans. Charles S. Singleton, 2 vols (London: Routledge and Kegan Paul, 1970), vol. 1, p. 369. Perhaps Dante echoes Ovid's *Metamorphoses* 2.254, where during the ride of Phaethon 'the Nile fled in terror to the ends of the earth, and hid its head, and it is hidden yet' (Nilus in extremum fugit perterritus orbem/ occuluitque caput, quod adhuc latet).

145 *Inferno*, 26.117. This line was interpreted by Pietro Alighieri in the second version of his commentary on the *Commedia* as Dante's affirmation of Augustine's rejection of antipodal inhabitation: see *Il 'Commentarium' di Pietro Alighieri nelle redazioni ashburnhamiana e ottoboniana*, ed. R. Della Vedova and M. T. Silvotti (Florence: Olschki, 1978) pp. 369–70. Consistent with this distinction between ecumene and *antoecumene* is the reference to the two hemispheres in Boccaccio's *Filocolo* 5.8: 'universo abitato e quello inhabitato'.

146 A few years earlier a text, attributed by modern scholarship to Bartholomew of Parma, proposed the inverse: at the end of time the zones would be realigned, and 'those who were in one zone will come into the other' (illi qui fuerint in una zona venient in alteram). This text is discussed, and relevant passages reproduced, in Bouloux, *Culture et savoirs géographiques*, pp. 32–4. For the attribution see *I primi due libri del 'Tractatus sphæræ' di Bartolomeo da Parma*, ed. Enrico Narducci (Rome: Bullettino di bibliografia e di storia delle scienze matematiche e fisiche, 1885), pp. 26–31.

147 *Purgatorio*, ed. Petrocchi, vol. 3, 1.22–7; *The Divine Comedy of Dante Alighieri*, ed. and trans. Robert M. Durling, 3 vols (Oxford: Oxford University Press, 2003), vol. 2, p. 19.

148 Pietro Alighieri, *Comentum super poema Comedie Dantis: A critical edition of the third and final draft of Pietro Alighieri's* Commentary *on Dante's* The Divine Comedy, ed. Massimiliano Chiamenti (Tempe: Arizona Center for Medieval and Renaissance Studies, 2002), p. 288.

149 Pietro, *Comentum*, p. 281, noted the passage in Augustine's commentary on Genesis; *Sancti Aureli Augustini De Genesi*

ad litteram libri duodecim, ed. Joseph Zycha (Vienna: Tempsky, 1894), 8.7, p. 242.

150 *Familiares* 6.3.57; 9.13.9; 20.1.12–13: *Le Familiari (Familiarium Rerum Libri)*, ed. Vittorio Rossi, 4 vols (Florence: Sansoni, 1933–42); *Lettere disperse* 9; 33: *Lettere disperse*, ed. Alessandro Pancheri (Parma: Fondazione Pietro Bembo, 1994); *Seniles* 11.17: *Opera quae extant omnia*, 4 vols (Basel, 1554), vol. 2, p. 989.

151 Bouloux, *Culture et savoirs géographiques*, pp. 41–2: 'un savoir partagé'.

152 Pierre de Nolhac, *Pétrarque et l'humanisme*, 2 vols (Paris: Champion, 1907), vol. 2, p. 80.

153 Francis Petrarch, *L'Africa*, ed. Nicola Festa (Florence: Sansoni, 1926), 2.399–400, p. 44. Compare *Africa* 8.1–3, where the sun is described in conventional terms as heading off to the antipodes at dusk, and see Petrarch's *Canzoniere* 22 and 50 for similar imagery: *Canzoniere*, ed. Marco Santagata (Milan: Mondadori, 1996).

154 Francis Petrarch, *Secretum*, in *Opere Latine*, ed. Antonietta Bufano, vol. 1 (Turin: Unione Tipografico-Editrice Torinese, 1975), 3.16.1–7, pp. 244–6.

155 *Francisci Petrarchae Opera quae extant omnia*, vol. 2, p. 989, slightly emended. Translation from Petrarch, *Letters of Old Age: Rerum senilium libri I–XVIII*, trans. Aldo S. Bernardo, Saul Levin, and Reta A. Bernardo, 2 vols (Baltimore: Johns Hopkins Press, 1992), vol. 2, pp. 433–4.

156 *Fam* 20.1.8, in *Le Familiari*, vol. 4, p. 5.

157 *Fam* 20.1.12–13, pp. 5–6.

158 On 'the paradox of exile as home' see Giuseppe Mazzotta, *The Worlds of Petrarch* (Durham: Duke University Press, 1993), pp. 181–92.

159 *Fam.* 20.1.20, p. 7.

160 *Fam.* 6.3.67–8, in *Le Familiari*, vol. 2, p. 76.

161 *Fam.* 6.3.57, pp. 73–4.

162 *Fam.* 6.3.62, p. 75.

163 *Fam.* 9.13.9, in *Le Familiari*, vol. 2, p. 248.

Chapter Six

Terra Recognita: *the Expansion of the Known World 1400–1493*

Around 1410 the Cardinal of Cambrai, Pierre d'Ailly, appended an 'Epilogus Mappe Mundi' to his geographical treatise, the *Imago Mundi.* D'Ailly, a prolific theologian and philosopher, intended the 'Epilogus' to be illustrated by eight figures, designed to provide visual elucidation of the theories discussed in the *Imago Mundi*, a digest that relied heavily on John of Sacrobosco's *De sphera*, Roger Bacon's *Opus Maius*, and Nicolas Oresme's *Traité de l'Espere.* The seventh figure in the 'Epilogus' is essentially a summation of geographical theories current in the early fifteenth century (fig. 19). In this image d'Ailly combined elements of three genres of medieval map: the T-O map; the zonal map; and the *climata* map. He divided the northern hemisphere not only between the three continents of Europe, Africa, and Asia (the standard division of T-O maps), but also into seven latitudinal zones from Ethiopia to the Arctic circle (the standard division of *climata* maps); in addition, the map shows vestiges of the division of the world into five zones: it extends from the north to the south pole, with Arctic and Antarctic circles, both tropics, and the equator marked. Consistent with the tradition of *climata* maps, d'Ailly has not so much drawn the map as written it: coastal outlines are not represented, but the positioning of the names of cities, nations, rivers, seas, and mountain ranges provides a fairly detailed if schematic representation of the northern hemisphere.[1] Particular prominence is given to the 'Mare mediterraneum vsque Asiam' (Mediterranean up to Asia), to the Nile, which extends through the cities of Syene and Meroë as far as the equator, to the Red Sea, represented by a diagonal line and caption in southern Asia, and to India, located at the world's easternmost extent. Four inscriptions on the map relate to the extension of land beyond the known world: two refer to inhabited land to the north and south of the *climata*; two others note the possible extent of India to the south (as far as the Tropic of Capricorn) and to the east.[2]

Consistent with his declared intention to rely on experience and probable histories rather than the imagination,[3] d'Ailly left the region beneath the Tropic of Capricorn blank. However, into this otherwise unmarked space he intruded an annotation that extends vertically from the northern hemisphere into the southern:

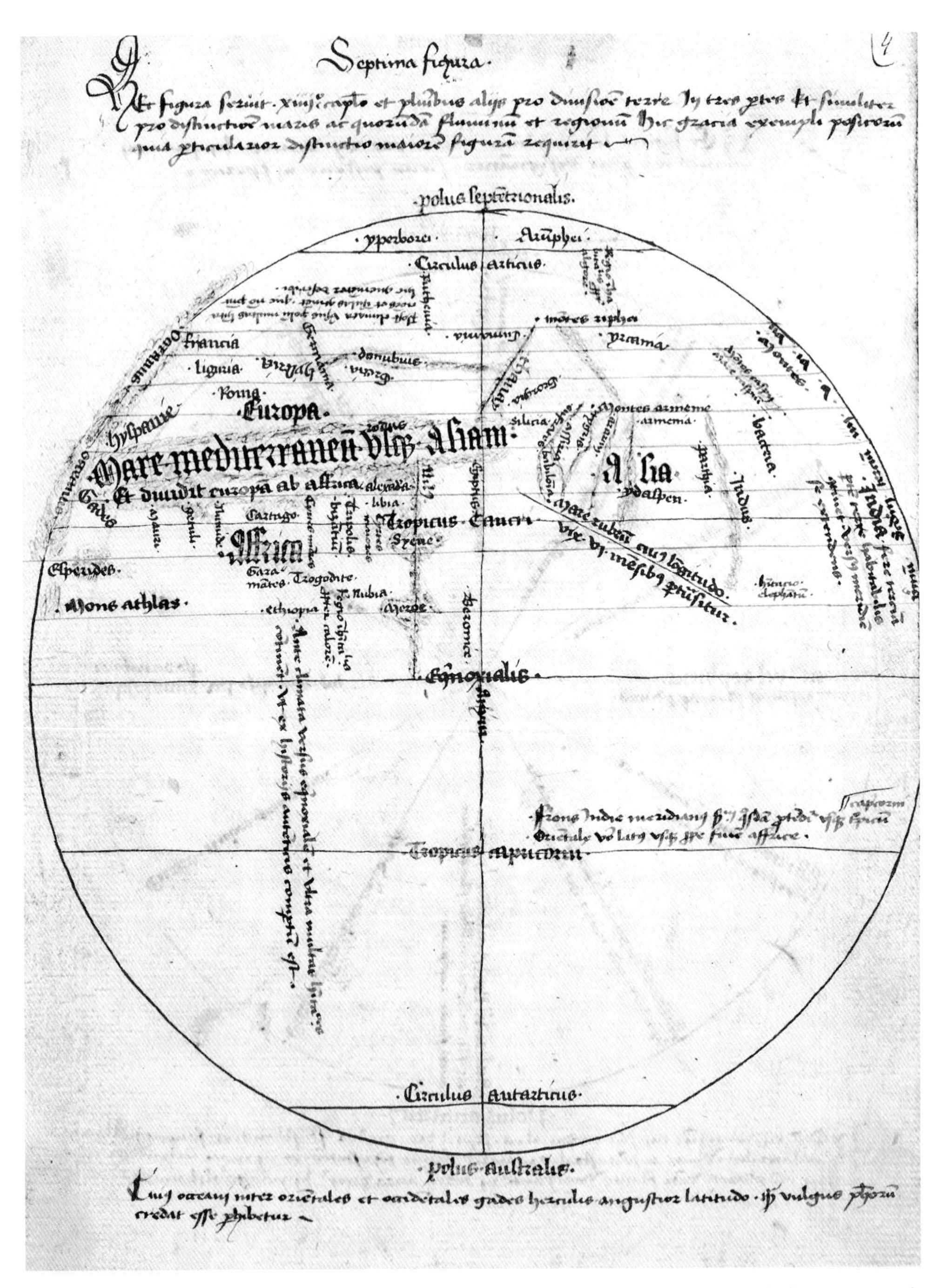

Fig. 19. Brussels, Bibliothèque royale, MS 21198–204, f. 4r. *Climata* diagram in Pierre d'Ailly, *Imago mundi*, fifteenth century. North at the top. The northern hemisphere is divided into seven *climata* from Ethiopia to the Arctic circle, and into the three continents of Europe, Africa, and Asia (the standard division of T-O maps). In addition to the Arctic and Antarctic circles, both tropics and the equator are marked. Prominent ecumenical features include the 'Mare mediterraneum vsque Asiam', the Nile, which extends from the north coast of Africa as far as the equator, the Red Sea, represented by a diagonal line and caption in southern Asia, and India, located at the world's easternmost extent. An inscription beneath the equator on the far right of the map notes the possible extent of India to the south (as far as the Tropic of Capricorn) and to the east; the inscription that extends vertically between the hemispheres states that 'before the *climata* towards the equator and beyond it are many habitations. As it is learnt from authoritative histories' (Ante climata versus equinoctialem et vltra multas habitationes continent. Vt ex historiis autenticis compertum est).

> Ante climata uersus equinoctialem et ultra multas habitationes continent. Ut ex historiis autenticis compertum est

> before the *climata* towards the equator and beyond it are many habitations. As it is learnt from authoritative histories.

The annotation's position and direction is no accident: it is a topographical reference, starting before the first *climata* (that of Meroë), and extending down to the equator and beyond ('et ultra'). The southward direction is made permissible by 'authoritative histories': the works of Pliny the Elder, Ambrose, Martianus Capella, Aristotle, Averroes, and the eleventh-century Arabic astrologer known as Haly ('Alī al-'Imrīn), all cited by d'Ailly in support of the idea of sub-equatorial habitation.[4] The words of the annotation suggest the habitability of the equatorial regions; its positioning represents graphically the tantalizing possibility of transit to the southern hemisphere.[5] The inscription across the equator mirrors a similar inscription in the far north,[6] but whereas that annotation heads across the page, contained within the surrounding toponyms of Germania, Hystria, and the Arctic circle, this one heads down, away from the place of the northern hemisphere, and into the speculative space of the south: it is a verbal exploration – tentative, limited – of unknown land.

D'Ailly's seventh figure cannot be said to have a date. It does not represent the world at a particular moment, even if it clearly was considered valid and informative in the first decade of the fifteenth century. Classical (Aristotle, Pliny), late antique (Ambrose, Martianus Capella), and Arab ('Haly', Averroes) authorities combine in the *Imago Mundi*, mediated by medieval Latin compilers of the thirteenth and fourteenth centuries (Roger Bacon, Nicolas Oresme). The image reveals, however, that for some time scholars had been engaged in rethinking and thereby reknowing the boundaries of the ecumene, and that such rethinking could find expression on the world image. This chapter considers the dramatic changes that occurred to the boundaries of the known world in the years that followed the compilation of d'Ailly's *Imago Mundi*. Change was driven by developments in two interrelated areas, intellectual and political. In the field of what was at the time termed 'cosmography', the translation of Claudius Ptolemy's *Geographia* from Greek into Latin in the first decade of the fifteenth century, and its relatively rapid dissemination throughout learned European circles, introduced a different mode of representing unknown land. As a result, the question of the antipodes was reformulated. Rather than being unreachable, unknown land was brought alongside the ecumene. It was, according to Ptolemy, at the southern, eastern, and northern borders of the known world. Yet it was quickly perceived that Ptolemy's image of the world required supplementation: certain areas unknown to him either had been known to his fifteenth-century European inheritors for some time, or were gradually coming to their attention.

Coterminous with the repopularization of Ptolemy was the expansion of Iberian power into Africa and the Atlantic, a process of exploration and colonization that culminated in the voyages of Columbus and the rounding of the Cape of Good Hope in the 1490s. *Terra incognita* became *terra inventa*, land discovered, brought not only into the European geographical consciousness but under its administrative practices. These two developments were interrelated because both demanded revision of the relationship between ecumene and *antoecumene*. Nowhere was that rethinking more evident than in the papal curia and the Church Councils of the first half of the fifteenth century. As scholars attempted to digest both the information contained in Ptolemy's *Geographia* and the reports of hitherto unknown land in Africa and Asia, the expansionary ambitions of the royal houses of Portugal, Aragon and Castile brought into dramatic focus the issue of the ownership of unknown, and recently discovered, land. The representation and the jurisdiction of *terra incognita* were linked by the fact that jurisdiction over land had to be assigned *through* acts of representation, whether on a map or a papal bull, or both. The process of the transformation of *terra incognita* to *terra inventa* therefore demanded both revision of places and images on the world map, and revision of the sources that informed the map. It is a process that has seemed to historians emblematic of many different things – the Renaissance, the birth of the modern world, a period of transition from the medieval to the modern. Rather than using these terms, however, I suggest that the changes to the world image that occurred in the fifteenth century are best understood as a process of reformulation, of land rethought and re-known: *terra recognita.*

The impact of Ptolemy

The basis for Ptolemy's geographical theories had been known in western Europe since the twelfth century, through his *Almagest,* and in the form of tables of geographic coordinates. However, the translation of Ptolemy's *Geographia* into Latin, a task commenced in Florence by Manuel Chrysoloras and completed during the first decade of the fifteenth century by Jacopo d'Angelo, added a practical cartographic guide to the theoretical basis provided by the *Almagest.*[7] With the *Geographia,* European geographers had the means of mapping the world and its regions according to a mathematically-determined system. Ptolemy sub-divided Europe, Africa, and Asia into regions, and provided co-ordinates in longitude and latitude so that the places in each region could be plotted on a global grid.[8] However, the image of the world transmitted by Ptolemy was based on the data available in second-century Alexandria – a considerable resource, but inevitably limited – and these limits were quickly realized by the *Geographia*'s fifteenth-century audience.

Ptolemy himself had acknowledged that the geographical information he presented was partial and in need of revision. In Book 1 of the *Geographia* he states that knowledge of portions of Europe, Africa, and Asia is incomplete because they have not been fully explored, or because they have been inaccurately described due to 'the carelessness of travellers' (ob peragrantium negligentiam), or because of mutations that have occurred to known places over time.[9] Ptolemy defined the word 'geōgraphia' ('cosmographia' in d'Angelo's translation) as an imitation of the entire *known* world.[10] Nevertheless, in the course of his descriptions of the regions of the three continents, Ptolemy noted the presence of unknown land beyond the limits of the known world: to the north of European Sarmatia (3.5), in Africa to the west and south of Ethiopia (4.8), and in Asia to the north of Asiatic Sarmatia (5.8), Scythia (6.14), and Serica (6.16), and to the east of Sina (7. 3). According to his summary of the inhabited earth, the known world was bounded not

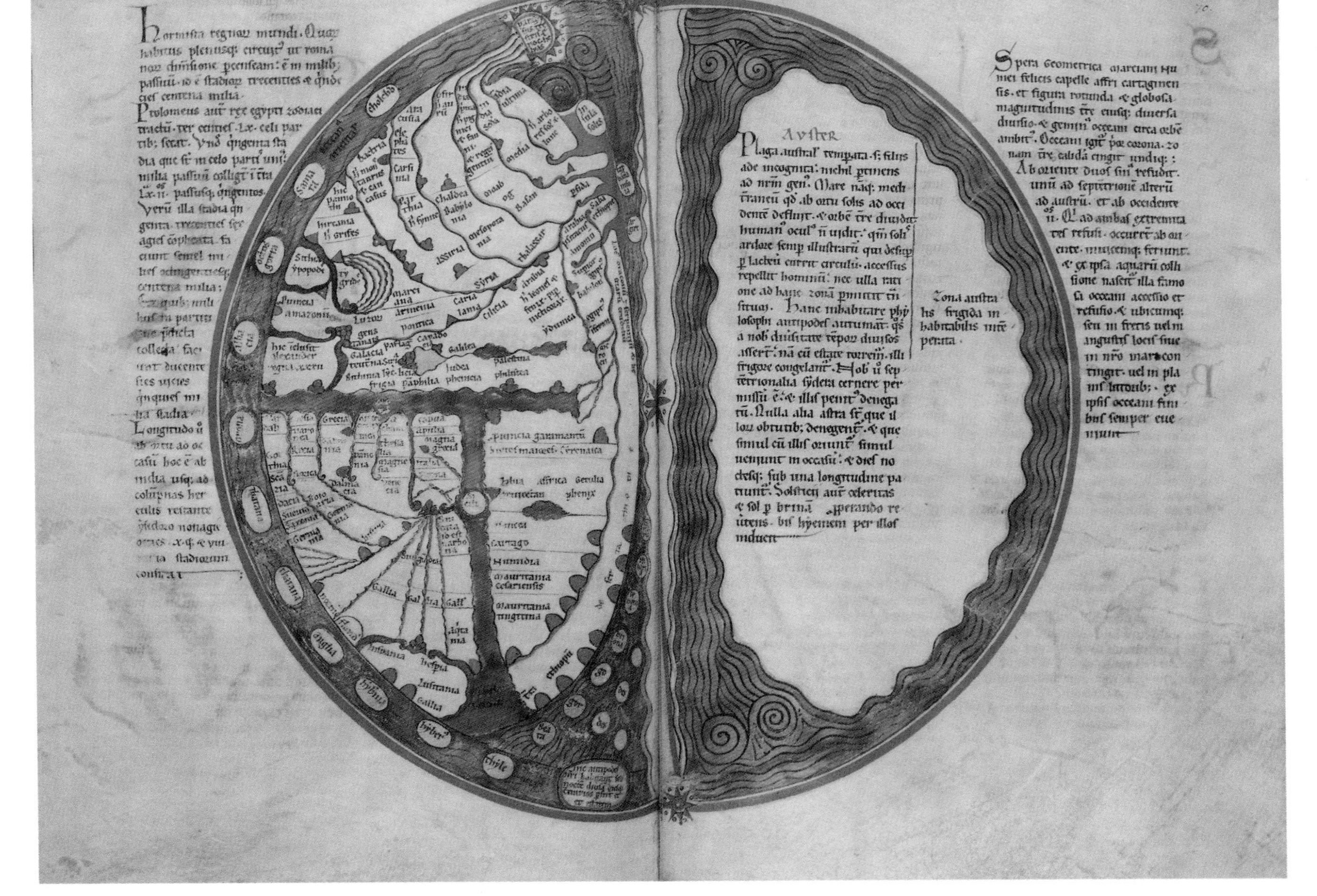

Plate 1. Wolfenbüttel, Herzog August Bibliothek, MS Gudeanus lat. 1, ff. 69v–70r. World map in a twelfth-century manuscript of Lambert of Saint-Omer, *Liber floridus*. East at the top. Northern and southern hemispheres are separated by the equatorial torrid zone. On the left, a fairly detailed ecumene, divided between Asia, Europe, and Africa, is surrounded by a series of islands. In far southern Africa a strip of land shows the 'Terra Ethiopum' and the 'Deserta Ethiopie'. At the far east of the ecumene the earthly paradise is represented as a peninsula, from which its four rivers (Tigris, Euphrates, Nile and Ganges) flow into the world. At the far west of the image an island represents the land of 'our antipodeans'. In the southern hemisphere a long text states that transit to the northern hemisphere is impossible (and vice-versa) but that philosophers assert the existence of people in the unknown region.

Plate 2. Reims, Bibliothèque municipale, MS 1321, f. 13r. World map of Guillaume Fillastre, illustrating his fifteenth-century manuscript of Pomponius Mela, *De chorographia*. East at the top. This world map is evidently very different in form from Fillastre's world map shown in figure 23, but it too shows clear signs of Ptolemaic influence, most notable in its repetition of 'terra incognita' in three places in the far north, north-east (to the east of the Caspian sea) and far south (south of Ethiopia).

Plate 3. Vatican City, BAV, MS Palat. lat. 1362b. World map of Andreas Walsperger, 1448. South at the top. The map is particularly notable for its conflation of the uninhabitable Antarctic region with the torrid central zone. Walsperger's map also expresses a clear sense of religious difference between Christianity and Islam by colouring cities according to their faith: red for Christian, black for non-Christian.

Plate 4. Venice, Biblioteca Nazionale Marciana. World map of Fra Mauro. South at the top. Fra Mauro's superb *mappamundi* was made at the Camaldolese monastery of San Michele di Murano, near Venice, probably between 1448 and 1450. Fra Mauro attempted to provide an image of the world that would account for the information derived from the expansion of European contacts with Asia and Africa, leading him to an often polemical interrogation of Ptolemy's *Geographia*.

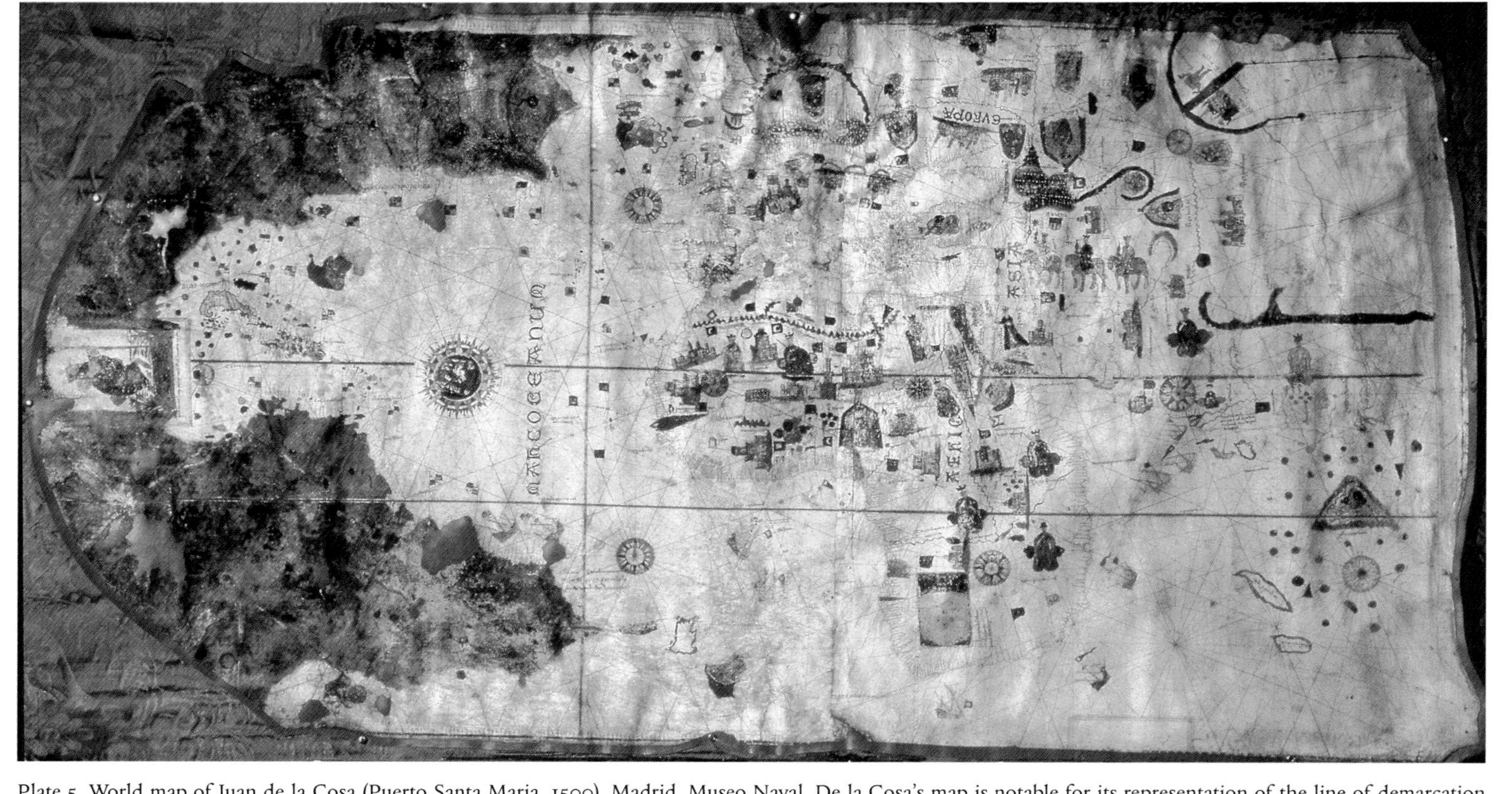

Plate 5. World map of Juan de la Cosa (Puerto Santa Maria, 1500). Madrid, Museo Naval. De la Cosa's map is notable for its representation of the line of demarcation between Portuguese and Spanish possessions in the New World. The area coloured green on the map signifies *terra incognita.* The unknown land contains a portrait of St Christopher bearing the Christ Child, probably a reference to Columbus, and the inscription 'Juan de la Cosa la fizo en el Puerto de S: ma [Santa Maria] en año de 1500'.

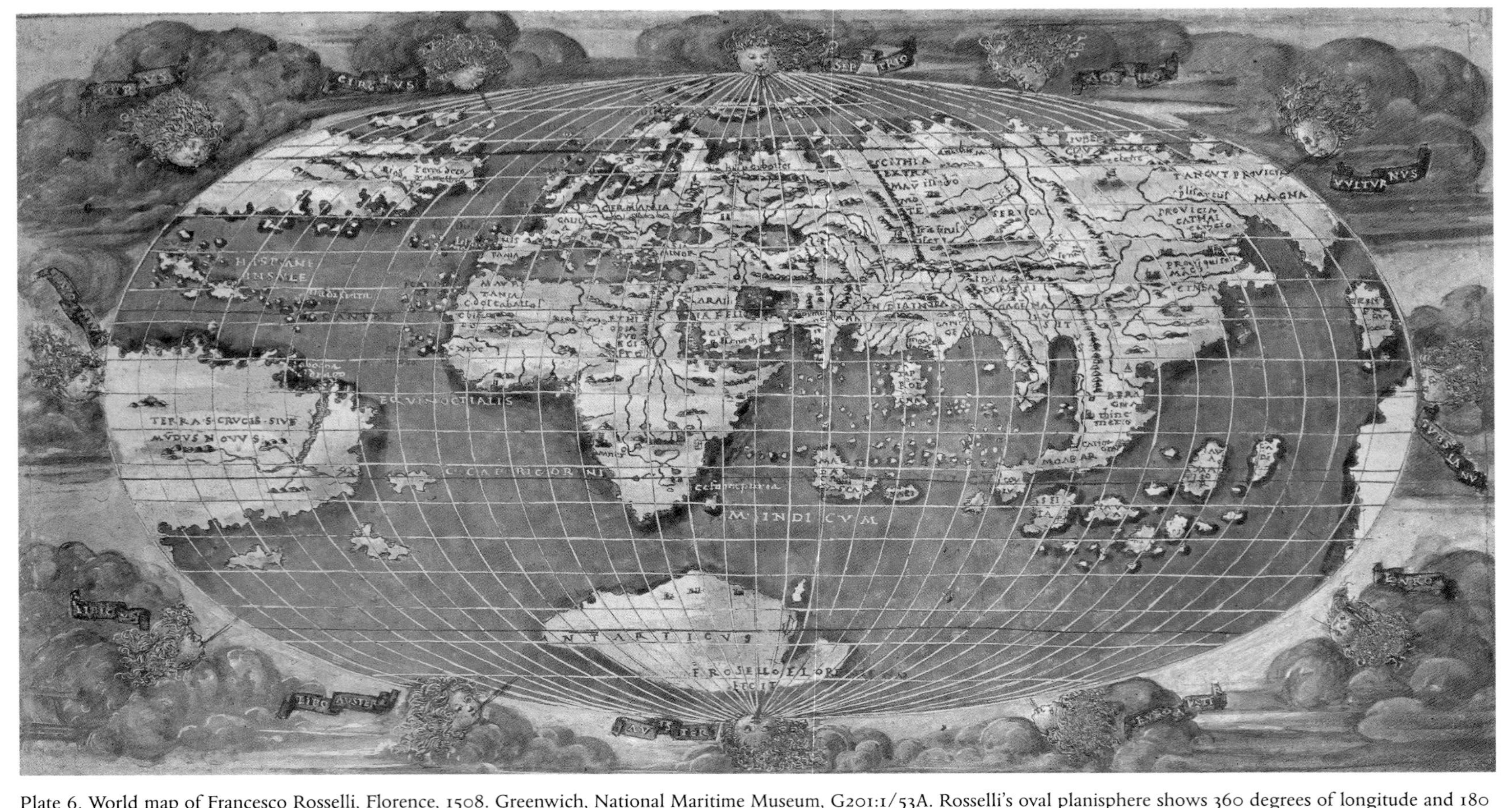

Plate 6. World map of Francesco Rosselli, Florence, 1508. Greenwich, National Maritime Museum, G201:1/53A. Rosselli's oval planisphere shows 360 degrees of longitude and 180 degrees of latitude. Rosselli clearly identifies the New World ('Terra S. Crucis sive Mundus Novus') as a separate continent, but he also represents Columbus' conviction that the places encountered on his fourth voyage were located on the east coast of Asia. At the Antarctic circle (marked 'Antarticvs'), Rosselli has included an unidentified and inchoate landmass, which he has used as the backdrop for his signature, 'F. Rosello Florentino Fecit'. © National Maritime Museum, Greenwich, London.

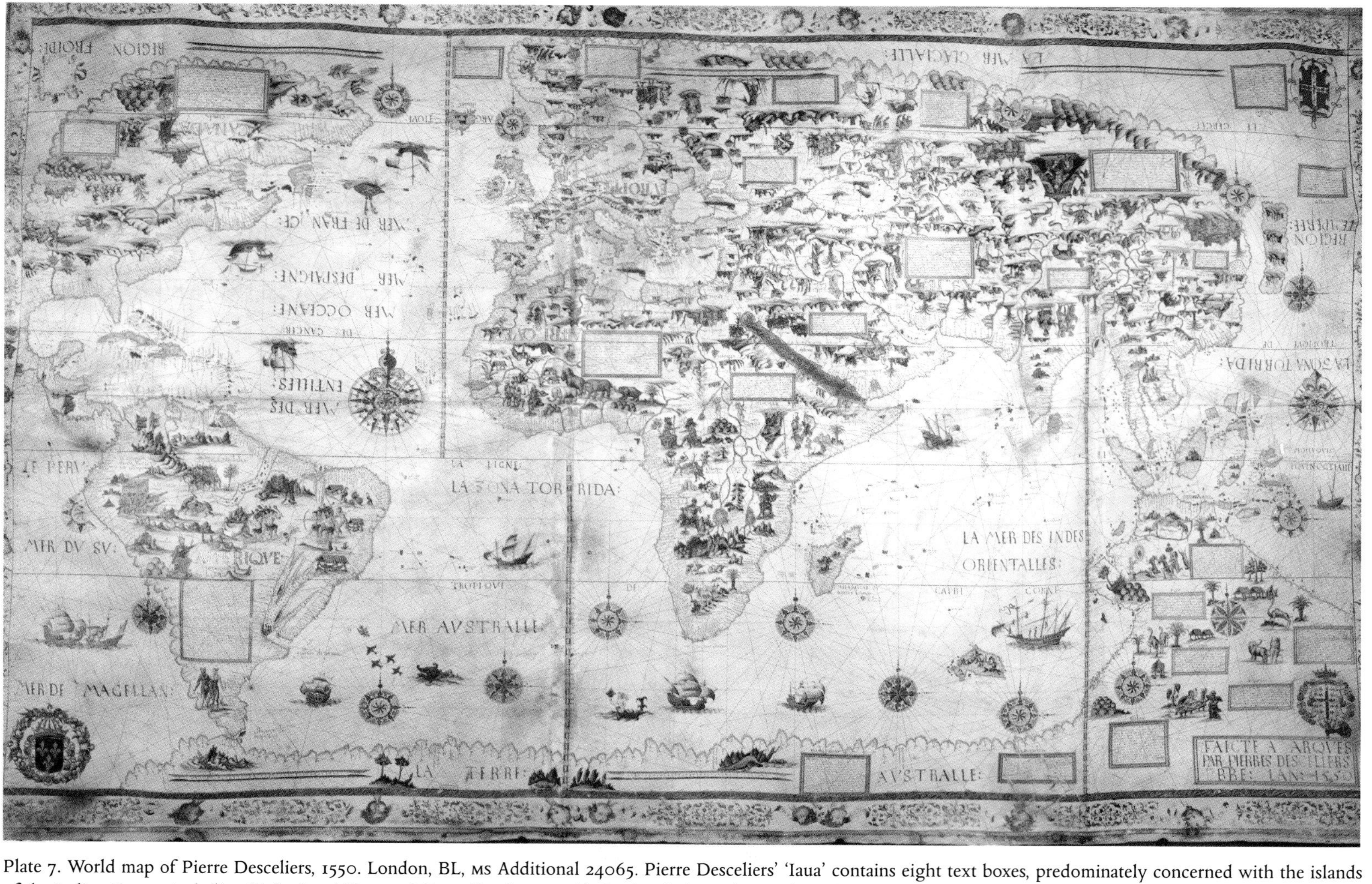

Plate 7. World map of Pierre Desceliers, 1550. London, BL, MS Additional 24065. Pierre Desceliers' 'Iaua' contains eight text boxes, predominately concerned with the islands of the Indian Ocean, including 'Seilan' and 'Samatra'. Desceliers has provided a detailed coastline and, in the interior, a series of vignettes, ranging from sun worship and dog-headed butchery to manual labour, along with an array of dwellings, animals, and foliage.

Plate 8. Jesuit world map, engraved by Bartholomew Kilian (1664). Private collection. An image showing the spread of the Society of Jesus throughout the world, in which St Ignatius Loyola appears prominently in the top right, receiving rays of light from Christ's side wound. His disciples and their converts are gathered around a heart-shaped map which depicts, at its base, *Terra Australis Incognita.*

only by Ocean, as most geographical theories current in the Middle Ages held, but also by *terrae incognitae*:

> Pars totius nostre habitabilis terminatur ab ortu Solis terra incognita que populis orientalibus Maioris Asie Sinarum scilicet atque Serum adiacet. A meridie similiter terra incognita que Indicum Pelagus cingit: queue amplectitur meridionalem Ethiopiam Lybie Regionem Agisymbam appellatam. Ab occasu etiam terra incognita que Sinum Africe Ethiopicum cingit et deinde occidentali Oceano qui Lybie et Europe ultimis occidentis partibus adiacet. A Septentrione oceano coniuncto qui insulas Brittanie complectitur ac partes Europe maxime Septentrionales claudit ... Preterea limites reliquos habet terram incognitam que partibus Asie maxime Septentrionalibus imminet[11]

> Our habitable part of the world is marked off to the east by the unknown land that adjoins the eastern peoples of Asia Major, that is the peoples of Sina and Serica. To the south similarly [it is marked off] by the unknown land that surrounds the Indian Sea and that embraces the southern Ethiopian region of Africa called Agisymba. To the west again [it is marked off] by the unknown land that surrounds the Ethiopian gulf of Africa, and finally by the western Ocean which adjoins the furthest parts of the west of Africa and Europe. To the north [it is marked off] by the continuation of Ocean which encompasses the islands of Britain and encloses the northernmost parts of Europe ... moreover [our habitable part] has as its remaining limits unknown land that borders upon the northernmost parts of Asia.

According to Ptolemy, the Indian Ocean was land-locked, so that Asia was joined to southern Africa 'by the unknown land that surrounds the Indian Sea' (per terram incognitam que Indicum pelagus circumplectitur) (see fig. 20). A further difference from standard medieval cartographic models was that Ptolemy located the southern boundary of the inhabited earth 16 degrees and 25 minutes south of the equator, a distance equivalent to that of the parallel of Meroë, the first of the *climata* north of the equator.

The gaps in Ptolemy's knowledge of the ends of the earth were thus spelled out to his fifteenth-century audience. Its response was gradually to supplement his world picture. The maps that accompanied Ptolemy's *Geographia* in the Latin editions (many of extreme opulence) that burgeoned over the course of the fifteenth century retained the basic form of those that accompanied the earlier Greek editions, but they often also revised Ptolemy's descriptions to incorporate features of which he was unaware.[12] Additional descriptions of islands discovered after Ptolemy's lifetime, and superior images of northern Europe, much of Asia, and the southern extent of Africa were provided, courtesy of exploration and the circulation of marine charts and geographic descriptions – the kind of reports cited by Ptolemy as such an important source of information.[13] The best example of such supplementation is the addition of the Danish cartographer Claudius Clavus' map of Scandinavia, including Greenland, to the *Geographia* compiled by the French cardinal Guillaume Fillastre c. 1418–27.[14] Fillastre notes in his annotations to the manuscript that Ptolemy made no mention, and seemed to have no information about 'that northern region [of Europe] up to unknown land' (illam plagam septentrionalem usque ad terram incognitam). Fillastre has consequently added to the standard ten maps of Europe in the *Geographia* an eleventh, since 'a certain Claudius Cymbricus described those northern parts and made of them a map' (quidam Claudius Cymbricus illas septentrionales partes descripsit et fecit de illis tabulam).[15] Further, Fillastre records on Clavus' witness that habitation was possible not

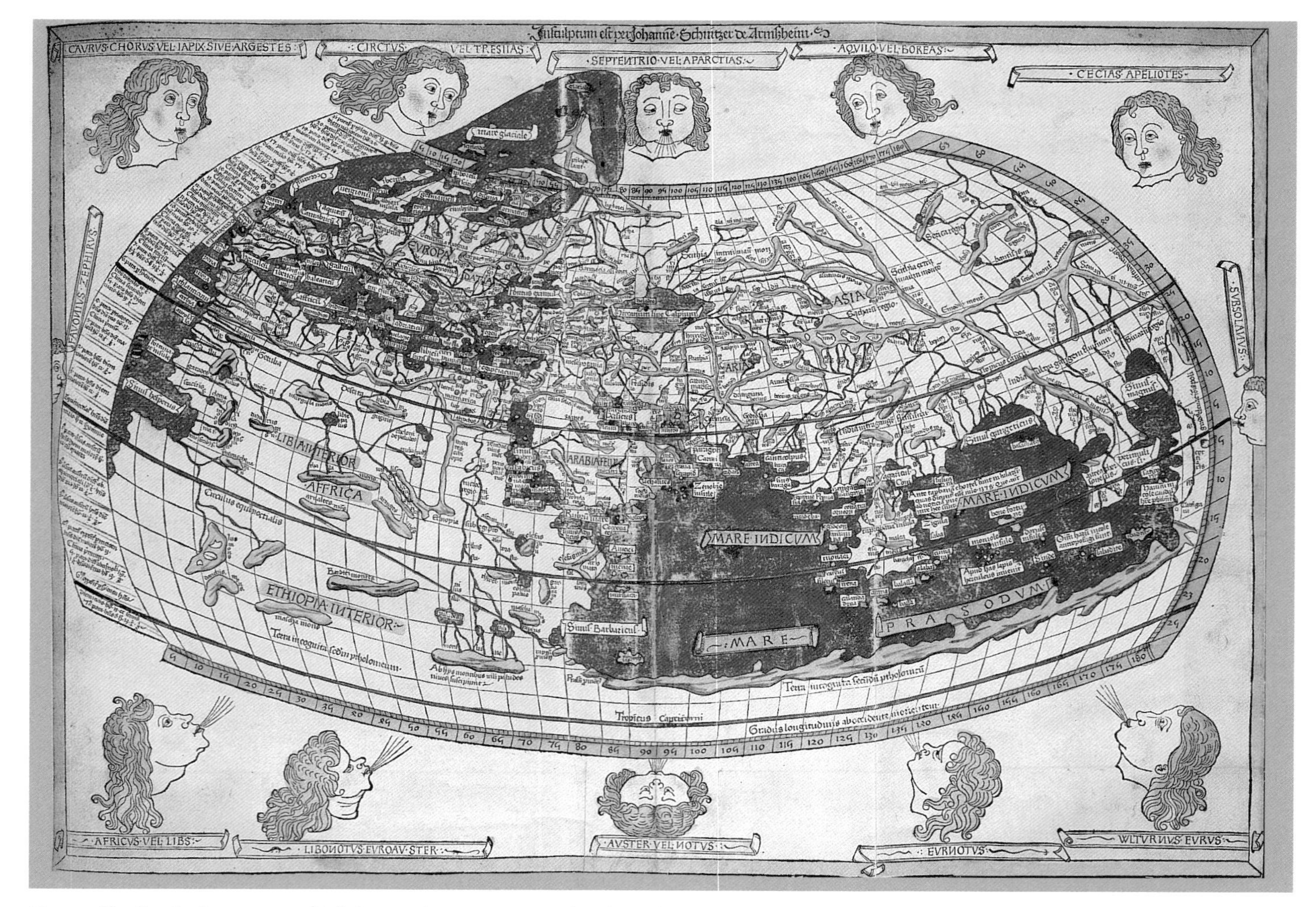

Fig. 20. Claudius Ptolemy, *Geographia* (Ulm, 1482), BL IC9306. Ptolemy's world map, second projection. This version of Ptolemy's world image shows a strip of 'terra incognita' connecting Asia to southern Africa beneath the Indian Sea and the island of Taprobana. *Terra incognita* is marked beneath Ethiopia in Africa, and beneath the Indian Sea, but both times with the added nuance 'secundum Ptholomeum' (according to Ptolemy). Note also the expansion of the world image to the north-west to accommodate a more accurate representation of Scandinavia. The Ulm Ptolemy was one of the earliest printed editions of the *Geographia*.

only at, but even beyond, the north pole. Clavus had testified to the invasions of Eskimos (Careli) into Greenland 'from the other side of the north pole' (ex altera parte poli septentrionali), leading Fillastre to posit a vast kingdom extending 'beneath the north pole towards the eastern extremities, so the pole that is north to us is southern to them' (sub polo septentrionali versus fines orientales, quare polus nobis septentrionalis est eis meridionalis).[16] The distinction between 'we' and 'they' here is not simply one of geography, but also of religion, since the infidel nature of the Eskimos is emphasized heavily in the notes that accompany Clavus' map.

Precisely because of this ongoing supplementation of the unknown or only partially known areas of Ptolemy's description, it is clear that the *Geographia* encouraged an altered perception and representation of *terra incognita.* Firstly, Ptolemy incorporated *terra incognita* in his theorization of cartographic practice, criticising maps that contracted areas because they were little known, and that expanded others because more was known about them.[17] Secondly, the possibility of a comprehensive set of regional maps made it harder to ignore those areas that were uncharted. Finally, the *Geographia* brought to the European image of the world a fairly specific identification of the point at which unknown areas began; since *terra incognita* was transmitted with Ptolemy's text it thereby became more precise, more clearly delineated, and less of an abstraction than it was in the classical philosophical tradition passed on through Macrobius and others. *Terra incognita* is not conceived of in the *Geographia* simply as space, existing beyond an impenetrable zone; rather, it is *placed,* located in proximity and relation to known places. So, for example, in describing the land of the 'Sine' (Sinae, modern-day Chinese), Ptolemy notes:

> terminantur a Septentrione parte Serum exposita. Ab ortu Solis atque meridie terra incognita. Ab occasu india extra gangem[18]
>
> it is marked off on the northern part by the part of Serica previously described. To the east and south by unknown land. To the west by India beyond the Ganges ...

Scythia within the Imaon mountain range, meanwhile,

> terminatur ab occasu Sarmatia Asiatica secundum latus expositum. A septentrione terra incognita. Ab oriente Imao monte ad septentrionem uergente secundum meridianam ferme lineam que a predicto oppido usque ad terram incognitam extenditur. A meridie ac etiam oriente Sacis quidem et sogdianis et Margiana[19]
>
> is marked off to the west by Asiatic Sarmatia on the side previously described. To the north by unknown land. To the east by the Imaon mountains inclining to the north closely following the meridian line which is extended from the foresaid town to unknown land. To the south and also the east by the Sacae and the Sogdians and by Margiana.

'Unknown land' in these descriptions fulfils substantially the same locative function as known places; like them, its boundaries are fixed in degrees of longitude and latitude. Many of the maps that illustrate fifteenth-century copies of Ptolemy's *Geographia* accordingly show areas of unknown land. On world maps the southernmost extension of Africa, and the band of land that stretches beneath the Indian Ocean are frequently marked 'Terra Incognita', sometimes with the added nuance 'according to Ptolemy' (secundum Ptolomeum) (fig. 20).[20] Less frequently marked

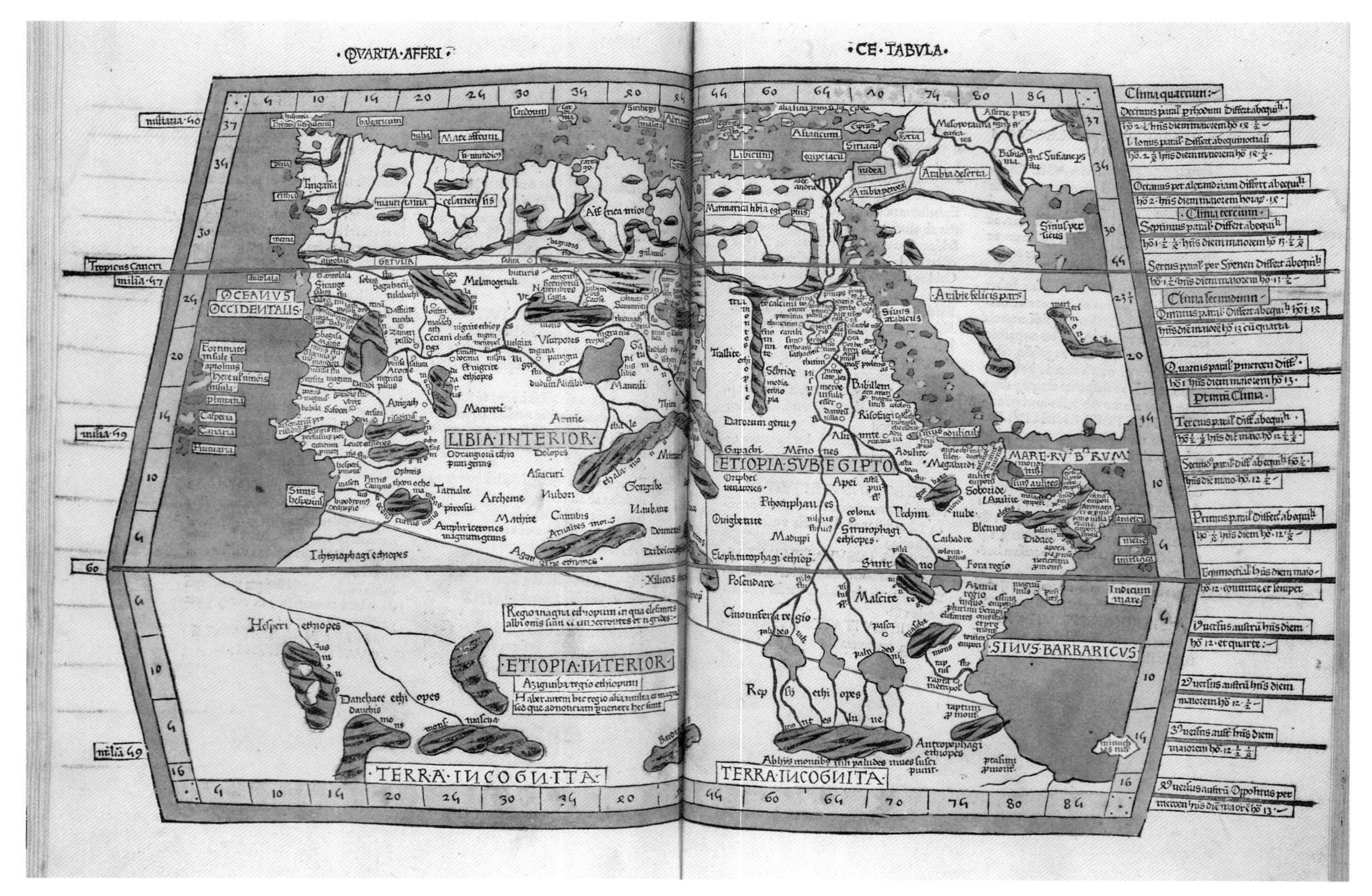

Fig. 21. Claudius Ptolemy, *Geographia* (Ulm, 1482). BL C.1.d.2. Fourth map of Africa (Quarta Africae Tabula). *Terra incognita* is marked at the far south of Africa beneath 'Etiopia Interior' and the putative source of the Nile.

are the areas of unknown land said by Ptolemy to lie at the far east and north of the world. This tendency persists on the regional maps that accompany the text: on the map of southern Africa (Africae Tabula IV) 'terra incognita' is fairly commonly shown (fig. 21), but it appears less frequently on the maps of northern Europe and northern and eastern Asia (Europae Tabula VIII, Asiae Tabulae II, IV, XI).[21] Unknown land, in short, is the site where the mediation of classical geographical inheritance is played out, where the authority of the cartographer is invoked ('secundum Ptolemeum') and perhaps subtly questioned, or where it is augmented by means of more accurate representations of formerly unknown land.

The process of negotiation and adaptation of Ptolemy's data is evident in an undated and anonymous manuscript of the *Geographia*, possibly of Venetian origin and probably dating from the second quarter of the fifteenth century, BL Harley 3686. The eighteen regional maps that appear in the manuscript seem to witness an attempt at a synthesis between the classical inheritance from Ptolemy and contemporary mapping practices.[22] So, for example, the place names that appear on the maps are a mixture of contemporary toponyms and those derived from Ptolemy; similarly, while those maps that show western Europe and north-west Africa accord with fifteenth-century marine charts, those of the north and east of the ecumene 'have Ptolemaic, schematic or even imaginary shapes'.[23] It is in these areas that *terrae incognitae* are marked, in accordance with Ptolemy's description.[24] The map of central and east Asia (ff. 98v–99) identifies areas of unknown land beyond the legendary 'caspey montes' and on either side of the Imaon mountain range ('terra Incognita de Sitia Intra Imaum montem'; 'terra Incognita de Sitia extra Imaum montem') (fig. 22).[25] Similarly, the map of the Baltic and Scandinavia and part of Germany (f. 100r) shows 'terra Incognita' above the 'Regnum Suetie', as well as a region of 'terra Incognita de sarmatia in Europe'. Here, unknown land becomes associated with the regions on which it borders, and thereby acquires a more specific and distinct geographic identity. It not always simply 'unknown', but 'unknown of Scythia', or of Sarmatia, or of Sarmatia in Europe; these are not wandering, conceptual blanks, but areas of ignorance tied with some firmness to a notion of place. This new naming practice foregrounds the contiguity of unknown land, and carries certain implications with it. At the edges of the known world, *terra incognita* has acquired a quasi-topographic identity. Certainly its size and significance varies, from its unobtrusive presence at the north of Europe to its more monumental manifestation at the south of Africa, but there is no question of it being unreachable, or unknowable. The reception of Ptolemy's *Geographia* in the fifteenth century encouraged, then, a reconfiguration of unknown land. But if this is the case, what were the implications of the alteration of the world picture for some of the central tenets of medieval world maps in the Macrobian tradition: the encircling ocean, the torrid zone, and the antipodes beyond it?

Guillaume Fillastre and the transmission of Ptolemy

At the Council of Constance (1414–18), Guillaume Fillastre, the Cardinal of St Mark and a close friend and ally of Pierre d'Ailly, composed a long introduction to Pomponius Mela's first-century AD treatise *De chorographia*. Even more profoundly than d'Ailly, Fillastre sought to

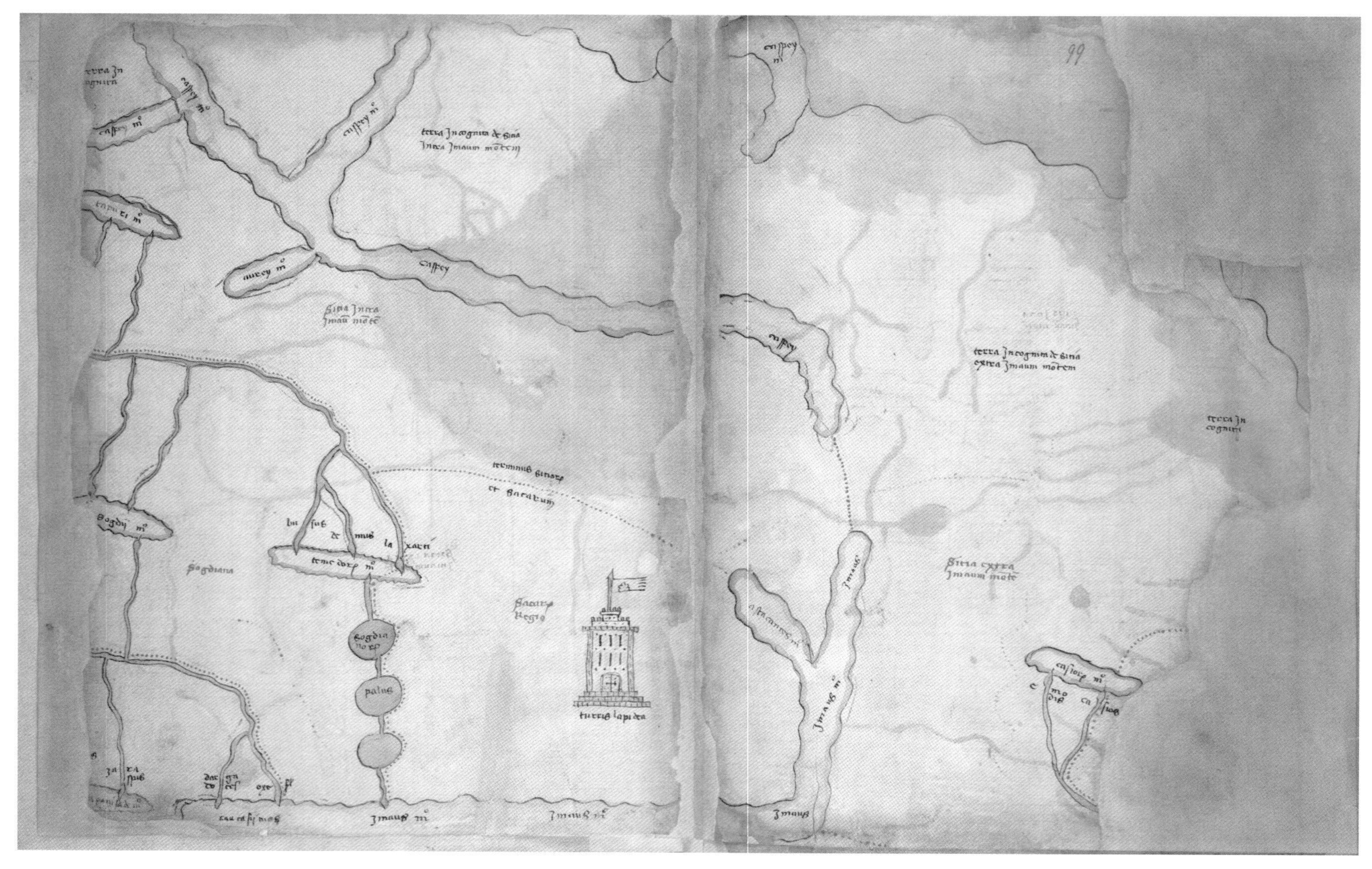

Fig. 22. London, BL, MS Harley 3686, ff. 98v–99r. Map of central and east Asia in a fifteenth-century manuscript of Claudius Ptolemy, *Geographia.* This unusual Venetian manuscript marks 'terra incognita' in several places in northern Europe and Asia, following Ptolemy's written description. 'Terra incognita' appears beyond the legendary 'caspey montes' and on either side of the Imaon mountain range ('terra Incognita de Sitia Intra Imaum montem'; 'terra Incognita de Sitia extra Imaum montem').

engage with a classical geographical legacy, to confront its contradictions, and to supplement it with information provided by geographers of his own time. Fillastre had commissioned manuscript copies of Mela's work while at Constance, and he wrote the introduction specifically to accompany them. As has only recently been fully appreciated, Fillastre's introduction is not only the earliest known commentary on Mela, but also an important geographical treatise in its own right, one that challenges Mela's geography by counterposing Ptolemy's *Geographia*, a copy of which Fillastre had also commissioned while at Constance.[26] Fillastre seems to have composed two versions of the commentary: one was intended for general circulation, although it was aimed particularly at the Italian humanist circles with which Fillastre was familiar, and which were well represented at the Council;[27] the other was written specifically for the canons of Reims cathedral, where Fillastre had been dean since 1392. This second version, the unique copy of which is contained in Reims, Bibliothèque municipale, MS 1321, contains numerous interpolations. It is difficult to state categorically which version was composed first, but it has been conjectured that Fillastre inserted the additions to the text in order to reassure the canons of the orthodoxy of his revised geography.[28] In this regard, it is not surprising to note that one of the additions concerns the orthodoxy of the Cardinal's views on the antipodes.

The canons of Reims were, it seems from references made by Fillastre, the inheritors of a rich cartographic legacy, including at least two *mappaemundi*.[29] And it was this legacy that Fillastre, disseminator of the ideas of Mela and Ptolemy, reworked, proposing in its place a significantly different world image. Like the compiler of Harley 3686, Fillastre attempted a synthesis of classical geographic texts with contemporary fifteenth-century knowledge of the world. Two maps seem to have been produced with the intention of accompanying Fillastre's copies of *De chorographia*: one, Ptolemaic in form, shows a detailed ecumene, with a land-locked Indian Ocean (fig. 23). This map extends as far as the boundaries of the known world specified by Ptolemy to the south – Agisymba in southern Africa – and to the east – Seres and Sina in east Asia.[30] It appears at the end of Fillastre's introduction, and seems designed to explicate the Ptolemaic conceptions discussed as a prelude to Mela's text. The other map, in the initial O of Orbis, the first word of *De chorographia*, appears only in the Reims manuscript, and was probably designed explicitly for the canons (Plate 2). Similar in form to standard circular representations of the known world, it diverges from this tradition of cartographic representation in at least one important respect. This is the representation of unknown land at the northern and southern edges of the map: the inscription 'terra incognita' appears three times, in northern Europe and northern Asia, and south of Ethiopia, an innovation that does not derive from Mela's description, but rather from the *Geographia* of Ptolemy.[31]

Fillastre was not only an important disseminator of geographical texts to northern humanist circles, but also, as is clear from his treatise, a commentator of some originality. As he recognized, a synthesis of the texts of Mela and Ptolemy with fifteenth-century geographical knowledge demanded a reconfiguration of the traditional description of the known world – and, crucially, a corresponding change in representational practices.[32] In his commentary on *De chorographia*, Fillastre quickly notes two vital issues: whether the Ocean encircles the earth (a theory denied by Ptolemy, but supported by Plato, Aristotle, Pliny and Mela), and whether habitable land is

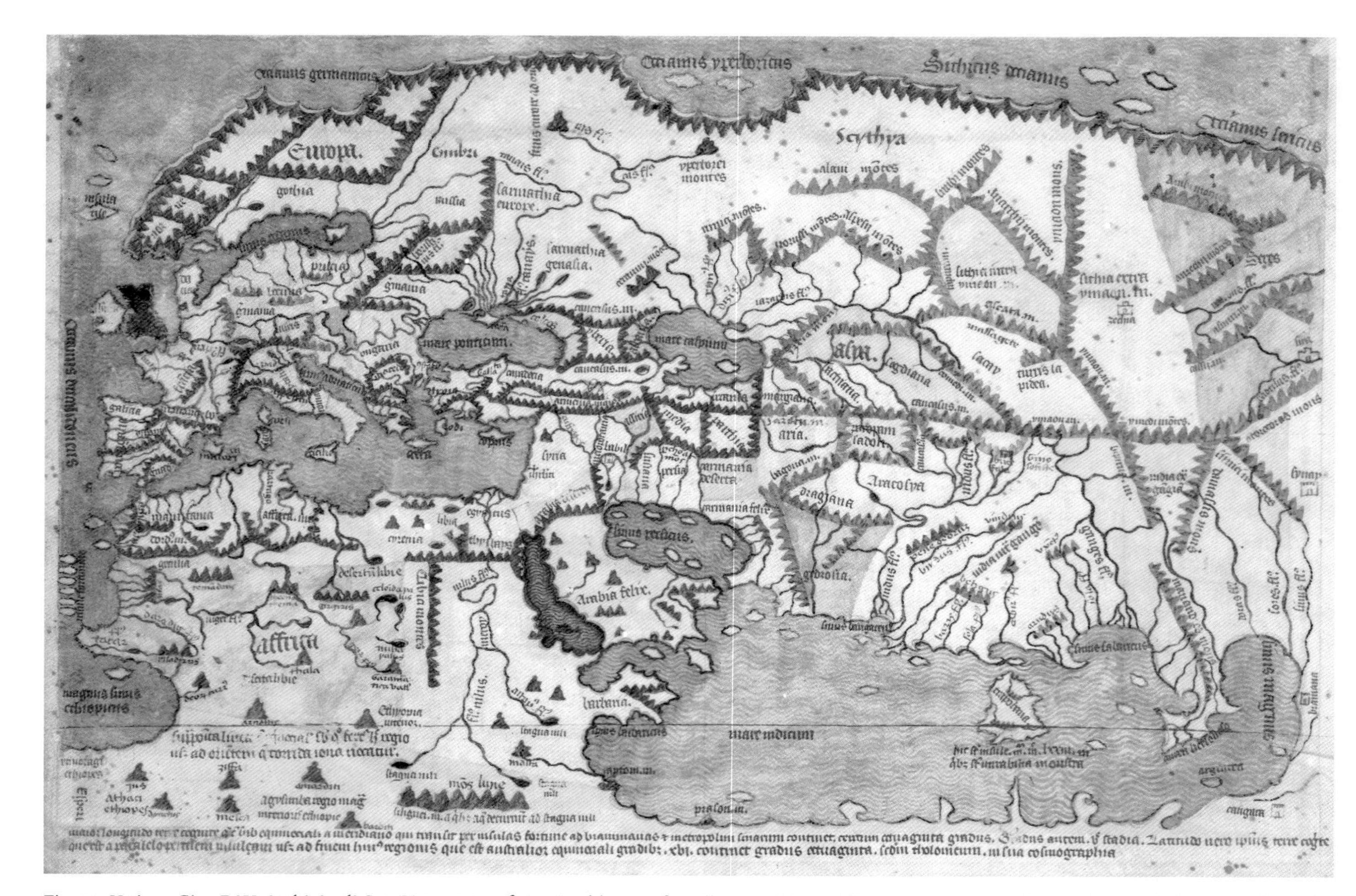

Fig. 23. Vatican City, BAV, Archivio di San Pietro, H 31, f. 8v. World map of Guillaume Fillastre, fifteenth-century. North is at the top. Ptolemaic in form, the map shows a landlocked Indian sea.

divided by a torrid zone running between the two hemispheres. The second issue raises the question of whether there exist peoples 'unknown to us'. In *De chorographia* Mela described both the standard tripartite division of the ecumene, and the division of the earth into zones. The two hemispheres ('latera') comprised five horizontal zones: the two outermost cold, the middle hot, and two temperate – one inhabited by 'us', the other by the Antikthones. The heat of the intervening zone prevented knowledge of the region beyond the known world ('illius situs ob ardorem intercedentis plagae incognitus'), but Mela went on to speculate that 'if there is another world, and if Antikthones are located directly opposite us in the south' (si est alter orbis suntque oppositi nobis a meridie antichthones), it is probable that the Nile originates in the unknown region, and emerges in Africa after passing through an unseen channel beneath Ocean, swelling at summer solstice when it is winter at its source.[33] In response to these claims Fillastre cites Lactantius' denial of the antipodes, and concurs with Augustine's statements in *De civitate dei*, adding in the 'Reims' version that he does so 'both from the authority of holy writ and from learned writings and even from the thinness of my own intelligence' (sensio tam ex scripture sacre auctoritate quam doctorum scriptis quam eciam ex mei exilitate ingenii).[34]

Fillastre challenges not only Mela's Antikthones, but also his division of the world into zones, and the corresponding limitation of the habitable part of the earth. Citing Ptolemy's description of the southern-most part of the ecumene, the Cardinal categorically denies the existence of a torrid, uninhabitable zone:

> Nunc autem de illa media zona dico quod non credo verum quod iste asserit illam propter calorem inhabitabilem primo auctoritate Tholomei in sua *Cosmographia* de qua supra, qui dicit quod Ethiopes omnes sunt sub illa zona, id est in illo terre spatio a solsticio in solsticium, et quod illi qui sunt sub equinoctiali sunt nigriores, plus elongati minus nigri, unde dicit quod eciam in Yndia sunt Ethiopes sicut in Affrica quia per Indiam et Affricam transit illa zona[35]

> Now about that middle zone I say that I do not believe to be true that which [Pomponius] claims, that it is uninhabitable on account of heat, firstly by authority of Ptolemy in his *Cosmographia* mentioned above, who says that all Ethiopians are under that zone, that is, in that region from one solstice to the other, and that they who are under the equator are more black, those further away are less black, wherefore he says that there are Ethiopians even in India just as in Africa, because that zone passes through India and Africa.

This evidence for habitation of the torrid zone throws into question the very notion of zones that are totally uninhabitable, and consequently leads Fillastre to posit a quite different, Ptolemaic, configuration. 'The furthest parts of the earth the same Ptolemy does not describe as uninhabitable, but says in both parts *the land is unknown*' (De extremis partibus terre idem Tholomeus non dicit inhabitabiles sed dicit in utraque parte *terram esse incognitam*):[36] the zones formerly held to be uninhabitable due to excessive heat or cold can in fact be shown, or at least conjectured, to be partially inhabited. To the north, Fillastre adduces as evidence reports of peoples living in the Hyperborean mountains 'in the furthest northern parts and almost under the Arctic pole ... in our hemisphere' (in ultimis septentrionis partibus et pene sub polo arthico ... in nostro emisperio).[37] To the south until the Antarctic pole the land may be habitable, because although Ptolemy records no city or people beyond the solstice, he describes the land as 'unknown beyond the last Ethiopians'.[38] Fillastre does not deny the existence of uninhabitable tracts of land, but he argues that they are uninhabitable because of their sterility and their habitation by beasts incompatible with humans, rather than the heat of the sun.[39] Any uninhabitable lands lie within areas of habitable land, and do not constitute zones of barrenness.

Having recast the division of land in this way, Fillastre returns to the question of water: the Ocean, he concludes – again concurring with Ptolemy – is at once continuous throughout its extent, and covers part of the earth on account of the earth's rotundity, but extends into some parts more and others less, depending on concavities in the earth. It may, he suggests, be possible to navigate from the west to the east 'through the part opposite to us, and in the opposite direction' (i.e. from western Europe, through the 'underside' of the northern hemisphere, to east Asia).[40] From these premises Fillastre returns to the question of the antipodeans, or Antikthones:

> dico cum correctione tantorum doctorum anthipodes esse, non per modum quem supponit Augustinus quod occeanus sit medius inter nos et ipsos propter quod dicit illos non esse, sed dico quod supposita figura terre sperica illi qui habitant in ultimis partibus orientis sunt antipodes illis qui habitant in ultimis partibus occidentis . . .[41]

> I say under correction of so many learned men that antipodeans do not exist in the way that Augustine supposes, that is, that Ocean is in the middle between us and them, on account of which he denies their existence, but I say that supposing the shape of the earth to be spherical they who live in the furthest parts of the east are antipodeans to those who live in the furthest parts of the west

With this statement, Fillastre completes his re-drawing of the boundaries of the world map. Having broken down the idea of a torrid zone, he turns his attention to the notion of a possibly inhabited southern temperate zone, and rejects that too. There is no habitable land that cannot be reached from the known world, because there is no intemperate hot zone restricting transit. There *is* unknown land, but it lies precisely at the furthest parts of the world, contiguous with the ecumene.

Fillastre's dismantling of the notion of the torrid zone echoed the comments of Albertus Magnus and Bacon on the subject;[42] in part too, his extension of the known world was the product of Ptolemy's calculation that the ecumene extended across 180 degrees of longitude.[43] The effect of Fillastre's revision of antipodal theory is to overturn the 'here/there' 'we/they' dichotomy of standard cosmological systems, and to articulate in more compelling form an idea previously mooted by Hermann of Carinthia and John Mandeville. 'They who live in the furthest parts of the east are antipodeans to those who live in the furthest parts of the west': 'they' live under 'those' (illi ... illis). That overturning of the dichotomy, the dramatic declaration that 'corrects' Augustine and so much classical authority, the temporary erasure of 'we', is predicated on a notion of the contiguity and reachability of all parts of the earth, and a consequent rejection of the idea of 'other worlds'. Augustine need not have worried: there is now no mystery about sons of Adam reaching the antipodes, because the antipodes are the furthest extent of the known world, and there is no mystery about the dispersal of peoples into Asia or western Europe. The emphasis on contiguity and reachability is, then, based upon readings of the hitherto little known classical geographical texts that were being promoted in the early fifteenth century, in conjunction with the debates about the extent of the known world, the possibility of 'other worlds' and their habitation, that had run throughout the Middle Ages. At the same time, the idea of contiguity was coterminous with a discourse of exploration and colonization.

Papal cartography

Reformulations of power and belief inform those of geography. It was no coincidence that Fillastre's revision of the world image took place in the context of the Council of Constance, a gathering charged with the resolution of the papal schism, the extirpation of heresy, and the reform of the Church in its head and members. The Council was kept informed of recent events of geopolitical significance, such as the Portuguese capture of the city of Ceuta on the Moroccan side of the Strait of Gibraltar in 1415,[44] and the occupation of two of the Canary Islands by Norman knights in the name of the king of Castile.[45] The extension of European sovereignty to non-European places and peoples brought about geographical revisions, such as the inclusion of the Canary Islands on maps, and their equation with the classical Fortunate Isles.[46] It also brought into sharp focus the vexed question of the rights of pagan peoples during and after military conquest by Christians. The papacy, as the body that could regulate and legitimise the conversion of conquered peoples within and without Europe, had an obvious motivation for overseeing such ventures. It had enunciated its claims to jurisdiction over territories conquered by Europeans at least as early as the thirteenth century,[47] and throughout the second half of the fifteenth century the theory and practice of papal jurisdiction was significantly developed in a series of papal bulls, culminating in the famous *Inter cetera* of Pope Alexander VI (1493), in which territories recently discovered, and to be discovered, throughout the world were divided between the kingdoms of Castile and Portugal.[48] The nexus between papal interest and *terra incognita* was grounded in biblical and patristic authority, since the head of the Church was charged with the apostolic duty of promoting the spread of the gospel to all peoples until the end of time.[49] If the inhabited world had expanded, so too should papal sovereignty. The purpose of such an expansion of sovereignty was, according to a long-held papal ambition, not only to convert, but also to unify Christendom, by rediscovering pockets of Christians far removed from Rome, such as the Nestorians rumoured to be located in Asia. Papal interest in the ends of the earth demanded acts of representation in a variety of forms, including the translation of geographical texts, the collection and commissioning of maps and illustrated geographical manuscripts, the writing and re-writing of the history of discoveries in the form of narratives that legitimated conquest, and the codification of such narratives in official documentation: in short, a papal cartography. Papal cartography was fully implicated in the administration of Christendom, a task that required not only the monitoring of Christian communities throughout the ecumene, but also the construction and propagation of a global vision to enable the adjudication of legal claims within and beyond the ecumene. These two facets of papal cartography belonged to the same geographical culture, one embedded in, though by no means limited to, humanist discourse. The old questions of classical and scholastic geography – what are the limits of the earth, what is the source of the Nile, which peoples dwell beneath the equator, do they know the name of Christ – were the same that preoccupied humanist geographical debate, only they were increasingly finding revised answers. Those answers demanded a swift practical response in the form of chart making – both the redrawing of maps, and the drawing-up of charters to assert possession: a cartography of the ends of the earth, directed from its ecclesiastical centre, Rome.

The geographic interests of fifteenth-century popes are attested by their receipt and commissioning of deluxe maps and editions of geographical works, most notably Ptolemy's *Geographia*. Jacopo d'Angelo's Latin translation of Ptolemy was dedicated to Pope Alexander V, and certain notable later editions of the *Geographia* found papal dedicatees, such as Nicolaus Germanus' Ptolemy, made for Paul II (1464–71).[50] Pius II (Aeneas Silvius Piccolomini) (1458–64) is also known to have commissioned maps from the Venetian cartographers, Antonio Leonardi and Girolamo Bellavista, in the course of his papacy; their work does not survive, but it may be these *picturae* to which Pius referred in his treatise *Asia* (1461).[51] Unfortunately, this evidence of geographic interests has too often been perceived simply as an expression of a bland humanist erudition on the part of fifteenth-century popes, and has all too rarely been placed in the context of papal administration. Instead, papal cartography was the product of and necessary companion to the exercise of papal power with regard to the occupation and evangelization of non-Christian land and peoples. It was the means by which the papacy maintained a stake in the dramatic realignment of European sovereignty beyond the boundaries of the known world.

Pope Nicholas V's bull of 1455 known as *Romanus pontifex* is one of the most striking of a series of papal documents that sought to define the rights of European sovereigns to newly detected land. The document amplifies the renewed papal interest in sponsoring crusading activity evident under the papacy of Eugenius IV in the late 1430s. It refines the position established in 1452 by the bull *Dum diversas*, in which Nicholas, a noted patron of humanism credited with founding the Vatican library, conceded to King Alfonso V of Portugal the faculty to attack Saracens, pagans, and other infidels, to occupy their territory and goods, and to submit them to perpetual servitude.[52] *Romanus pontifex* witnesses the transformation of *terra incognita* into land not only known, but actually appropriated. In so doing, it shows that the presence of unknown lands at the ends of the ecumene demanded a response from the papal administration that displayed legal acuity, as well as an appreciation of the momentous symbolism of penetration into hitherto unknown regions.

Addressed to Alfonso V, *Romanus pontifex* begins by praising the Catholic kings and princes who have not only taken up arms against Saracens and other enemies of the faith, but have also subdued to their temporal dominion those kingdoms and places 'most remote and unknown to us' (in longissimis nobisque incognitis partibus).[53] The bull then notes, more specifically, the achievements of Alfonso's son, Henry: he has revealed, extolled and venerated the name of Christ throughout the world, even in 'the most remote and unknown places'; he peopled islands in the Atlantic with the faithful and constructed churches there, converting many of the inhabitants; most remarkably, he had navigated 'through the Atlantic towards the southern and eastern shores' (per huiusmodi Oceanum mare versus meridionales et orientales plagas).[54] In the late 1440s the Florentine humanist and papal secretary Poggio Bracciolini had written a letter to Henry in which he explicitly compared 'the Navigator' with luminous figures of classical antiquity. In going beyond the limits of southern Africa to parts 'neither known nor easily reached' and conquering 'unknown and savage peoples' (incognitas atque efferas nationes) Henry had, Poggio reckoned, surpassed even Alexander and Caesar.[55] That sense of Henry as a latter-day Alexander, although never explicit, can be detected in *Romanus Pontifex*'s increasingly breathless account of his feats: he travelled 'in very fast ships, called caravels ... towards the southern parts *and the Antarctic pole*' (in velocissimis navibus, caravellis nuncupatis ... versus meridionales partes *et polum antarticum*); he reached the province of Guinea, occupied some adjacent islands, and navigating further 'they reached the mouth of a certain great river, commonly thought to be the Nile' (ad ostium cuiusdam magni fluminis, Nili communiter reputati, pervenerunt).[56] Henry is a transgressor of geographic barriers: heading towards the antipodes, he comes upon the Nile, and penetrates so far into Africa that he encounters 'those who worship the name of Christ', but who have hitherto been separated from the Church of Rome. This penetration to lands 'unknown to us in the West' offers the possibility of uniting Christians in the southernmost part of Africa with those in Europe against Saracens and other enemies of the faith. Moreover, those pagans not of the 'sect of the most wicked Machomet' but living amongst it may now be subdued, and learn of the 'most holy name of Christ, unknown to them' (eisque incognitum sacratissimum Christi nomen).[57] The conflation between unknown land and the unknown name of the saviour is deliberate: as European geographic ignorance is pushed back ever further, it encounters its spiritual equivalent.

In this bull a powerful discourse of colonization is, to say the least, expressed with clarity. Papal authorization of the dispossession of non-Christian peoples follows at the heels of the expansion of geographical knowledge. Nicholas concedes to Alfonso the full and free power:

> quoscumque saracenos ac paganos, aliosque Christi inimicos ubicumque constitutos ... invadendi, conquirendi, expugnandi, debellandi et subiugandi, illorumque personas in perpetuam servitutem redigendi, ac regna, ducatus, comitatus, principatus, dominia, possessiones et bona sibi et successoribus suis applicandi, appropriandi, ac in suos successorumque usus et utilitatem convertendi[58]
>
> of invading, conquering, subduing, making war, and subjugating any Saracens and pagans, and any other enemies of Christ wherever they fix their abode ... of forcing their persons into perpetual servitude, of annexing and appropriating kingdoms, duchies, counties, principalities, dominions, possessions and goods for him and his successors, and for converting them to the use and utility of himself and his successors.

Romanus pontifex bears witness to the process of transformation from *terra incognita* – land at the boundaries of knowledge – into *terra inventa*, discovered land in which the rights of the inhabitants may be extinguished by conquest.[59] It authorizes progression from the 'discovery' of land to its 'annexing and appropriation'.

Papal cartography was not a process limited to the acts of popes; instead, I suggest, it can only be understood as an expression – and more, an instrumentalization – of the scholarly culture that surrounded and informed the papacy. The works of two humanist scholars, Poggio Bracciolini and Flavio Biondo, who were closely attached to the curia and who participated in the geographical debates of the middle of the fifteenth century, provide particularly valuable testimony about the ways in which intellectual, political, and ecclesiastical concerns intersected. Both men attest in particular to the continuing importance of Church councils as sites for the reception and dissemination of information, and for the confrontation between classical text and contemporary report.[60] In the first place, the Councils of Basel and Ferrara-Florence that followed on from Constance in the 1430s and 1440s continued to provide a forum for the transmission of geographical texts lost to the Latin west. Of particular significance was the introduction of the geography of Strabo to Italian humanist circles following the visit of Georgius Gemistus Pletho to Italy in 1439 as part of the Orthodox delegation to the Council of Florence, convened with the purpose of negotiating the unification of the Roman and Greek churches.[61] Around this time Pletho composed a brief geographical treatise in which he compared the world image of Strabo with that of Ptolemy, devoting particular attention to the limits of the ecumene.[62] Amongst other errors, Pletho criticized Strabo for regarding the torrid zone as uninhabitable; he admitted, however, that very little was known about land on the underside of the earth (ἀντίχθων).[63]

At the same time, the Council of Florence heard about regions at the ends of the known world from people who had actually lived there. During the course of the Council, Pope Eugenius IV received a group of Ethiopian Christians along with a delegation of Egyptian Copts,[64] and an Italian merchant who had travelled to east and south-east Asia, Niccolò de' Conti. The appearance at the Council of ambassadors and informants from the far south and far east of the known world was noted by Poggio Bracciolini in book four of his *De varietate fortunae* (1448), a work dedicated to Nicholas V.[65] According to Poggio, Niccolò de' Conti was able to give detailed descriptions of parts of Asia not known to Ptolemy, including Sumatra and Java; the Ethiopians were interrogated as to their knowledge of the sources of the Nile, and the reason for its flooding.[66] The Ethiopians, Poggio emphasizes, were seen as providing an opportunity to resolve matters on which 'ancient writers and philosophers and Ptolemy' were ignorant.[67] Book four of *De varietate fortunae* enjoyed considerable short-term influence: Niccolò's account was mentioned (albeit with scepticism) by Pius II in his description of Asia, and was without doubt also a source for the 'Genoese' portolan chart of 1457 and the world map of Fra Mauro.[68] It was first printed in 1492 by the Milanese diplomat Christoforo da Bollate under the title *India recognita*:[69] the sense conveyed is of land both recognized and re-known, recovered from ignorance; the duality of the title testifies to the work's function as a revision and expansion of classical geographical knowledge by means of reportage.

Book four of *De varietate fortunae* opens with an attempt to connect the voyage of Niccolò with the theme of fortune: 'the power of fortune may be viewed not a little in this also, that it caused a man to be tossed from the extreme ends of the world through so many seas and lands for twenty-five years and returned safe to Italy' (Quanuis et in hoc quoque uim fortune haud paruam licet conspicere, que hominem ab extremis orbis finibus per tot maria ac terras quinque et uiginti annos iactatum sospitem in Italiam reducem fecerit).[70] The crucial word in this sentence appears to be 'reducem'. The force of fortune manifests itself in the act of bringing back Niccolò. Admittedly the great extent of his travels is emphasized, as well as its length, but of central importance is the reintroduction of this far-flung traveller to his place of origin within Europe. This *reductio* figures the broader intellectual function that the book will perform: the revision – that is, better comprehension and supplementation – of classical knowledge of the ends of the earth. For, as Poggio notes in the next sentence, 'many things were said about the Indies both by ancient writers, and by common repute. Of these the certain knowledge brought to us argues that some were closer to fables than to truth' (Multa tum a ueteribus scriptoribus, tum communi fama de Indis feruntur, quorum certa cognitio ad nos perlata arguit quedam ex eis fabulis quam uero esse similiora).[71] At stake, then, is the fortune of ancient writers, more than the fate of the traveller: the task of the contemporary travel narrative is to help distinguish fables from verisimilitude. This is not to say that Niccolò is used to supplant classical authority. The genre remains that of Pliny's *Naturalis historia*: the traveller provides information about peoples, about the place and customs of the Indians, about animals and trees, about aromas. Indeed, even Niccolò's achievement in crossing the Ganges and travelling 'far beyond the island of Taprobana' only matches the exploits of two others, a commander of the fleet of Alexander the Great, and a storm-tossed Roman from the reign of Tiberius.[72] Niccolò does not go beyond the classical world, but reenacts, rearticulates, and reinstates it.

As narrated in the fourth decade (book two) of Flavio Biondo's *Decades*, an ambitious history from the fall of the Roman empire to his own day, the episode of the Ethiopians at the Council of Florence resulted in a similar interrogation of classical geography. Biondo, who worked within the papal bureaucracy in various capacities from 1433 until his death in 1463, explains that the audience with the Ethiopians arose from Eugenius' desire to clasp to the papal bosom those who had deviated from the Catholic faith in Asia and Africa. Preoccupation with Church union led to the need for information and contacts with those Christian communities at the ends of the earth, a qualification amply filled by the Ethiopians, since as they themselves put it, 'we believe no-one to have come from a more remote part of the world than we have, since we inhabit not only the furthest part of the world, but Ethiopia, placed almost beyond that very world' (neminem credimus remotiori ab orbis parte huc se conferre quam nos, qui non extremam modo orbis partem, sed paene extra ipsum orbem positam, incolimus Aethiopiam).[73] Biondo's account indicates that, as well as their religious practices, and their attitude to the Roman church, the Ethiopians who appeared before Eugenius and the curia in 1441 were quizzed with particular intensity by three cardinals – Giuliano Cesarini, Jean de Térouanne, and Juan de Torquemada – on the habitation, geography, and political organization of Ethiopia and surrounding areas ('quanti popoli, quasque regiones, quantumque potentiae'), all of which was up to that point 'unknown in Italy' (hactenus ... in Italia ignoratum).[74] The Ethiopians confirmed that their country contained rivers, which discharged either into a lake of such magnitude that it appeared to be a sea, or into the Nile. More controversially, they claimed that their country bordered on India. This statement, their questioners gave them to understand, contradicted the authority of Ptolemy, an expert in the measurement of land and sky held in the greatest esteem by the Greeks and Romans, who placed a sea and several provinces between Ethiopia and India. In response, the Ethiopians replied that either Ptolemy was wrong, or there was more than one India (a faulty translation may also have been to blame, since, according to Biondo, the Ethiopians produced reports in Arabic script, which were read first in Arabic and then translated into Latin).[75]

In Biondo's account the confrontation between Ptolemy's 'terra Ethyopiae incognita' and the Ethiopians' account of a populous, even irrigated, region works in two ways. Intitially, the authority of Ptolemy is put to the Ethiopians, who insist on the validity of their own knowledge, but offer a means of resolving the difficulty (by positing a second India). Shortly after this passage, however, Biondo turns the confrontation back upon his readers. Apparently sensing that those of his audience learned in Greek and Latin might dismiss the evidence of the Ethiopians as fabulous precisely because of the greater reliability of the *Geographia*, Biondo points out that Ptolemy was ignorant of a great many things. Underneath the Arctic, beyond Britain, Ptolemy's *terra incognita* is now known to hold fifty islands inhabited by Christians, to whom the Pope had assigned a bishop, and whose envoys had appeared before the curia. Further, to shore up his case for the supplementation of Ptolemy's description of Africa, Biondo adduces the evidence of Old Testament texts, Paralippomenon book 2, and Esther 1.1, where over 120 provinces are mentioned between Ethiopia and India, as opposed to Ptolemy's four.[76] We see in this account an exploration of limits, textual and geographical. Even as the Ethiopians, dwellers of a region 'almost beyond the world', are quizzed about the ends of the earth, the limits of Ptolemy are made abundantly clear. The Alexandrian's regions of *terra incognita* turn out to offer the keen reader the possibility of some leverage against his authority. At this stage, however, the task remained to update the *Geographia*, rather than, as it would become for sixteenth-century geographers, to restore it to its pristine antiquity.

Biondo's ongoing concern with exploration of Africa, and its historical significance, is evident in a letter he wrote to King Alfonso of Portugal in 1459. The letter concerns a proposed history of Portuguese conquests in Africa, to be composed by Biondo in Latin. Biondo praises the conquests of the King and his transformation of *terra incognita* to land not only known, but colonized. Alfonso's success in invading and subjugating Africans and Moors is put into a classical context. He has sought out African peoples:

> ut ignotos semper hactenus Romanis, olim orbis domitoribus et omnis doctrinae ac rerum peritiae scientia ornatissimis, populos gentes ac nationes sub ipso ardenti ac flagrantissimo meridie degentes, christianum nomen et eius dignitatem et simul Europae populorum mores edoceas.[77]

> so that you might instruct in the name and dignity of Christ and the customs of the peoples of Europe those peoples, races, and nations living beneath that hot and most blazing south, hitherto unknown to the Romans, once rulers of the world and most well endowed with knowledge of all branches of learning.

Indeed Alfonso has succeeded in surpassing the Romans by establishing colonies – not, as the ancients had done, in known and highly cultivated areas, but rather 'you have begun to bring forth your colonies beneath skies always unknown to our geographers' (tu tuas sub caelo nostris semper geographis incognito colonias deducere incepisti).[78] The space of unknown land here becomes the site for moving beyond the reach of classical empire. At the end of his letter, Biondo draws Alfonso back into the Roman orbit, comparing him to the emperors Trajan, Hadrian, and Theodosius (the first two Spanish-born boundary pushers).[79] The empire is on the move once again, now being driven from the Iberian peninsula rather than the Tiber – but still, Biondo emphasizes, in need of record in Latin by a properly qualified historian. Crucially, empire still establishes its colonies and conquers native peoples for the glory of Rome – only this time it is the Rome of the papacy, with its plans for church union. The crusading impetus is never too far away from the surface of accounts of European exploration in this period: when the Ethiopians at Florence were quizzed whether they knew the legendary Christian king Prester John they replied promisingly that one 'Zareiacob' (Zara Ya'qob) ruled over 100 lesser kings, and was prepared to to use his armies against the Saracens in order to liberate Jerusalem and the Holy Land.[80] These dreams made manifest the reward for mission to, and conquest of, peoples beyond the known world, the potential gain to Christendom of those Ptolemy-challenging Ethiopians. Church, state, and classicism intersected in European intervention in unknown lands, driven by dreams of filling lacunae, of refashioning a legacy, of making the divided whole.

The extension of the ecumene and the world image

The connections between mapping and diplomatic are evident in the fairly frequent references to cartographic representation in fifteenth-century documents relating to European expansion. In *Romanus pontifex* Nicholas V describes himself not only as the vicar of Christ, the successor of Peter, but also one who extends paternal consideration over 'all the *climata* of the world and the state of every nation inhabiting them' (cuncta mundi climata omniumque nationum in illis degentium qualitates).[81] Such reference to 'climata' was by no means unique to Nicholas' pontificate: Eugenius IV's bull of 1436, *Romanus pontifex,* which concedes to the king of Portugal the right of appropriating those Canary Islands not yet possessed by Christian rulers, declares that the Pope 'out of the duty of pastoral office extends the cutting edge of his consideration to each of the *climata* of the world' (ad singula mundi climata ex pastoralis officii debito aciem sue considerationis extendit).[82] Although 'climata' could be understood simply to mean 'climates', it could also be understood to refer to the division of the known world into climatic zones, as in a *climata* map. Jurisdiction, papal and regal, was thereby expressed in terms that could be given cartographic form. Given the implication of maps in fifteenth-century administrative practice, it is important to ask how cartographers of the period represented *terrae incognitae*, lands at the ends of Europe's expansionary ambitions.

While interest in unknown regions was considerable throughout the fifteenth century, it would be misleading to suggest that *terra incognita* became an integral or even a common feature on fifteenth-century world maps. In fact, the inscription of unknown land was a standard feature only on the world map of Ptolemy's *Geographia*, and even then only on the map's southernmost latitude. However, a rethinking of ecumenical boundaries is evident on many surviving fifteenth-century world maps, including those that incorporated Ptolemaic features within pre-fifteenth-century representational traditions. In particular, mapmakers of the period showed an increased willingness to depict both of the polar regions, a departure from the *mappamundi* tradition which had traditionally left the poles to zonal maps. Several Italian maps from the fourteenth and fifteenth centuries oppose an Arctic region uninhabitable due to excessive cold to a vaguely delineated southern region – at times as far south as the Antarctic pole – uninhabitable due to extreme heat.[83] In these instances, the expansion of the ecumene seems to have led to a conflation of the equatorial torrid zone with the frigid zone of the southern polar region. The precise reasons for such a conflation are not obvious, but it seems to have emerged from a desire to represent the world from pole to pole at the same time as accounting for the extension of the known world beyond the equator. Evidence for the further transformation of cartographic tradition around the middle of the fifteenth century can be found in the world map compiled in Constance in 1448 by the Benedictine monk Andreas Walsperger (Plate 3).[84] On his map Walsperger explicitly cites the *Geographia* in his representation of the southernmost part of the earth, noting Ptolemy's estimate of the extent of habitable land.[85] Walsperger, however, extends the land beyond this inscription to the frame of the map. Although the southern *plaga* is thus incomplete, Walsperger locates the Antarctic pole there, recording that 'there land is uninhabitable. And around this pole are most marvellous monsters not only beasts but even men' (ibi terra est inhabitabilis. Et circa hunc polem sunt mirabilissima monstra non solum in feris sed etiam in hominibus). The Arctic pole is similarly described as uninhabitable, because of the cold that causes 'perpetual congelations'. The extension of the southernmost part of the earth to the Antarctic pole makes the standard association of the southern Africa with extreme heat difficult to maintain. Nevertheless, Walsperger perpetuated the traditional associations of southern Africa with monstrous races, expressly noting the presence of 'extremely fast monopods' near to the south polar region. There is a kind of mingling, then, between the tradition of the location of monstrous races in southernmost Africa on *mappaemundi* (such as the Hereford and Ebstorf world maps), Ptolemy's description of Africa, bounded to the south by *terra incognita*, and the frigid Antarctic region represented on zonal maps. The proximity of southern Africa to the southern polar region on Walsperger's map provides a visual counterpart to the statement of Nicholas V in *Romanus pontifex* that the Portuguese had searched the sea and maritime provinces in their fast boats 'towards the southern parts and the Antarctic pole' (versus meridionales partes et polum antarticum). Both texts dispense with the representation of the southern hemisphere as irrevocably separated from the north by an intervening and impenetrable zone. On the contrary, northern and southern hemispheres are represented as contiguous, and the implication follows that both poles are within European reach.

Walsperger's map and Nicholas' bull also share an interest in the spread of Christianity throughout the known world. The sense of a territorializing Christianity that runs through Nicholas' bull, the tension it expresses between the advance of Islam and the contrary penetration of Christianity into regions unknown, is echoed in possibly the most remarkable feature of Walsperger's map. For Walsperger coloured the circles that represent cities on his map according to their religion: red for Christian cities, black for non-Christian.[86] This action may help to explain the map's exclusion of an unknown, and unknowable, antipodal continent. The spread of the world from one pole to the other, without the barrier of an excessively hot equatorial zone, is consistent with Walsperger's aim to map the geographic extent of the faith, showing either anxiety at Christianity's contraction, or possibilities for its expansion – or both.

The vigorous debate about the extent of the known world and the optimum manner of its representation, evident in Walsperger's map, was given particular clarity of verbal as well as visual expression on the renowned *mappamundi* of his contemporary, Fra Mauro (Plate 4). Fra Mauro's map was produced between 1448 and 1450[87] at the Camaldolese monastery of San Michele di Murano, near Venice. An inscription on the extant version of the map makes it clear that it was constructed for the Venetian Signoria, but another version of the map appears to have been made by Fra Mauro for the court of King Alfonso V of Portugal between 1457 and 1459.[88] In either case, the map's audience was clearly educated and interested, in all senses, in a comprehensive image of the world that took account of the information brought back by recent explorations. Amongst the more extraordinary aspects of Fra Mauro's *mappamundi* are its lengthy vernacular inscriptions, thirty-two of which are constructed in the first-person singular: using this means the mapmaker is able to address his readers frequently, anticipating their objections, justifying his decisions, and declaring his sources.[89] The effect is to open the map out, to lay at least some of its workings bare. The mapmaker's subjectivity is most apparent at the ends of the earth. Here Fra Mauro appears to continue the interrogation of Ptolemy's *terrae incognitae* commenced by Fillastre and d'Ailly. But the real impetus for the rethinking of *terra incognita* is navigation: in several inscriptions the mapmaker emphasizes the importance of navigators' reports to his image. Where did this leave the classical and medieval heritage? In a state of revision – but not rejection:

> Io non credo derogar a Tolomeo se io non seguito la sua cosmographia, perché se havesse voluto observar i sui meridiani over paralleli over gradi era necessario quanto a la demonstration de le parte note de questa cicumferentia lassar molte provincie de le qual Tolomeo non ne fa mention, ma per tuto maxime in latitudine çoè tra ostro e tramontana dice terra incognita, e questo perché al suo tempo non li era nota.
>
> I do not think that I am being unfaithful to Ptolemy if I do not follow his *Cosmography*, because if I had wanted to observe his meridians, parallels and degrees, I would have had to omit many provinces within the known part of the world that Ptolemy does not give: everywhere in his account, but especially to the north and south, he gives areas as *terra incognita* because in his day they were not known.[90]

This passage clarifies a significant problem facing the mapmaker: the supplementation of Ptolemy's world image required its expansion, and that in turn threatened the integrity of his projections – the organization of the map by means of meridians, parallels, and degrees (fig. 24). Fra Mauro's map is usually described as the terminus of medieval cartography – its high point, and its end – precisely because of this rejection of Ptolemaic projection, because of the retention of a circular form, the vestiges of the T-O schema.[91] However, the mapmaker's statement shows us that the Ptolemaic model has been considered and found wanting, precisely because the map aspires to show the things 'of which Ptolemy does not make mention'. This attitude to traditions medieval and classical is by no means dismissive, but neither is it slavish – both are assessed in terms of their compatibility with information at the mapmaker's disposal. That information, we are repeatedly told, includes eyewitness testimony: 'so I say that in my own day I have been careful to verify the texts by practical experience, investigating for many years and frequenting persons worthy of faith, who have seen with their own eyes what I faithfully report above' (Per tanto dico che io nel tempo mio ho solicitado verificar la scriptura cum la experientia,

Fig. 24. Venice, Biblioteca Nazionale Marciana. World map of Fra Mauro, detail showing northern Europe and Asia. Fra Mauro's lengthy and argumentative inscriptions indicate his critical reception of Ptolemy's *Geographia*, with the issue of *terrae incognitae* and the extent of the known world of particular concern.

investigando per molti anni e praticando cum persone degne de fede, le qual hano veduto ad ochio quelo che qui suso fedelmente demostro).[92]

The contrast that emerges from this and other inscriptions on the map historicizes Ptolemy and at the same time *terra incognita*. What was unknown for Ptolemy is known to the mapmaker. And as an expanded ecumene begins to make unsustainable a strict division into habitable and uninhabitable zones, as the images of southern Africa and northern Europe and Asia swell towards their respective poles, as the expanded range of source material makes it difficult to maintain complete ignorance, so Ptolemy becomes a shadow in those areas he had deemed unknown. The southern part of Africa was 'almost unknown to the ancients' (quasi està incognita a li antichi); the province of Serica in Asia 'Ptolemy labels as unknown land' (Tholomeo fa terra ignota); next to the toponym 'Europa' the mapmaker notes his amazement that Ptolemy named and located 'Scandinaria' in his fourth table, but that 'all this area of Norway and Sweden was unknown to him' (tuta questa parte de norvegia e svetia li sia sta' ignota).[93] A transition is made not so much from unknown to known but from unrecorded to noted: the emphasis is on what has been seen, named, and marked on the map, and reported back to Europe's political and cultural centres. Re-cognition – rewriting, reassessment – is here a rethinking not only of places, but also of sources. However, Ptolemy is not left behind, for the trace of his (and his time's) ignorance lingers, not erased, itself duly noted, a provocation to be answered.

The mode of rethinking established traditions made explicit on Fra Mauro's map was not the last gasp of the medieval mind, nor a halfway house to modernity: it was no different to the method employed at the end of the fifteenth century in the *postille* (annotations) made by Christopher Columbus to the geographical works of Pierre d'Ailly and Pius II. These *postille* carefully record classical theories of uninhabitable and unknown land; they note even more carefully the evidence adduced by d'Ailly and Pius to contradict such theories. Above all, though, they supply evidence of contemporary eyewitness accounts of travel up to and beyond the boundaries of classical and medieval knowledge. So, as an annotation to the second chapter of Pius' *Asia*, and to d'Ailly's 'Epilogus Mappe Mundi', Columbus notes:

> Serenissimo regi Portugallie renunciatum fuit ab uno suo capitaneo, anno de .88., quem miserat ad tentandum terram in Guinea, quod navigavit ultra equinocialem gradus .45.[94]

> In the year '88 the King of Portugal was informed by one of his captains, whom he had sent to explore land in Guinea, that he sailed 45 degrees beyond the equator.

It would be easy to see this statement as a straightforward victory of empiricism over the dead hand of tradition – and it would be mistaken. For just a few chapters later Columbus notes that 'Ptolemy puts habitation beneath, and far beyond, the equator' (Ptholomeus sub equinociali et ultra multa ponit habitacio).[95] As it was for Fra Mauro, *terra incognita* was for Columbus something to be interrogated, frequently from personal experience, but it does not follow that he sought to undermine geographical authorities. Instead, he sought a concordance, in which the authority of Ptolemy, the arguments of d'Ailly and Pius, and the most up-to-date reports from the royal courts could be brought into some kind of harmony. It should be noted that the questioning of *terra incognita* came not simply from those who had most recently travelled through it, but from classical authors and their medieval adaptors. Ptolemy's extension of habitable land beyond the equator finds its confirmation in the discoveries of Portuguese navigators, but we should not lose sight of the fact that it was transmitted to Columbus through the commentaries of d'Ailly and Pius, where it is compared with the theories and data of Pliny, Pomponius Mela, Strabo, Eratosthenes, Avicenna, Albertus Magnus and numerous other classical and medieval authors. Distinctions between authorities on the basis of period (classical, medieval, and contemporary) are not made by Columbus in the *postille*. The sense of separation between ancient and modern geography was to grow markedly in the sixteenth century, but even in the late fifteenth century the prevailing attitude seems similar to that espoused by Fillastre and Fra Mauro: debate, reassessment, revision, and a capacity to accommodate diverse sources of information.

The process of sifting authorities leads to a conclusion of great practical significance, one that is repeated several times in the *postille*: a region of great heat does not impede navigation to the south ('concordat quod mare sit tot navigabilem, nec impedit maximum ardorem').[96] That capacity to travel beyond theorised boundaries, the newfound confidence that people 'without number' were dwelling in the far north and south of the known world, encouraged global vision and with it the possibility of global navigation. 'From pole to pole', Columbus noted, echoing Roger Bacon, 'water runs in the body of the sea, and between the end of Spain and the beginning of India not a great length extends' (A polo in polo decurrit aqua in corpus maris, et extenditur inter finem Hispanie et inter principium Indie non magne latitudinis).[97] The next step, of course, was navigation from west to east, the idea described in the Florentine humanist Paolo dal Pozzo Toscanelli's famous letter of 1474 to Fernão Martins as 'subterranean navigation':

> et non miremini si voco occidentales partes vbi sunt aromata, cum communiter dicantur orientales quia nauigantibus ad occidentem semper ille partes inveniuntur per subterraneas nauigaciones.[98]

> and do not be amazed if I call western those parts commonly called eastern where there are spices, because by sailing to the west those parts are found by subterranean navigations.

The putative western sea route to the furthest ends of Asia appears to be understood by Toscanelli as a voyage across the underside of the earth. As subsequently expressed by Columbus, the collapse of the distinction between east and west is coterminous with a vision that extends from pole to pole. But where did this revision of the possibilities of navigation, and of the relationship between the ends of the earth, leave the antipodes?

The notion of extending the boundaries of the known world to the south pole, into the regions traditionally regarded as antipodal, was expressed with great clarity in a remarkable map that accompanied Antoine de la Sale's rambling, 'indigestible' compilation, 'La Salade' (fig. 25). La Sale, a tutor to the house of Anjou from 1434 and to the sons of Louis of Luxembourg from 1448, composed 'La Salade' between 1437 and 1444 for the son of René of Anjou.[99] The work includes a geographical description of the three known continents of the world, which survives in three redactions, none of which discusses the southern hemisphere, or any theory of antipodal regions. The 1521 and 1527 printed editions of 'La Salade', however, contain a woodcut world map with significant antipodal content. The form and features of the map indicate that it dates from the fifteenth century, and that the woodcut was derived from a lost manuscript version.[100]

At first appearances La Sale's map is a most unusual sight.[101] Here more than in any other fifteenth-century map the known world can be seen to have burst its boundaries. The tripartite form is consistent with that of a T-O map: circular, with a division into one half and two quarters. However, what is shown is a striking elaboration on this familiar model: a fairly standard representation of the northern hemisphere in the top half of the map (including the terrestrial paradise in the far east), is set against a highly speculative representation of its extension into the southern hemisphere in the lower half of the map. Africa pushes well beyond the equator, and even beyond the tropic of Capricorn, into 'terra Incognita et deserta'. Asia, meanwhile, has extended its reach almost to the south pole thanks to the sprawl of 'Patalis regio'. This vast peninsula, whose name was subsequently incorporated into the sixteenth-century's *Terra Australis Incognita*, stretches from the easternmost part of Asia into the sub-equatorial sea marked 'Mare Antipodes et Incognitum'.[102]

The manner of representation of the southern hemisphere – in which the known world has expanded beyond the equator – is certainly a unique feature of the map. In many respects, though, the 'La Salade' map resembles other world maps of the mid to late fifteenth century. Like the maps of Walsperger and Fra Mauro, it is an attempt to create a synthesis of Ptolemy's geography with certain traditions of medieval cartography. Its location of 'terra incognita' in sub-equatorial Africa and its depiction of 'Patalis regio' seem likely to have been derived from a combination of the *Geographia*, Pliny's *Naturalis historia*, and perhaps d'Ailly's *Imago Mundi*.[103]

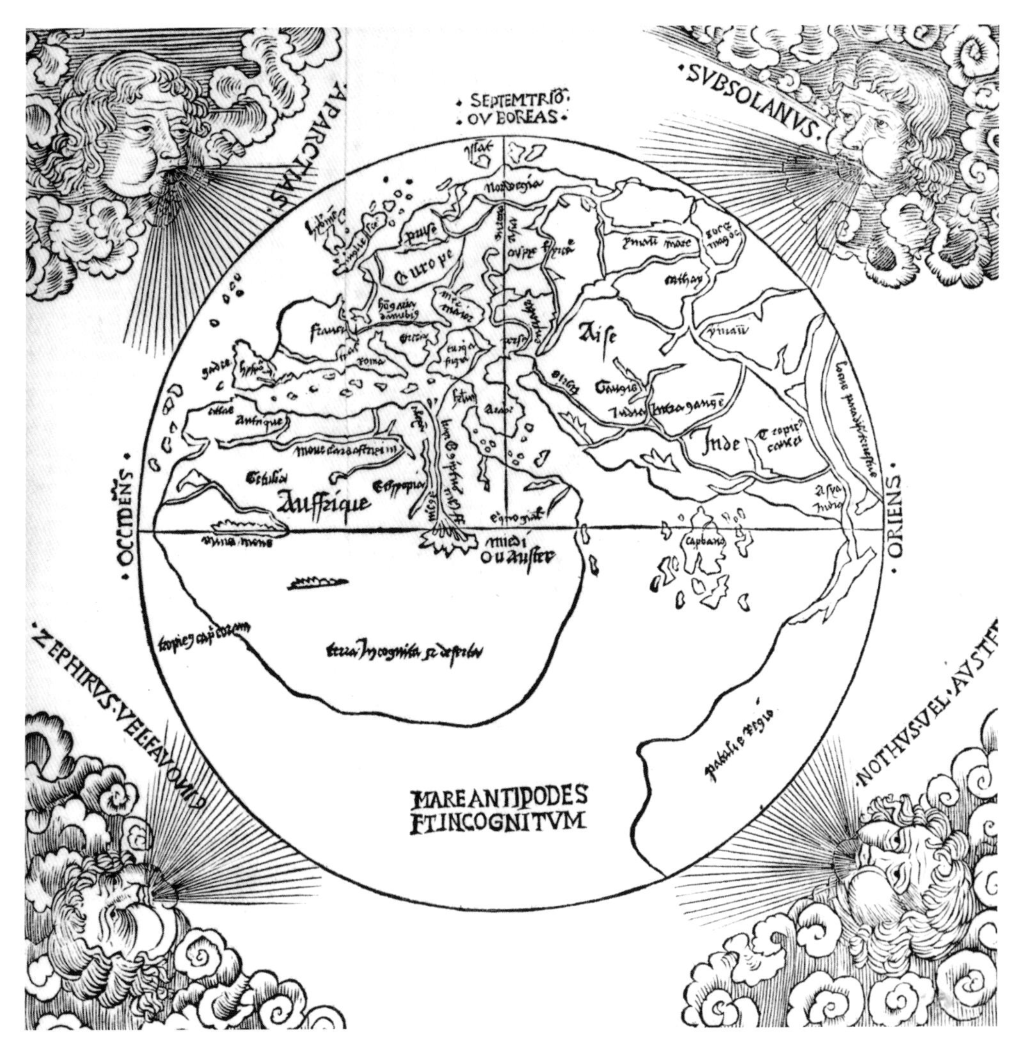

Fig. 25. London, BL, 216.a.8. World map in Antoine de La Sale, *La salade* (Paris, 1521). The tripartite form of this image is consistent with that of a T-O map: circular, with a division into one half and two quarters. The fairly standard representation of the northern hemisphere in the top half of the map (including the terrestrial paradise in the far east), is set against a highly speculative representation of its extension into the southern hemisphere in the lower half of the map. Africa pushes beyond the equator into 'terra Incognita et deserta'. Asia extends almost to the south pole thanks to 'Patalis regio'. The sub-equatorial sea is marked 'Mare Antipodes et Incognitum'.

However, other features, such as the map's circular shape and the representation of the terrestrial paradise and Gog and Magog in the far east of Asia, indicate its retention of non-Ptolemaic influences. Interesting in this regard too is the fact that Ptolemy's land-locked Indian Ocean has been discarded, and that the island of 'Capbano' (Taprobana) is represented as straddling both hemispheres. Classical authors, such as Pliny, had long associated Taprobana with antipodal regions, noting that shadows fell to the south there, a fact that seemed to indicate its sub-equatorial position.[104] Here it is shown on the threshold of the lower hemisphere, extending across the equator along with the continents of Africa and Asia. This southern expanse of Asia and Africa shows the mapmaker's somewhat crude conception of the 'antipodal' lands as merely an extension of the ecumene. Here, as in Fillastre's introduction to Pomponius Mela, the antipodes have been subsumed into the known world. Perhaps, then, La Sale's map represents the ultimate fantasy of contiguity, the transformation of one quarter of the world into the whole by means of the steady pushing back of boundaries, the erasure of unreachable areas of *terra incognita*, to be replaced by the seamless construction of more accessible formations: areas that invite, rather than repel, the passage of men.

'By making and constructing a line ...'

The idea that antipodal places could be reached from the known world continued to raise endemic questions about the inhabitants of such regions and their capacity to be incorporated into the Catholic Church. The dream of converting antipodal peoples was, in the last two decades of the fifteenth century, both a literary fantasy and a reality with serious political and spiritual ramifications. Two texts, ostensibly very different, illustrate these different manifestations of antipodal space, and convey the particularly charged significance of spaces beyond, but increasingly within reach of, the known world. One, Alexander VI's adjudication on New World discoveries, in which papal ambition and oversight struggle to keep pace with the expansion of European sovereignty, was both improvization and consolidation of power. The other, an antipodal digression in Luigi Pulci's chivalric epic *Morgante*, was an energetic reworking of classical and medieval literary tradition. What both texts have in common is that they engage in the process of *recognitio* – a rethinking and re-knowing of antipodal space, based on traditional modes of representation, but in the knowledge that those modes required reformulation. The antipodes, as both authors realized, was a space with a past, which had functioned as a sign of intellection, and a topic for debate; but it was also, evidently, a space with a future.

The *Morgante*, first printed in its entirety in 1483, but under composition from the time it was commissioned by Lucrezia de' Medici in the early 1460s, consists of two essentially separate poems. The first (conventionally printed as Cantos 1–23) tells of the adventures of Orlando and Rinaldo, knights of the court of Charlemagne. Picaresque and eccentric, the poem includes Rinaldo's encounters with demons and giants, including the eponymous Morgante. The theme of the antipodes is introduced in the second poem, 'La rotta di Roncisvalle' (Cantos 24–8 of the *Morgante*), written between 1479 and 1483, which features frequent theological, cabbalistic, and zoological digressions from its central theme, the climactic battle and defeat of the Christian army led by Charlemagne at Roncevaux.[105] One such digression occurs in Canto 25, when the demon Astarotte oversees the retrieval from the antipodes of a herb that renders Rinaldo invisible. Astarotte explains that the columns of Hercules do not mark the ends of the earth, and that the other hemisphere is not only populated, but can boast cities, castles, and empires.[106] The people there worship Jupiter and Mars, have plants and animals as in the 'upper' hemisphere, and 'often fight great battles against each other' (spesso insieme gran battaglie fanno). They are,

Astarotte explains, descended from Adam and can, he suggests, be saved, since 'all were saved by the cross', and because their ignorance of Christ does not put them in the same category as those, namely Muslims and Jews, that know the truth but do not follow it. Rinaldo subsequently expresses the desire to travel beyond the columns of Hercules 'because down there they have wars' (perché di' colaggiù si fa guerra).[107] There is a suggestion that when he takes leave of Charlemagne's court in the poem's final canto Rinaldo's destination may in fact be the 'other hemisphere' – and that his purpose may be its evangelization:

> E perché ancor di lui quell'angel disse:
> — Ogni cosa esser può, quando Iddio vuole —,
> acciò che quelle gente convertisse
> ch'adoravan pianeti e vane fole,
> e se ancor vivo un giorno e' rïuscisse
> dall'altra parte ove si lieva il sole
> (come molte miracoli si vede)
> qual maraviglia? Chi più sa, men crede.[108]
>
> Besides, it was an angel said of him,
> 'All things are possible if God so wills.'
> Now, if he wanted to convert those throngs
> that worshipped planets and vain fables still,
> and, safe and sound, at last one day he reached
> the other hemisphere where the sun rises
> (there have been other miracles before),
> what then? He less believes who knows much more.[109]

Critics have tended to regard the passages on the antipodes in 'La rotta di Roncisvalle' as a remarkable intimation, just a few years before Columbus' voyages, of European exploration across the Atlantic. In particular, Rinaldo's evangelizing impetus seems similar to the motivation for exploration and conquest professed by the Castilian and Portuguese royal houses.[110] The significance of the matter of the antipodes in 'La rotta' can also be explained as part of a continued (perhaps revived) interest in Medicean Florence in the 'underside of the earth'. Pulci may have been influenced by the public lectures of Lorenzo Buonincontri, who printed the Stoic author Manilius' *Astronomica* in 1484, and commented specifically on its passages on the antipodes.[111] In its consideration of the vexed question of the religious beliefs, and descent, of the people of the 'other hemisphere', 'La rotta' attests to the continued resonance of Augustine's comments on the antipodes. Astarotte's account of the world below, and his response to Rinaldo's questions about the origins of the people there, constitute, as Gabriella Moretti puts it, 'a fully-fledged theological disquisition, in which a possible future evangelization and redemption of the antipodes is actually prophesied'.[112] However, what is at stake here is not so much the future of the antipodeans, as the past of their potential redeemers. The underside of the earth is, I would suggest, the site for Pulci's playing out of Europe's pagan antiquity. Its inhabitants are said to be a warlike people who worship not just any pagan deities, but Jupiter and Mars. The reasons given for their possible redemption recall, moreover, the trope of the virtuous pagan, a familiar and repeated concern of Christian authors of the Middle Ages, and one frequently applied to virtuous Romans, such as Virgil, or the emperor Trajan.[113] The final suggestion that Rinaldo will head beyond the columns of Hercules is not, then, so much a 'prevision' of evangelization of the New World, as a post-vision of the evangelization of the Roman empire. It is no coincidence that the underside of the earth is said to contain 'imperio' as well as cities and castles; it is the underside – the ancestor – of Charlemagne's aggressively Christian empire. Rather than departing to convert Saracens or Jews – already aware of the truth, and therefore damned for not converting when they could have – Rinaldo is made to head back towards a people whose belief is still pure, but misguided, in order to inform them of their error in time for judgement. His journey is inward, an attempt to align Christian present and future with classical past.

In 1493, a decade after the completion of the *Morgante*, the political and juridical significance of land on the underside of the earth was given unprecedented clarity of definition. The response of Alexander VI to reports of Columbus' discovery of remote and previously unknown land to the west of Europe was to issue a series of bulls that sought to regulate the possession and occupation of what was about to become the New World. The most important of this series, the two bulls entitled *Inter cetera*, dated 3 May and 4 May 1493, acknowledge news of the discoveries, and draw upon the plenitude of papal authority to mediate the claims of the Spanish and Portuguese royal houses to land in the Atlantic, and along the west coast of Africa. The two versions of *Inter cetera* differ in one crucial respect: into the latter a clause was inserted which provides for the construction of the line of demarcation in the Atlantic, one hundred leagues to the west and south of the Azores and Cape Verde islands, running from the Arctic to the Antarctic pole. Part of the power of Alexander's donation lies in its determination of the direction of future exploration: all islands and *terrae firmae* discovered and *to be discovered* to the west and south of the line were to be possessed by the monarchs of Castile and León, Ferdinand and Isabella.[114] This line of demarcation was moved the following year a further 270 leagues to the west of the Cape Verde islands, when the Spanish and Portuguese negotiated the Treaty of Tordesillas, but the principle of global division remained the same.[115]

There is a remarkable air of simplicity to the May 4 *Inter cetera*: the document explains that land remote and hitherto unknown has been discovered in the Ocean by a worthy and commendable man, Christopher Columbus, and his fellow sailors. This land contains peaceable, naked savages who are not cannibals. Ferdinand and Isabella propose to subdue these lands and lead their inhabitants to the Catholic faith; Alexander therefore constructs a line from one pole to the other, and concedes to them any lands to the west and south of it not already in possession of a Christian king before 25 December 1492. No person may possess, or trade in, these lands without the permission of these sovereigns.[116]

The response to Columbus' landfall contained in this document represents the culmination of fifteenth-century geographical and legal culture. It is the culmination of a century's interest in unknown land, and of the reconceptualization of the reachability of unknown regions. The document repeats the legal formulas used in papal bulls earlier in the century to justify the papacy's concession of remote and unknown lands and islands. And, in drawing a line pole-to-pole – 'fabricando et constituendo unam lineam a polo artico, scilicet septentrione, ad polum antarcticum, scilicet meridiem' – the bull reveals the impact of changes in fifteenth-century cartography. The expanded frame of ecumenical mapping is evident here: the maps of d'Ailly, Walsperger, Fra Mauro, even la Sale – maps that extend the known world below the equator towards the Antarctic pole – make the idea of drawing a line that divides the world in two possible. Such lines had been drawn many times before, but this line was a cartographic act of practical, rather than theoretical, significance. The apparent ease with which lands and peoples could be assigned to these monarchs was, as many historians have pointed out, the product of long-running debates among theorists of papal power about the extent of papal authority and responsibility for non-Christians.[117] The discourse of colonization, emphatically preoccupied with habitation, and the response to inhabitants, recognized the naked, non-cannibalistic natives of previously unknown lands as descendants of the sons of Adam. Hence Alexander's concern that they be lead back to the Christian faith: this is his condition for sanctioning the subjection of their land.[118]

It is no more than a commonplace to say that acts of colonization are accompanied by drawing lines on maps, by a process of 'making and constructing' territories. In this instance, though, the line of demarcation was actually an attempt to retain papal control over expansion, an assertion of jurisdiction over European powers as much as the peoples of the New World; the papal construction of lines from one end of the world to the other was a means of keeping pace with secular rulers, and thereby shaping colonization. What is perhaps not so well noted is the ancient presence in the European scientific and literary imagination – and hence in its cartographic imagination – of unknown lands. *Terra incognita*, land glimpsed only in a visionary moment, could encourage dreams of pushing colonization beyond the boundaries of empire – dreams like that of the evangelizing Rinaldo in Pulci's *Morgante*. Or it could curtail such expansionary impulses: as in the 'Somnium Scipionis', *terra incognita* could be used to confine *imperium* within its established boundaries through an awareness of the vastness of what lies beyond and the danger of losing sight of duty to the state. That ideological malleability, or ambiguity, of the presence of unknown land – its signification of critique or incitement to empire – seems during the fifteenth century to have tilted decisively in favour of expansion. The Alexandrine bull transforms lands unknown, lands 'by others thus far unreported', to lands found. The transformation is ongoing, since some land is acknowledged to remain *as yet* unknown, outside the sovereigns' reach. As boundaries are rolled back, *terra incognita* is increasingly conjoined with land newly found, 'terra inventa' or 'terra reperta'. Unknown land is still noted, but its figuration as unreachable or uninhabitable is shattered: it is now *terra nondum cognita*, 'not yet' land that will be found, and that has already been claimed.

Notes

1 For a detailed discussion, and transcription, of the seventh figure of d'Ailly's 'Epilogus' see Anna-Dorothee von den Brincken, '*Occeani Angustior Latitudo*: Die Ökumene auf der Klimatenkarte des Pierre d'Ailly', in *Studien zum 15. Jahrhundert: Festschrift für Erich Meuthen*, ed. Heribert Müller, 2 vols (Munich: Oldenbourg, 1994), vol. 1, pp. 565–82. The seventh figure survives in eight fifteenth-century manuscripts: Destombes, *Mappemondes*, pp. 161–3, although one of these (Basel, Universitätsbibliothek, MS F.IV.24) differs notably from the others (it does not show a southern hemisphere). The *Imago Mundi* was printed in Louvain between 1480 and 1483 with the accompanying figures from the 'Epilogus'. The printed 'septima figura' differs in certain details from those in the manuscripts, including the direction of the inscription concerning land beyond the *climata* (horizontal in the printed version, rather than vertical).

2 In the easternmost part of the figure: 'extending towards the south, India contains nearly one-third of habitable land' (India fere terciam partem terre habitabilis continet, versus meridiem se extendens); in the southeast between the equator and the tropic of Capricorn: 'according to some the southern coast of India extends beyond the tropic of Capricorn; and its eastern side almost to the African coast' (Frons Indie meridianus secundum quosdam protenditur usque tropicum capricorni; Orientale vero latus usque prope finem affrice): von den Brincken, Die Ökumene auf der Klimatenkarte des Pierre d'Ailly', p. 572.

3 'In hac igitur opinionum varietate ymaginarias rationes hinc et inde apparentes non recito que in hiis rebus non tam ymaginationibus quam experientiis et probabilibus historiis reputo certitudinaliter adherendum': Pierre d'Ailly, *Ymago Mundi*, ed. Edmond Buron, 3 vols (Paris: Maisonneuve Frères, 1930), vol. 1, p. 202.

4 *Ymago Mundi*, vol. 2, pp. 230–4; 524. D'Ailly's later work, the 'Compendium cosmographiae', pits these authorities against Ptolemy's *Geographia*: *Ymago Mundi*, vol. 3, pp. 658–86.

5 Although d'Ailly adduced the argument of Augustine against the possibility of a temperate, inhabited antipodal region, he consistently emphasized the extent of land and habitation as far as the Arctic circle in the north and the equator in the south: *Ymago Mundi*, vol. 1, pp. 198–202, 230–4; vol. 2, pp. 522–4.

6 'After the *climata* towards the pole there are many habitations and islands, which cannot be conveniently described here' (Post climata versus polum multas habitationes et insulas continent, que non possunt hic convenienter descibi): von den Brincken, 'Die Ökumene auf der Klimatenkarte des Pierre d'Ailly', p. 572.

7 The *Geographia* had, it seems, been rediscovered in Constantinople only in the thirteenth century. The date of the completion of the Latin translation is disputed: it was certainly in progress in 1406 and was dedicated to Pope Alexander V (1409–10). On the contribution of Chrysoloras see Sebastiano Gentile, 'Emanuele Crisolora e la "Geographia" di Tolomeo', in *Dotti bizantini e libri greci nell'Italia del secolo XV*, ed. Mariarosa Cortesi and Enrico V. Maltese (Naples: D'Auria, 1992), pp. 291–308. For an overview of the *Geographia*'s reception see Patrick Gautier Dalché, 'The Reception of Ptolemy's *Geography* (End of the Fourteenth to Beginning of the Sixteenth Century)', in *The History of Cartography*, vol. 3: *Cartography in the European Renaissance*, ed. David Woodward (Chicago: University of Chicago Press, 2007), pp. 285–364.

8 O. A. W. Dilke, 'The Culmination of Greek Cartography in Ptolemy', in *The History of Cartography*, vol. 1, pp. 177–200, esp. pp. 190–9.

9 *Cosmographia* 1.5. My discussion of Ptolemy in this chapter refers to the Latin translation of Jacopo d'Angelo contained in the 1478 Rome edition of the *Cosmographia* (reprinted Amsterdam: Theatrum Orbis Terrarum, 1966), by book and chapter number.

10 'Cosmographia designatrix imitatio est totius *cogniti* orbis cum iis que fere uniuersaliter sibi iunguntur' (1.1). My emphasis.

11 *Cosmographia* 7.5.

12 It should be noted, however, that not all copies of the *Geographia* contained maps; those not produced for royal, aristocratic, or papal patrons, for example, often contained either no maps, or maps that lacked the rich colours and decoration found in some of the more splendid manuscripts of the *Geographia*: see Marica Milanesi, 'A Forgotten Ptolemy: Harley Codex 3686 in the British Library', *Imago Mundi* 48 (1996), 43–4.

13 For alterations to Ptolemy's description of southern Africa made in a mid-fifteenth-century codex see O. A. W.

and Margaret S. Dilke, 'The Wilczek-Brown Codex of Ptolemy Maps', and Susan L. Danforth, 'Notes on the Scientific Examination of the Wilczek-Brown Codex', *Imago Mundi* 40 (1988), 118–25. Ptolemy's fourth map of Africa had been extended to the south by the first compiler of this codex (no areas of *terra incognita* are represented), and a subsequent owner painted over this extension so as to represent the continent's south coast.

14 Nancy, Bibliothèque municipale, MS 441. For Clavus see Axel Anthon Björnbo and Carl S. Petersen, *Der Däne Claudius Claussøn Swart, der älteste Kartograph des Nordens, der erste Ptolemäusepigon der Renaissance* (Innsbruck: Wagner'schen Universitäts Buchhandlung, 1909).

15 Patrick Gautier Dalché, 'L'œuvre géographique du cardinal Fillastre', *Archives d'histoire doctrinale et littéraire du Moyen Âge* 59 (1992), 319–83: p. 374. Gautier Dalché provides an edition of Fillastre's annotations to his Ptolemy edition (pp. 372–83). See also Joseph Fischer, *Claudii Ptolemaei Geographiae Codex Urbinas Graecus 82*, 2 vols in 4 (Leiden: Brill, 1932), vol. 1, p. 303.

16 Björnbo and Petersen, *Der Däne Claudius Claussøn*, p. 145. Discussed by Christiane Deluz, 'L'Europe selon Pierre d'Ailly ou selon Guillaume Fillastre? De l'*Ymago Mundi* aux légendes de la carte de Nancy', in *Humanisme et culture géographique à l'époque du concile de Constance autour de Guillaume Fillastre*, ed. Didier Marcotte (Turnhout: Brepols, 2002), pp. 151–60: p. 158.

17 *Cosmographia* 8.1.

18 *Cosmographia* 7.3.

19 *Cosmographia* 6.14. The Sacae were thought to be a nomadic North Asian race of cave-dwellers; the Sogdians inhabited the region north of Baktria in central Asia; Margiana was a district of the Parthian empire, usually depicted on fifteenth-century Ptolemaic maps to the south-east of the Caspian Sea. The town referred to in this passage was represented as a nodal point on the Imaon mountains on Ptolemy's Septima Asiae Tabula.

20 By 1503 the Freiburg edition of Gregor Reisch's *Margarita philosophica* had amended this inscription to 'this is not land but sea: in which there are islands of great size, but they were unknown to Ptolemy' (hic non terra sed mare est: in quo mirae magnitudinis Insulae, sed Ptolomeo fuerunt incognitae).

21 See the maps reproduced in Fischer, *Claudii Ptolemaei Geographiae Codex Urbinas Graecus 82*, vol. 2. As it explicitly does on Fillastre's Ptolemy, Clavus' map effectively removed an area of *terra incognita* in northern Europe from many subsequent editions of Ptolemy. On the Florentine reception and adaptation of Ptolemy's *Geographia* over the course of the fifteenth century, see Sebastiano Gentile, 'L'ambiente umanistico fiorentino e lo studio della *Geografia* nel secolo XV', in *Amerigo Vespucci. La vita e i viaggi*, ed. L. Formisano *et al.* (Florence: Banca di Toscana, 1992), pp. 9–63, esp. pp. 12–45.

22 Milanesi, 'Forgotten Ptolemy', p. 56: 'the author of Harley MS 3686 tried to redraft essentially portolanic outlines to accord with the dictates of Ptolemy and, unlike anybody else, he did it on a regional, almost Ptolemaic, scale'. Milanesi suggests that the manuscript was compiled between 1436 and 1450 (p. 55), based on its orthography and non-humanistic provenance (restricting it to a date before 1450), and its possible use of Andrea Bianco's atlas of 1436 as a source.

23 Milanesi, 'Forgotten Ptolemy', p. 45.

24 Milanesi contends that stippling in the map showing north-western Africa (f. 99v) represents not the Sahara desert, but *terra incognita*: 'Forgotten Ptolemy', pp. 53–4. If so, this represents a departure from Ptolemy's description, a fact which may explain the non-verbal form of representation.

25 On the 'caspey montes' see Milanesi, 'Forgotten Ptolemy', p. 61 *n*27. This mountain range is not mentioned in the *Geographia*, but was associated with the Alexander romance tradition, and appears on several fifteenth-century *mappaemundi*.

26 Gautier Dalché, 'L'œuvre géographique du cardinal Fillastre', pp. 331–44. Gautier Dalché describes Fillastre's introduction as 'a commentary on Ptolemaic conceptions' (p. 334), and characterizes this commentary as a clash between the representational schemes used by descriptive and scientific geography (i.e. T-O maps v. zonal maps), the first of its kind to address the contradictions of the two schemas (p. 330). See also Gautier Dalché's subsequent modification of his position on the compatibility of 'traditional' schemes with Ptolemy, in the 'Avertissement' to the reprinting of 'L'œuvre géographique du cardinal Fillastre' in *Humanisme et culture géographique à l'époque du*

concile de Constance, p. 293. Fillastre is credited with introducing the Latin text of Ptolemy's *Geographia* to northern Europe.

27 There are three extant manuscripts of this version of Fillastre's introduction to the *Chorographia*: Vatican City, BAV, Archivio di San Pietro, H 31; Florence, Biblioteca Laurenziana, Gaddi 91 inf. 7; and Rennes, Bibliothèque municipale, MS 256. One of the three (BAV, Archivio di San Pietro, H 31) was copied for Cardinal Giordano Orsini by Pirrus de Noha, the same scribe that copied Orsini's manuscript of the Latin translation of Ptolemy (BAV, Archivio di San Pietro, H 32): Catherine M. Gormley, Mary A. Rouse and Richard H. Rouse, 'The Medieval Circulation of the *De chorographia* of Pomponius Mela', *Mediaeval Studies* 46 (1984), 266–320: p. 316. The influence of *De chorographia* only began as a result of its dissemination within humanist circles, following its introduction to Italy by Petrarch in the fourteenth century. Its influence prior to this was extremely limited, having survived in only a few copies: G. Billanovich, 'Dall'antica Ravenna alle biblioteche umanistiche', in *Annuario della Università Cattolica del Sacro Cuore 1955–57* (Milan, 1957), pp. 73–107; Gormley, Rouse and Rouse, 'Medieval Circulation of the *De chorographia*', pp. 269–311. On the fourteenth-century humanist reception of Pomponius Mela see Bouloux, *Culture et savoirs géographiques*, pp. 159–67.

28 Gautier Dalché, 'L'œuvre géographique du cardinal Fillastre', p. 334. The introduction is anonymous in the Vatican, Florence, and Rennes manuscripts, but a note in Fillastre's own hand asserts his authorship in Reims, Bibliothèque municipale, MS 1321; the textual evidence supports the conclusion that Fillastre was the author of both introductions: Gormley, Rouse and Rouse, 'Medieval Circulation of the *De chorographia*', p. 317; Gautier Dalché, 'L'œuvre géographique du cardinal Fillastre', pp. 353–5.

29 Gautier Dalché, 'L'œuvre géographique du cardinal Fillastre', p. 332.

30 This map survives only in Giordano Orsini's manuscript, BAV, Archivio di San Pietro, H 31; it seems also to have appeared in the Reims manuscript and in Florence, Biblioteca Laurenziana, Gaddi 91 inf. 7, but to have been subsequently removed from both: Gormley, Rouse and Rouse, 'Medieval Circulation of the *De chorographia*', p. 316.

31 Gautier Dalché, 'L'œuvre géographique du cardinal Fillastre', p. 335.

32 Gautier Dalché, 'L'œuvre géographique du cardinal Fillastre', pp. 330–42.

33 Pomponius Mela, *De chorographia*, ed. Parroni, 1.4 and 1.54, pp. 112, 120; English translation from F. E. Romer, *Pomponius Mela's Description of the World* (Ann Arbor: University of Michigan Press, 1998), pp. 34, 50. Mela also speculated that the island of Taprobana may be the 'first part of the second world' (3.70): *De chorographia*, p. 167; Romer, *Pomponius Mela's Description of the World*, p. 122.

34 Fillastre, *Introductio*, p. 357. All references are to Gautier Dalché's edition of Fillastre's *introductio* in 'L'œuvre géographique du cardinal Fillastre', pp. 355–65, by page number.

35 Fillastre, *Introductio*, pp. 357–8. Fillastre here refers to the *Geographia* 1.9.

36 Fillastre, *Introductio*, p. 358. My emphasis.

37 Fillastre, *Introductio*, p. 358. Discussion of the Hyperboreans, marking the northernmost extent of human habitation, was commonplace in classical Greek and Latin geographical commentary from Herodotus onwards; they are, for example, mentioned in Martianus Capella's *De nuptiis*, Macrobius' *Commentarii*, and Mela's *De chorographia*. On the legend of the Hyperboreans in classical Greek ethnography see Romm, *Edges of the Earth in Ancient Thought*, pp. 60–7.

38 Fillastre, *Introductio*, pp. 358–9.

39 Fillastre, *Introductio*, p. 359.

40 Fillastre, *Introductio*, p. 359. This suggestion had already been made by Roger Bacon, and subsequently repeated by Nicolas Oresme and d'Ailly: Gautier Dalché, 'L'œuvre géographique du cardinal Fillastre', p. 337.

41 Fillastre, *Introductio*, p. 359.

42 It is likely that Fillastre would have known both Albertus' *De natura loci* and Bacon's *Opus Maius*; he would probably have been aware of d'Ailly's comments in chapter 6 of the *Imago Mundi* in which he argues for the habitability of the torrid zone.

43 For a similar subsumption of the idea of the antipodes within an expanded ecumene see Marica Milanesi, 'Il commento al "Dittamondo" di Guglielmo Capello (1435–37)', in *Alla corte degli Estensi: Filosofia, arte e cultura a Ferrara nei secoli XV e XVI*, ed. Marco Bertozzi (Ferrara: Università degli Studi, 1994), pp. 365–88.

44 *Acta Concilii Constanciensis*, ed. Henrich Finke, 4 vols (Münster i. W.: Regensbergsche Buchhandlung, 1898–1928), vol. 2, pp. 300–1.

45 Fillastre, *Introductio*, p. 360. On the Canaries and their particular role in the history of European (specifically Castilian) colonizing practices see Felipe Fernández-Armesto, *The Canary Islands after the Conquest: The Making of a Colonial Society in the Early Sixteenth Century* (Oxford: Clarendon Press, 1982).

46 The Canaries appeared on European marine charts as early as the second quarter of the fourteenth century: London, BL, MS Additional 25691, a map produced in the workshop of Angelino Dulcert on Mallorca, c. 1340, shows the islands of Lanzarote and Fuerteventura, as well as 'Insula Canaria'.

47 James Muldoon, *Popes, Lawyers, and Infidels* (Philadelphia: University of Pennsylvania Press, 1979), pp. 6–18; 56–9. See also Antonio García y García, 'Las donaciones pontificias de territorios y su repercusión en las relaciones entre Castilla y Portugal', in *Las relaciones entre Portugal y Castilla en la época de los descubrimientos y la expansión colonial*, ed. Ana María Carabias Torres (Salamanca: Ediciones Universidad de Salamanca, 1994), pp. 293–310.

48 In a series of articles Charles-Martial de Witte considered the papal *acta* that both legitimated and to some extent moderated fifteenth-century Portuguese exploration and colonization: 'Les bulles pontificales et l'expansion portugaise au XVe siècle', *Revue d'histoire ecclésiastique* 48 (1953), 683–718; 49 (1954), 438–61; 51 (1956), 413–53, 809–35; 53 (1958), 5–46, 443–71.

49 Matthew 28:19–20.

50 Vatican City, BAV, MS Urb. lat. 274. See Fischer, *Claudii Ptolemaei Geographiae Codex Urbinas Graecus 82,*, vol. 1, pp. 351–6.

51 Nicola Casella, 'Pio II tra geografia e storia', *Archivio della Società romana di storia patria* 95 (1972), 35–112: pp. 78–80.

52 On the differences between *Dum diversas* and *Romanus pontifex* see de Witte, 'Les bulles pontificales et l'expansion portugaise', 51 (1956), pp. 425–41. Essentially Nicholas concedes similar powers to the Portuguese in *Romanus pontifex*, but limits them within a more concise – if still generous – geographical formulation than appeared in the earlier bull.

53 Nicholas V, 'Romanus Pontifex', in *Bullarium Romanum*, ed. F. Gaude and A. Tomassetti, 27 vols (Turin: Augustae Taurinorum, 1857–85), vol. 5, pp. 110–15: p. 111.

54 'Romanus Pontifex', p. 111. The reference to activity in the Atlantic islands specifically invokes Portuguese interests in Madeira, the Azores, and the attempted evangelization of the Canaries.

55 Poggio Bracciolini, *Lettere*, ed. Helene Harth, 3 vols (Florence: Olschki, 1984–7), vol. 3, pp. 88–90.

56 'Romanus Pontifex', p. 112 (my emphasis).

57 'Romanus Pontifex', p. 112.

58 'Romanus Pontifex', p. 113.

59 On this issue see James Muldoon, 'The struggle for justice in the conquest of the New World', in *Proceedings of the Eighth International Congress of Medieval Canon Law*, ed. Stanley Chodorow (Vatican City: Biblioteca Apostolica Vaticana, 1992), pp. 707–20. See also the discussion of V. Y. Mudimbe, *The Idea of Africa* (Bloomington: Indiana University Press, 1994), pp. 31–7.

60 See Thomas Goldstein, 'Geography in Fifteenth-Century Florence', in *Merchants and Scholars: Essays in the History of Exploration and Trade* (Minneapolis: University of Minnesota Press, 1965), pp. 9–32, who regards the Council of Florence as the culmination of the geographical discussions of a quasi-academy – 'an apparently sizeable group of scholars and dilettantes' – based at Florence from c. 1410 (p. 16). For more recent accounts of the study of geography in fifteenth-century Florence see Gentile, 'L'ambiente umanistico fiorentino', esp. pp. 27–44, and idem, 'Toscanelli, Traversari, Niccoli e la geografia', *Rivista Geografica Italiana* 100 (1993), 113–31.

61 See Sebastiano Gentile, 'Giorgio Gemisto Pletone e la sua influenza sull'umanesimo fiorentino', in *Firenze e il concilio del 1439*, ed. Paolo Viti, 2 vols (Florence: Olschki, 1994) vol. 2, pp. 813–32, esp. pp. 825–32.

62 Aubrey Diller, 'A Geographical Treatise by Georgius Gemistus Pletho', *Isis* 27 (1937), 441–51.

63 Diller, 'Geographical Treatise', pp. 445–6.

64 The Ethiopian delegation is likely to have been formed in Cairo, where in 1440 the papal envoy Alberto da Sarteano held negotiations with the Copts on their adhesion to the Council, before introducing the two delegations to the Council in the following year: Salvatore Tedeschi, 'Etiopi e Copti al Concilio di Firenze', *Annuarium historiae conciliorum* 21 (1989), 380–407. See also Gautier Dalché, 'Reception of Ptolemy's *Geography*', pp. 309–12.

65 Poggio was himself the redactor of Eugenius' bull of concession of the conquest of the Canary Islands to the King of Portugal (15 September, 1436): de Witte, 'Les bulles pontificales', 48 (1953), p. 717

66 Poggio Bracciolini, *De varietate fortunae*, ed. Outi Merisalo (Helsinki: Suomalainen Tiedeakatemia, 1993), Book 4, pp. 153–75. For a discussion of the motives of Poggio for appending de' Conti's travel narratives to the *De varietate fortunae*, and for his use of classical models, especially Pliny, see Thomas Christian Schmidt, 'Die Entdeckung des Ostens und der Humanismus: Niccolò de' Conti und Poggio Bracciolinis *Historia de Varietate Fortunae*', *Mitteilungen des Instituts für österreichische Geschichtsforschung* 103 (1995), 392–418.

67 *De varietate fortunae*, p. 174. See further Gentile, 'L'ambiente umanistico fiorentino', pp. 32–3.

68 Bracciolini, *De varietate fortunae*, ed. Outi Merisalo, pp. 21–2; Schmidt, 'Entdeckung des Ostens', pp. 416–18; Angelo Cattaneo, 'Scritture di viaggio e scrittura cartografica. La *mappamundi* di Fra Mauro e i racconti di Marco Polo e Niccolò de' Conti', *Itineraria* 3–4 (2004–5), 157–202: pp. 200–1.

69 See Outi Merisalo, 'Le prime edizioni stampate del *De varietate fortunae* di Poggio Bracciolini', *Arctos* 19 (1985), 81–102.

70 *De varietate fortunae*, p. 153. The fourth book of Poggio's work is a marked diversion from the preceding three, which illustrate his theme of the mutability of fortune through exemplary material drawn from ancient Rome (book I) and fourteenth- and fifteenth-century history up to and including the pontificate of Eugenius (books II and III). The fourth book, which consists solely of geographical and ethnographical material, was written later (post-1442) than the other three (which were composed 1432–5), and was subsequently the book most often transmitted separately: Schmidt, 'Entdeckung des Ostens', pp. 409–13.

71 *De varietate fortunae*, p. 153.

72 *De varietate fortunae*, p. 153.

73 'Quartae Decadis Liber II', in *Scritti inediti e rari di Biondo Flavio*, ed. Bartolomeo Nogara (Rome: Tipografia poliglotta vaticana, 1927), p. 20. The second book of the fourth decade appears to have been composed in 1442, but it did not appear in early editions of the work: *Scritti inediti e rari di Biondo Flavio*, pp. lxxxvi–viii. An early copy of the work was sent by Biondo to the English bishop Thomas Bekynton: *The Official Correspondence of Thomas Bekynton*, ed. George Williams, 2 vols, Rolls Series 56 (London: Longman, 1872), vol. 2, pp. 327–38.

74 'Quartae Decadis Liber II', p. 19. The precise number of Ethiopians who appeared at the Council is uncertain. Biondo refers to eight, but it seems likely that four of this number were in fact part of the delegation of Copts: Tedeschi, 'Etiopi e Copti al Concilio di Firenze', p. 397.

75 'Quartae Decadis Liber II', p. 22, where Biondo comments on the difficulties of communication. See also Tedeschi, 'Etiopi e Copti al Concilio di Firenze', pp. 393–4.

76 'Quartae Decadis Liber II', p. 24.

77 *Scritti inediti e rari di Biondo Flavio*, p. 190.

78 *Scritti inediti e rari di Biondo Flavio*, p. 190.

79 *Scritti inediti e rari di Biondo Flavio*, p. 191–2.

80 'Quartae Decadis Liber II', p. 23

81 'Romanus pontifex', p. 111.

82 de Witte, 'Les bulles pontificales', 48 (1953), p. 718. De Witte provides an edition of Eugenius' bull and the letter from the King of Portugal on the subject of the rights of the Portuguese to colonize the Canary Islands that provoked it (pp. 715–18). The formula describing the extent of papal power occurs also in Martin V's *Romanus Pontifex*, a bull of 1418.

83 See, for example, the map of Pietro Vesconte in Marin Sanudo's *Liber secretum fidelium Crucis*; Giovanni Leardo's three maps of 1442–53; the fifteenth-century copy of Johann of Udine's c. 1344 world map; and the map in a 1447 manuscript of Guglielmo Capello's commentary on the 'Dittamondo': Tony Campbell, 'Portolan Charts from the Late Thirteenth Century to 1500', in *History of Cartography*, vol. 1, pp. 371–463: pp. 406–7; Destombes, *Mappemondes*, pp. 208–12, nos 52.7, 52.8, and 52.9; Scott D. Westrem, 'Against Gog and Magog', in *Text and Territory: Geographical Imagination in the European Middle Ages*, ed. Sylvia Tomasch and Sealy Gilles (Philadelphia: University of Pennsylvania Press, 1998), pp. 54–75: p. 63; Milanesi, 'Il commento al "Dittamondo" di Guglielmo Capello', p. 375.

84 Destombes, *Mappemondes*, pp. 212–14, no. 52.10.

85 Dana Bennett Durand maintained that Walsperger's circular map was derived from a world map produced c. 1425 by members of the scriptorium at the monastery of Klosterneuburg, near Vienna, and that it contained a synthesis of a variety of cartographic forms, including Ptolemaic geography, Arabic cartographic practice, marine charts, and T-O maps: *The Vienna-Klosterneuburg Map Corpus of the Fifteenth Century* (Leiden: Brill, 1952), pp. 209–13. But see Patrick Gautier Dalché's doubts about the nature and originality of the 'Klosterneuburg school': 'Pour une histoire du regard géographique: conception et usage de la carte au XVe siècle', *Micrologus* 4 (1996), 77–103: pp. 93–4.

86 Walsperger placed a gloss underneath the map, which includes an explanation of this scheme: 'rubra puncta christianorum civitates nigra vero infedelium in terra marique existentium'. See Bertrand Hirsch, 'L'espace nubien et éthiopien sur les cartes portulans du XIVe siècle', *Médiévales* 18 (1990), 69–92 for a discussion of fourteenth-century mapping of Christian-Muslim conflict.

87 Susy Marcon, 'Leonardo Bellini and Fra Mauro's World Map: the *Earthly Paradise*', in *Fra Mauro's World Map*, ed. and trans. Piero Falchetta (Turnhout: Brepols, 2006), pp. 135–61, p. 143. Falchetta argues that the map 'was certainly drawn up before 1450 and probably before 1448': *Fra Mauro's World Map*, p. 58. Previous scholarship tended to date the completion of the map to 1459, the year of Fra Mauro's death: Destombes, *Mappemondes*, pp. 223–6, no. 52.14; Woodward, 'Medieval *Mappaemundi*', p. 315. Monastery records indicate that Fra Mauro and two co-workers had begun a map by 1448, and that between 1457–9 another map, the (now lost) copy for the Portuguese royal court, was in production; a date of 26 August 1460 on the reverse of the surviving map may commemorate its final framing. Roberto Almagià suggested that the map's failure to note the fall of Constantinople indicates a date before 1453: Almagià, 'Presentazione', in *Il Mappamondo di Fra Mauro*, ed. Tullia Gasparrini Leporace (Venice: Istituto Poligrafico dello Stato, 1956), pp. 6–7; more recently Angelo Cattaneo has proposed a date of 1448–50 for the geographical part of the map, and 1455–60 for its depiction of the terrestrial paradise, now attributed to Leonardo Bellini: 'Scritture di viaggio e scrittura cartografica', p. 159 *n*5.

88 Angelo Cattaneo, 'Fra Mauro Cosmographus incomparabilis and His Mappamundi: Documents, Sources, and Protocols for Mapping', in *La cartografia europea tra primo Rinascimento e fine dell'Illuminismo*, ed. Diogo Ramada Curto *et al.* (Florence: Olschki, 2003), pp. 19–48: p. 30.

89 For valuable discussions of the narrative form of Fra Mauro's inscriptions see Cattaneo, 'Scritture di viaggio e scrittura cartografica', pp. 190–7, and Falchetta, *Fra Mauro's World Map*, pp. 120–2; see also Wojciech Iwańczak, 'Entre l'espace ptolémaïque et l'empire: les cartes de Fra Mauro', *Médiévales* 18 (1990), 53–68: p. 56.

90 *Fra Mauro's World Map*, ed. and trans. Falchetta, p. 711, inscription number 2892.

91 Bagrow called it 'the last characteristic mediaeval map, the summit of Church cartography': Leo Bagrow, *History of Cartography*, trans. D. L. Paisey, rev. R. A. Skelton (London: Watts, 1964), p. 72; Woodward described it as 'stand[ing] at the culmination of the age of medieval cartography', but also 'transitional', because of its use of portolan charts, information from Ptolemy's *Geography* and contemporary exploration: 'Medieval *Mappaemundi*', p. 315; cf. Iwańczak, 'Entre l'espace ptolémaïque', p. 68. Cattaneo, 'Scritture di viaggio e scrittura cartografica', makes an important argument to the contrary, seeing the map in terms of 'synthesis' rather than transition (p. 160

and *passim*). Falchetta makes a similar argument in his discussion of Fra Mauro's use of Ptolemy, although he somewhat unhelpfully reiterates the notion of the *mappamundi* as 'a "closed" work that did not admit of the possibility of future renewal': *Fra Mauro's World Map*, p. 55.

92 *Fra Mauro's World Map*, ed. and trans. Falchetta, pp. 699–700, inscription no. 2834. On this caption and the interjection of vernacular oral and written culture into high culture texts see Cattaneo, 'Fra Mauro Cosmographus incomparabilis and His Mappamundi', p. 44.

93 *Fra Mauro's World Map*, ed. and trans. Falchetta, pp. 201–2, 585, 715. Inscription nos 98, 2243, 2904.

94 *Raccolta di documenti e studi pubblicati dalla R. Commissione Colombiana*, ed. Cesare De Lollis (Rome: Ministero della pubblica istruzione, 1894), 1, vol. 2, p. 291.

95 *Raccolta*, vol. 2, p. 294.

96 *Raccolta*, vol. 2, p. 377.

97 *Raccolta*, vol. 2, p. 398.

98 N. Sumien, *La correspondance du savant florentin Paolo dal Pozzo Toscanelli avec Christophe Colomb* (Paris: Société d'Éditions, 1927), pp. 10–11. Sumien argued on the grounds of chronology and style (pp. 37–8) that 'per subterraneas nauigaciones' was an insertion made by Columbus, who preserved Toscanelli's letter in his volume of Pius II's *Historia rerum ubique gestarum*. Henry Vignaud famously asserted the inauthenticity of Toscanelli's letter in *The Letter and Chart of Toscanelli* (London: Sands, 1902). For a more recent overview of this vexed question, see Ilaria Luzzana Caraci, *Colombo vero e falso* (Genoa: Sagep Editrice, 1989). For a thoughtful reassessment of Toscanelli's interest in geography see Gentile, 'Toscanelli, Traversari, Niccoli e la geografia', esp. pp. 120–8.

99 In addition to eclectic works such as 'La Salade' and 'La Sale' (1451), La Sale wrote successful chivalric romances and treatises: *Le Reconford de Madame du Fresne* (1447/48), *Des anciens tournois et faicts d'armes* (1459), and his most famous work, *Le Petit Jehan de Saintré* (1456).

100 Rodney W. Shirley, *The Mapping of the World: Early Printed World Maps 1472–1700*, 2nd edn (Riverside: Early World Press, 2001), p. 56 (no. 50). Shirley gives the date of the first printed edition as 1522. See also Manuel de Santarem, *Essai sur l'histoire de la cosmographie et de la cartographie*, 3 vols (Paris: Maulde et Renou, 1849–52), vol. 3, pp. 450–9, who describes a different state of the map to those in extant printed versions. The map does not survive in the two extant manuscripts that contain part or all of the work: Chantilly, Bibliothèque Condé, MS 924, and Brussels, Bibliothèque royale, MS 18.210–15. The Paris 1521 and 1527 editions were printed by Michel Le Noir, and Philippe Le Noir respectively, and dedicated to François I. For a discussion of the manuscripts and the printed editions see Antoine de la Sale, *Œuvres complètes*, ed. Fernand Desonay, 3 vols (Paris: Les Belles Lettres, 1935), vol. 1, pp. xxvi–xlii, and la Sale, *Paradis de la Reine Sibylle*, ed. Desonay (Paris: Droz, 1930).

101 The map exists in at least two variant forms. Reproductions can be found in Santarem's *Atlas* (Paris, 1842–53), A. E. Nordenskïold, *Facsimile-Atlas till Kartografiens äldsta historia* (Stockholm, 1889), p. 100, and Shirley, *Mapping of the World*.

102 Three islands dot the sea in the redaction printed by Santarem.

103 Pierre d'Ailly had mentioned the 'regio Pathalis' in his *Imago Mundi*, noting that 'the southern coast of India is extended to the tropic of Capricorn on account of the region of Pathalis and nearby lands which the great arm of the sea surrounds descending from the ocean sea which is between India and lower Spain or Africa' (frons Indiae meridianus pellitur ad Tropicum Capricorni propter regionem Pathalis et terrarum vicinarum quas ambit brachium maris magnum descendens a mari oceano quod est inter Indiam et Hyspaniam inferiorem seu Africam ...): *Ymago Mundi*, vol. 1, pp. 260–2. The passage was copied from Bacon's *Opus Maius*, pp. 308–10. The name 'Regio Pathalis' seems to derive from Pliny, *Naturalis Historia*, 6.23 (book six is cited by Bacon/d'Ailly immediately before this remark). 'Regio Pathalis' extends well beyond the tropic of Capricorn on the map in *La Salade*, of course. Ptolemaic toponyms in the northern hemisphere include 'Ymaus', 'India intra gangem', and 'nuna mons' (for Montes Lune?). The presence of four windheads in the corners of a square outer frame could be seen as another Ptolemaic element.

104 Pliny, *Naturalis Historia*, 6.81–91.

105 Mark Davie, *Half-Serious Rhymes: The Narrative Poetry of Luigi Pulci* (Dublin: Irish Academic Press, 1998), pp. 19–22.

106 Luigi Pulci, *Morgante*, ed. Franca Ageno (Milan-Naples: Ricciardi, 1955), 25.230 (p. 899).

107 Pulci, *Morgante*, 25.245 (p. 904).

108 Pulci, *Morgante*, 28.34 (p. 1068). The fullest treatment of the antipodal matter in the Morgante appears in Dieter Kremers, *Rinaldo und Odysseus: Zur Frage der Diesseitserkenntnis bei Luigi Pulci und Dante Alighieri* (Heidelberg: Winter, 1966).

109 Luigi Pulci, *Morgante: The Epic Adventures of Orlando and His Giant Friend Morgante*, trans. Joseph Tusiani (Bloomington: Indiana University Press, 1998), p. 737.

110 E.g. Moretti, *Gli antipodi*, pp. 117–18, and Rainaud, *Le continent austral*, pp. 134–5.

111 G. Uzielli, *Paolo dal Pozzo Toscanelli, iniziatore della scoperta d'America* (Florence, 1892), pp. 535–43. See also Gentile, 'L'ambiente umanistico', p. 42, for evidence of Medicean interest in Ptolemy.

112 Moretti, *Gli antipodi*, p. 115. For a detailed discussion of the nature of Pulci's theology see Kremers, *Rinaldo und Odysseus*, esp. pp. 41–8; 77–89. For a contrasting view see Constance Jordan, *Pulci's* Morgante*: Poetry and History in Fifteenth-Century Florence* (Washington: Folger Books, 1986), pp. 148–55. For both Kremers and Jordan, Pulci's treatment of the antipodes is a response to (and rejection of) Dante's treatment of Ulysses in *Inferno* 26.94–9. The ongoing nature of debate about the antipodes at the end of the fifteenth century is seen in Zacharia Lilius' curious tract *Contra Antipodes* (1496), which relies only in part on the authority of Augustine and Lactantius to disappoint those who desire to know whether antipodeans exist. Lilius also gives an account of exploration from Noah onwards, including the feats of Alexander and Julius Caesar, and the preaching of the apostles, to show that 'neither in the upper or lower part of our habitable earth, nor in islands of Ocean, or the Mediterranean are antipodeans found' (neque in superiori aut inferiori parte habitabilis nostre neque in insulis tam oceani quam mediterannei maris Antipodes inuentiantur): *Contra Antipodes* (Florence, 1496), sig. f. v. verso.

113 On the virtuous pagan in medieval literature generally see Alistair Minnis, *Chaucer and Pagan Antiquity* (Cambridge: Cambridge University Press, 1982), pp. 31–60. On this issue in the *Morgante* see Kremers, *Rinaldo und Odysseus*, pp. 85–9.

114 A bull beginning *Eximiae devotionis* was also dated 3 May, but seems to have been composed after the first *Inter cetera*, and appears to be in greater harmony with the second (4 May) *Inter cetera*; a subsequent bull entitled *Dudum siquidem*, of 25 September 1493, confirmed and extended the second *Inter cetera*. All four bulls are printed in *America Pontificia*, ed. Josef Metzler, 3 vols (Vatican City: Libreria editrice vaticana, 1991–5), vol. 1, pp. 71–89, and in *Italian Reports on America 1493–1522: Letters, Dispatches, and Papal Bulls*, ed. Geoffrey Symcox and Giovanna Rabitti, trans. Peter D. Diehl, Repertorium Columbianum vol. 10 (Turnhout: Brepols, 2001), pp. 93–100; 104–5.

115 For the Treaty see *European Treaties bearing on the History of the United States and its Dependencies to 1648*, ed. Frances Gardiner Davenport (Washington: Carnegie Institution, 1917), pp. 84–100.

116 Alexander VI, 'Inter cetera', in *America Pontificia*, vol. 1, pp. 79–83.

117 These debates extended at least as far back as the *Decretum* of Gratian, but were given definitive expression by Innocent IV (1243–54) and his student Hostiensis: Muldoon, *Popes, Lawyers, and Infidels*, pp. 3–48. On the practical papal response to discovery and colonization see de Witte, 'Les Bulles Pontificales'.

118 A point emphasized by James Muldoon, 'Papal Responsibility for the Infidel: Another Look at Alexander VI's *Inter Cetera*', *The Catholic Historical Review* 64 (1978), pp. 168–84: '*Inter cetera* … is a statement on the nature of Christian-infidel relations and on the responsibility of the pope to protect the infidels and to convert them to Christianity' (p. 169).

Chapter Seven

Nondum Cognita: *the Antipodes and the New World 1493–1530*

Heinrich Bünting's marvellous map of 1581, entitled 'The entire world in a clover leaf' (Die gantze Welt in ein Kleberblat), shows precisely the way in which the 'New World' could be added to the old as a supplement.[1] The fully formed tripartite world of the medieval ecumene is divided between the elegant leaves of Europe, Asia, and Africa, with Jerusalem the centre of the trefoil (fig. 26). At the bottom left of the map emerges the inchoate mass of … something new. America unsettles the sense of perfection represented by the natural form of the clover, the delicate balance of the three continents of the Old World; it disrupts in particular their convergence onto a shared centre, Jerusalem. The pattern of the old world, redolent with civic heraldry, since the clover leaf represented the arms of Hanover,[2] cannot, now, be seen without the rude emergence of a new and unassimilable space. The New World supplements, adds to and changes, the old.[3] Such supplementation is no less significant because undertaken unwillingly: the primary purpose of the maps in Bünting's *Itinerarium Sacrae Scripturae* was to illustrate biblical narrative and to show the relation (including distances in German miles) of the three Old World continents to Jerusalem. In a note to another, more conventional, world map Bünting explained that he had been compelled to represent America in the interests of cartographic accuracy even though the New World 'was not conceived of in holy Scripture' (in heiliger Schrifft nicht gedacht).[4] Yet perhaps the genius of Bünting's image lies in the way he reminds the viewer of a history of supplementation. To the north of the 'entire world' appear England, an insular blob, and a kind of promontory representing Denmark and Sweden. These are the old ends of the earth, whose function of standing outside the continental world is displaced by the new colossus of externality, America. Bünting appears to be conscious of the fairly common location of Britain on the very edge, even the frame, of many medieval world maps.[5] On medieval *mappaemundi* the islands at the end of Europe offered an alternative, a distraction, from the centripetal bulk of the three principal parts of the earth. On Bünting's map, the meaning of England and Scandinavia is altered by the new landmass, a mass that is not the end of the earth, but, it would seem, the beginning of a new one. At the same time, the new presence changes the dimensions of the world, such that the former ends of the earth are now unequivocally a part of the Old World. Their days on the edge are numbered.

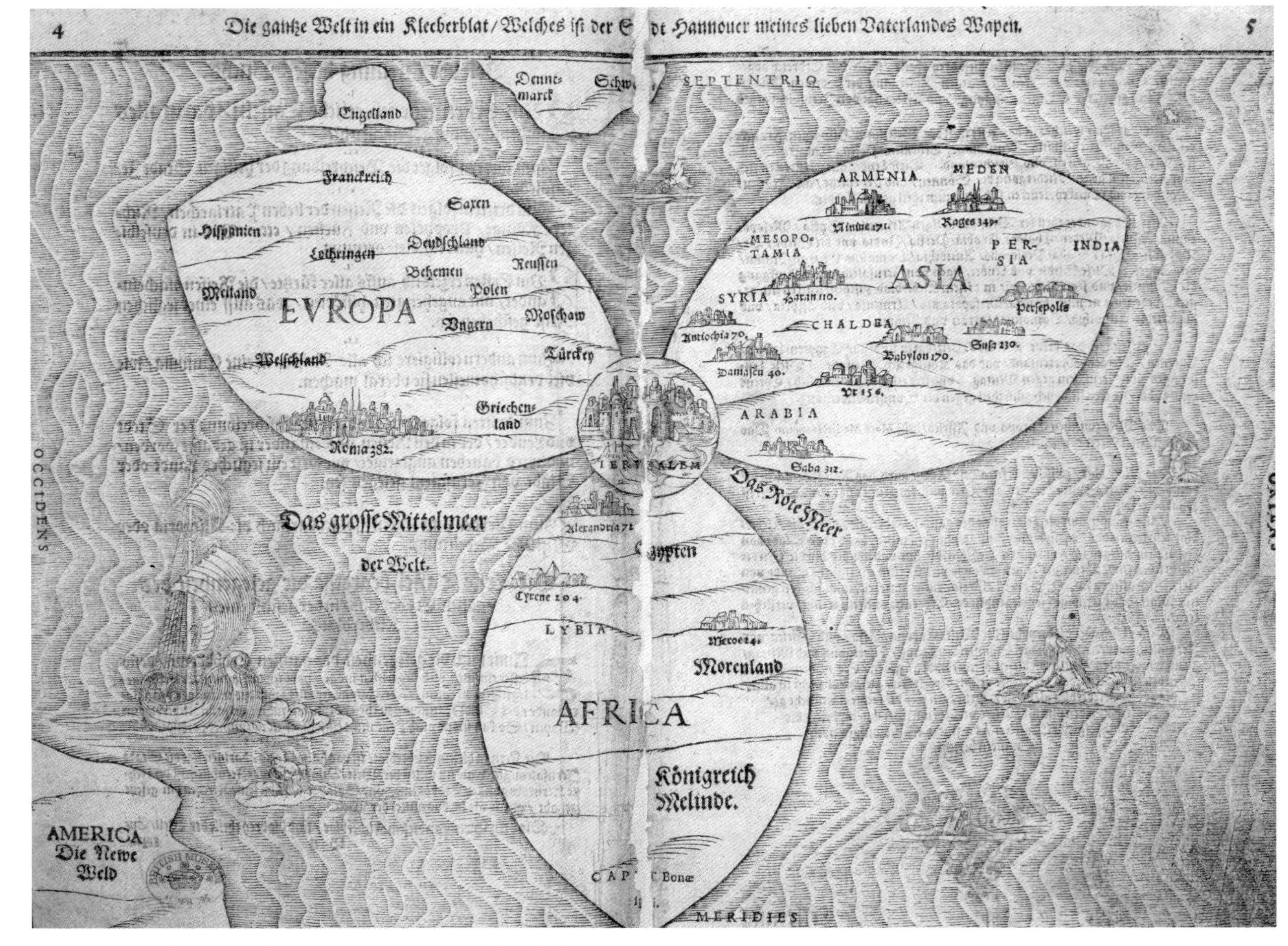

Fig. 26. Heinrich Bünting, 'Die gantze Welt in ein Kleberblat', in *Itinerarium Sacrae Scripturae* (Magdeburg, 1585). London, BL 3105.a.14. Bünting's display of cartographic virtuosity refashioned the map of the world into the form of a clover leaf, centred on Jerusalem. But not all the world could be accommodated within this shape: England and Scandinavia remain outside the leaves, as does the New World.

In the years immediately following European discovery of land in the western ocean much remained unresolved. Was the land discovered insular or part of a continent; was it part of Asia or a new, 'fourth' part of the world; and by whom was it inhabited? Resolution of these and other matters arising (was there a North–West passage from Europe to Asia around the north coast of the Americas?) was slow, but while debates continued the world image changed in order to register different possibilities and to incorporate recent information, reliable or not, from the New World. In the midst of such revision, what role did *terra incognita* play in the emerging space of the New World? On one level, clearly, cartographic representations of unknown lands drew attention to the frontiers of European knowledge. By showing uncharted alongside recently discovered land, maps also presaged the rolling-back of the frontier, and the eventual subsumption of unknown land within European possession. The representation of *terra incognita* on maps thus made manifest a *process* of territorial expansion. At another – but obviously related – level, the acknowledgement of *terra incognita* made manifest the process of map-making. Uncharted space offered, as was the case in medieval cartography, a space for various kinds of writing, encompassing not only commentary but also self-inscription, and anecdote. However, as a result of the conjunction of *terra incognita* with the New World, some sixteenth-century mapmakers resorted to a new technique: they began to fill unknown lands with pseudo-topographical representation, imitating the forms of the known world, fabricating coastlines and interiors, even place-names.

In the first three decades of the sixteenth century a discrete southern continent began to be represented on world maps. Initially anonymous, and then subject to a variety of nomenclature, by around 1530 this continent had been named 'Terra Australis'. The invention of 'Terra Australis', no less than the invention of America, was the product of a cartography of exploration because the world image responded to a significant extent to the reports and maps of explorers, evangelists, merchants, and travellers. But those who produced images of the world did not jettison the classical and medieval inheritance that formed its very foundation. Instead, a variety of practices took place to accommodate discovery on the map. These included confrontation between old and new knowledge, but also compromise and addition, as well as replacement. In this chapter I characterize this phase, at least in its initial stages, as one of supplementation. As in Bünting's clover leaf, supplementation involved not simply the addition of new geographical formations to pre-existing cartographic models, but also the revision and sometimes purification of inherited structures. Accordingly, the classical and medieval concept of the antipodes was not discarded; instead, unknown antipodal space was reinvented in the form of *Terra Australis.*

The southern land mimicked the process of the New World's construction. Even in the second quarter of the sixteenth century both America and *Terra Australis* could be described, as several maps put it, as 'recently discovered but not yet fully explored' (nuper inventa sed nondum plene examinata). However, there was a critical difference between America and the southern land: one was subject to a gradual process of discovery, the other to the elaboration of a fantasy. Map-makers constructed an antipodean world from a variety of sources contemporary, medieval and classical, assigned it a number of toponyms, a tentative coastline, even garbled accounts of its habitation. And as the New World and its dimensions were charted, the tendencies of the representation of the southern continent towards speculation and fabrication seemed to become more extreme. *Terra Australis* was the shadow to New World cartography, or the supplement to the supplement – a kind of colonial overspill that had to be explored by the European imagination before it was explored by its navigators.

'As a supplement to Ptolemy': new maps, New World

The response of many European scholars to the news of discoveries was not only to seek out and to produce maps of newly-discovered land; it was also to revise the classical sources of geography in the light of new information. To a large extent such revision was a continuation of fifteenth-century practices. Fifteenth-century printed editions of Ptolemy's *Geographia* such as the Ulm editions of 1482 and 1486, and the Rome edition of 1490, included maps that were explicitly termed 'modern'. Indeed, as suggested in the previous chapter, from its outset at the beginning of the fifteenth century the reception of the Latin Ptolemy was characterized by attempts to supplement the world and regional images contained in the *Geographia*. What began to change in the sixteenth century was the manner and extent to which revision occurred. Discoveries meant that it was necessary to provide an increasingly large number of 'modern' maps. Partly as a consequence, a more rigid line came to be drawn between ancient and modern geography. The aim ceased to be to update Ptolemy's world map, in the manner of a Fra Mauro, or a Nicolaus Germanus, by adding parts of the world unknown to the Alexandrian. Instead it was to create a 'Tabula Moderna' distinct from the 'Tabula Antiqua' of Ptolemy. The process of distinguishing ancient from modern involved the assimilation of the discoveries of navigators to construct up-to-date images of the world, but it also involved a more thorough examination of the text of the *Geographia*, expressed above all in a mounting dissatisfaction with Jacopo d'Angelo's Latin translation, and a desire to recuperate the original Greek text. What began in the fifteenth century as a process of supplementation ended in the sixteenth as supplantation, with the text and world image of Ptolemy the sign of ancient knowledge, the ancestor, but not the parent, of the modern map.

The principle of cartographic supplementation is well expressed by the first major revision of Ptolemy's *Geographia* to appear in the sixteenth century, the edition printed in Strasbourg in 1513. The defining feature of the 1513 Ptolemy is its series of twenty maps, added explicitly 'in Claudii Ptolemei Supplementum'. The work was arguably the greatest product of the 'Gymnasium vosagense', a collaboration of savants based at Saint-Dié in the Vosges mountains, initially assembled in the early 1500s under the auspices of René, Duke of Lorraine, and titular King of Jerusalem.[6] The precise nature of the genesis of this group is unclear, but it seems to have been founded by Gauthier Lud, chaplain and secretary to René; it included his brother or nephew Nicholas Lud, Jean Pélerin (Viator), miniaturist and author of the treatise *De artificiali perspectiva* (1505), the Latinist Jean Basin de Sandaucourt, and two young German scholars: the cosmographer Martin Waldseemüller (also known by the Hellenized version of his name as Hylacomilus), and the poet Matthias Ringmann (Philesius Vogesigena), a former student of Gregor Reisch at Heidelberg, and Jacques Lefèvre d'Etaples at Paris. The geographical output of this group seems to have been driven by Waldseemüller and Ringmann, and while the exact division of their labours is unclear, in broad terms Waldseemüller seems to have taken responsibility for the production of maps and charts to accompany the texts prepared by Ringmann.[7] The pair aimed to co-ordinate classical geographical knowledge with that provided by recent explorations, and particularly by early reports of the New World. The new edition of Ptolemy that appeared in 1513 was evidently underway as early as 1505; work on it required scrutiny of Greek manuscripts of the *Geographia*, and the production of an updated series of maps. Several texts produced by Ringmann and Waldseemüller before 1513 testify to the aim of reconstituting Ptolemaic geography on the basis of new information about previously unknown land: Ringmann's *De ora antarctica per regem Portugalliae pridem inventa* (1505), a translation of the *Mundus Novus* attributed to the explorer Amerigo Vespucci;[8] Waldseemüller's *Cosmographiae introductio* (1507), which included further accounts of Vespucci's four voyages to the New World; and a globe and world map produced by Waldseemüller in 1507. In the *Cosmographiae introductio*, Waldseemüller explained that he had prepared the globe and world map by using a Greek manuscript of Ptolemy's books 'and *supplementing them from the description of Amerigo Vespucci's four voyages*' (vt me libros Ptholomaei ad exemplar Graecum quorundam ope pro virili recognoscente et quatuor Americi Vespucii nauigationum lustrationes adiiciente: totius orbis typum tam in solido quam plano … parauerim).[9] Waldseemüller's explanation draws attention to two kinds of reception: the revision of the text of Ptolemy through the recovery of apparently more accurate manuscript sources (Ringmann is known to have made at least two trips to Italy in search of such manuscripts), and the addition of geographical information furnished by contemporary accounts of exploration.

The first section of the 1513 *Geographia* contains twenty-seven Ptolemaic maps, printed, the reader is assured, 'so that their antiquity might stand separated and more pure' (ut incorruptior et selecta stet antiquitas sua); these maps are accompanied by the Latin text of the *Geographia*, revised on the basis of Greek manuscript sources.[10] The twenty maps added as a 'supplement' to Ptolemy consist of one world map, one map of the Atlantic, and eighteen regional maps – thirteen of European regions or islands, two of Africa, and three of Asian regions – drawn from a variety of sources, including marine charts, contemporary world maps, and 'modern' maps in fifteenth-century printed Ptolemy editions.[11] The most striking of the supplementary maps are the first two to appear: the world map, produced according to the 'tradition/practice of hydrographers' (iuxta hydrographorum traditionem) (fig. 27); and the 'Tabula terre nove', which displays the Atlantic littoral – the west coasts of Britain, the Iberian peninsula, and Africa, and the islands and east coasts of the emerging continent of the New World (fig. 28). A caption records the discovery of this mainland and its adjacent islands by the Genovese Columbus on the orders of the King of Castile;[12] its interior remains 'terra incognita'.

The stated purpose of the maps drawn as a supplement to Ptolemy was to provide a 'more modern illumination of land and sea', their greater accuracy based on the explorations of 'our age' (modernior lustratio terrae marisque singula positionibus certissimis regulatius tradens ad saeculi nostri peragrationes).[13] At the same time, however, the function of the supplement was to purify Ptolemy, to strip away a process of corruption, to reveal his work 'in its antiquity'. Even

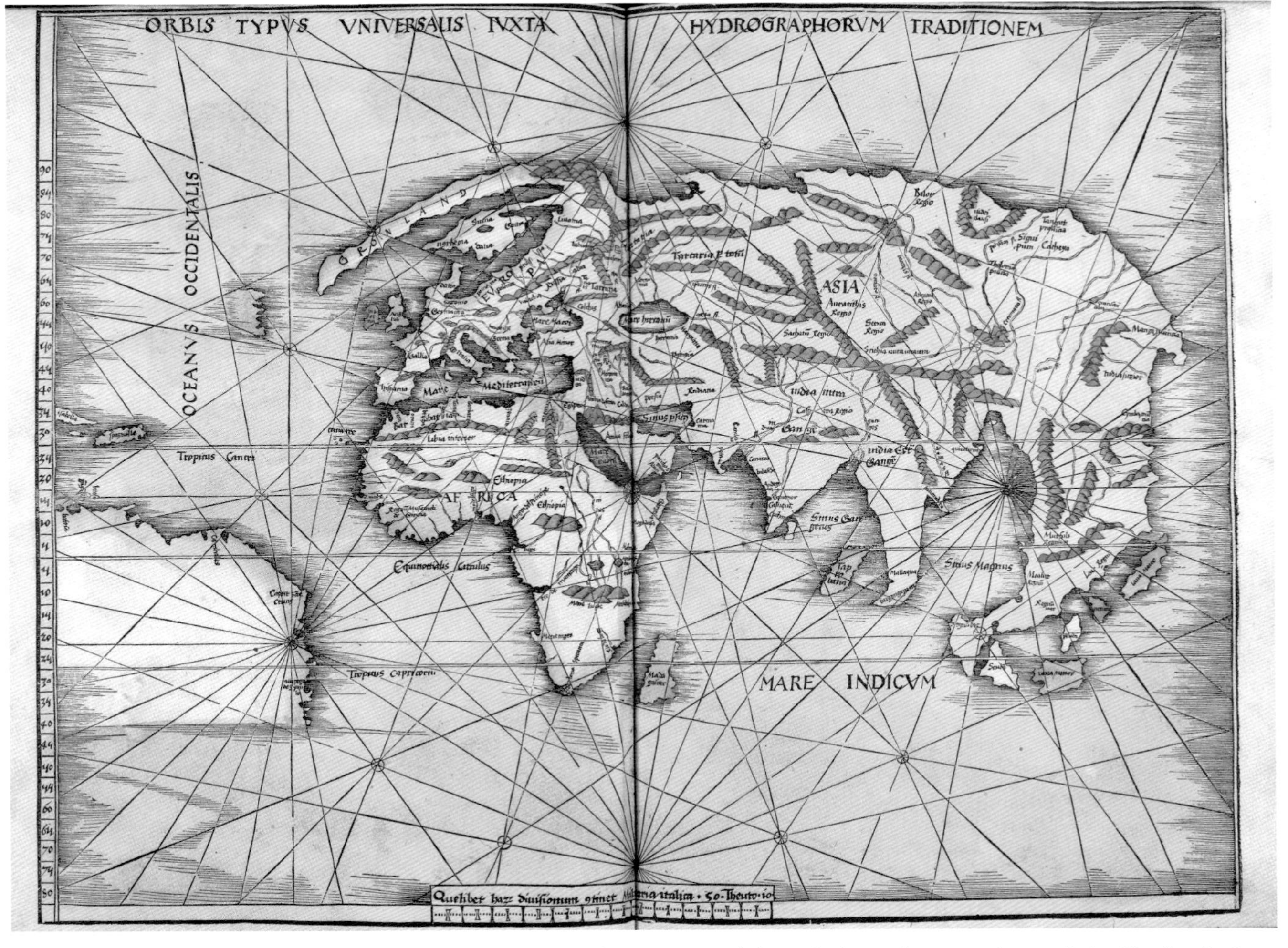

Fig. 27. Supplementary world map 'according to the tradition of hydrographers' (iuxta hydrographorum traditionem), in Claudius Ptolemy, *Geographię opus nouissima traductione e Gręcorum archetypis castigatissime pressum* (Strasbourg, 1513). London, BL, Maps.C.1.d.9. One of the 'modern maps' added by Martin Waldseemüller to supplement the *Geographia*, this image includes parts of the world unknown to Ptolemy.

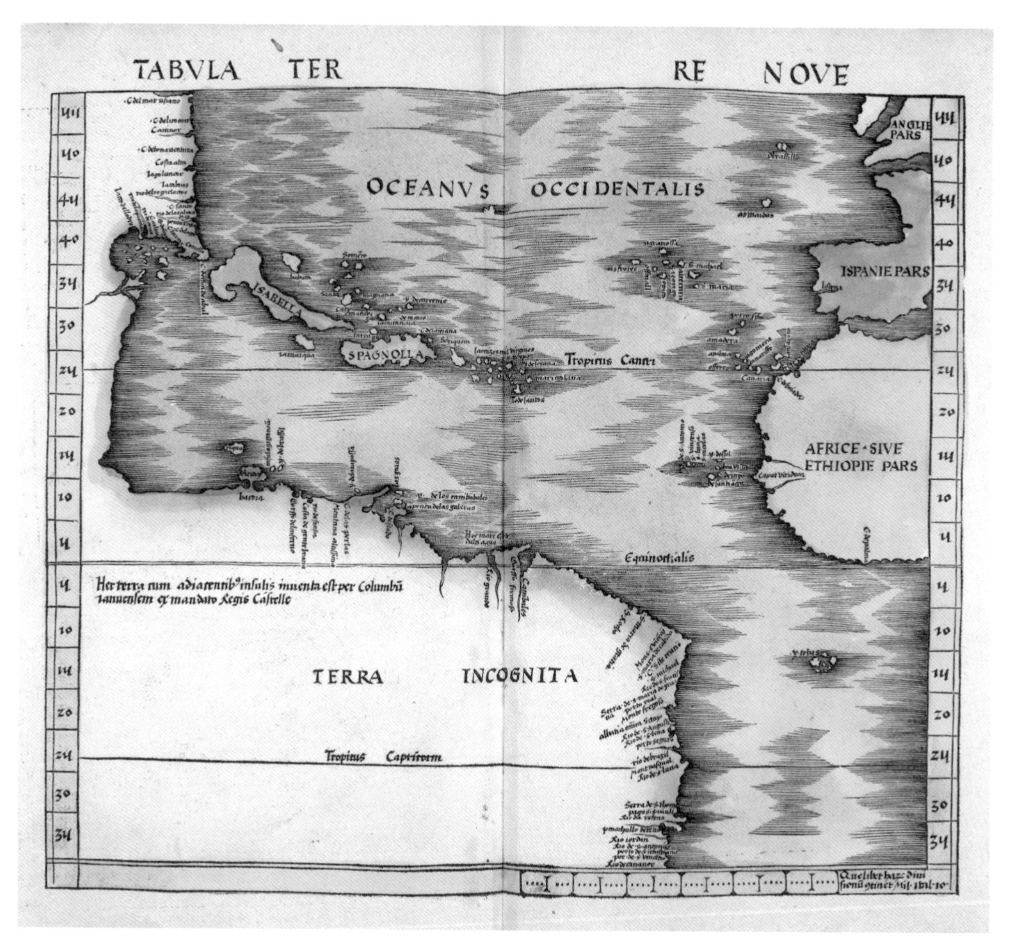

Fig. 28. 'Tabula terre nove', in Claudius Ptolemy, *Geographię opus nouissima traductione e Gręcorum archetypis castigatissime pressum* (Strasbourg, 1513). London, BL, Maps.C.1.d.9. Another of the 'modern maps' used to supplement the *Geographia*, the 'Tabula terre nove' displays the Atlantic littoral: the west coasts of Britain, the Iberian peninsula and Africa, and the islands and east coasts of the New World. A caption records the discovery of this mainland and adjacent islands by Columbus on the orders of the King of Castile; its interior remains 'terra incognita'.

as it offers the reader a text improved by consultation of Greek archetypes, the 1513 edition emphasizes the deviance of the author from 'more modern' geographers. On the verso of the title page 'In Supplementum', the reader is told that toponyms, especially, have changed since Ptolemy's day: so, for example, the regions now called Hungary and Austria were once Pannonia, and cities on the Rhine known to Ptolemy as Canodorum, Augusta Rauricum, Elcebus and Berthomagus will now be familiar under very different names, which the reader can see on the supplement. A revealing comparison can be made between the 1513 Ptolemy and the edition of the *Geographia* produced by the Italian scholar Bernardus Sylvanus in 1511, in which Ptolemy's place names were retained, but the coastlines of his maps redrawn on the basis of marine charts.[14] In his dedicatory letter to the Duke of Atri, Sylvanus expressed his astonishment at the difference between Ptolemy's maps and 'the navigations of our time'. He noted, in addition, that the numbers denoting distances between places in the standard text of Ptolemy were inconsistent with the words; to remedy this discrepancy, he 'corrected the numbers' by using the data of contemporary sea-voyages (rather than consulting Greek manuscript sources, as the editors of the 1513 Ptolemy did).[15] For Sylvanus the *Geographia* was an organic text, subject to a process of refinement and adjustment. By contrast, the supplement of the 1513 Ptolemy denies the notion of the *Geographia* as a fluid resource that lends itself to constant updating; the act of supplementation is an attempt to return to the integral Ptolemy, a Ptolemy purged of accretions and mutations, a work arrested in its antiquity. The role of *terra incognita* is crucial to this process, since unknown land is now no longer contiguous with the Old World, but located, in the 'Tabula terre nove', across the physical boundary of the Atlantic ocean, a passage which is in some sense also temporal, the boundary between ancient and modern. The addition of supplementary lands is in this way an updating that drags Ptolemy back to his antique setting, and memorializes him there.

We, They

To understand more fully the process of the reception of New World explorations, and the changing role of *terra incognita*, it is necessary to look more closely at the works that did the supplementing, in the case of the 'Gymnasium vosagense' the narratives of Amerigo Vespucci's voyages. In a curious sense Vespucci was Christopher Columbus' supplement: a follower, and a supplanter. Like the Genovese Columbus, Vespucci was an educated Italian who found employment in the service of Castile and Portugal; the two were, apparently, on good terms, since Columbus recommended Vespucci with some warmth to his son, Diego.[16] Unlike Columbus, Vespucci did not agitate for, nor lead, voyages of exploration in the Atlantic. Nor did he found or assume command of settlements in the New World on behalf of European powers. Rather, he appears to have taken part in at least, but perhaps no more than, two voyages: one in the service of Castile, 1499–1500; another, in the service of Portugal, 1501–2; a third may have been undertaken in 1503–4.[17] Vespucci, again like Columbus, recorded the voyages and the places and peoples encountered on them, reports that were quickly translated, printed, and disseminated throughout Europe. Indeed Vespucci's texts are in parts so 'like' Columbus' that critics have suspected him of plagiarism, or, less dramatically, argued that certain of the texts attributed to Vespucci may contain a degree of conflation of sources, or may be forgeries.[18] For one thing, the number of voyages supposedly made by Vespucci seems to mimic Columbus' own four voyages. Secondly, certain passages bear marked similarities to those of Columbus' letters.[19] All these similarities raise the vexed question of the authorship of Vespucci's writing.

Six Vespuccian texts survive: three 'familiar' letters to Lorenzo di Pierfrancesco de' Medici as well as a fragment of Vespucci's reply to a lost letter from a Florentine correspondent, all of which circulated only in manuscript form; there are in addition two 'public' texts (that is to say, printed within Vespucci's lifetime): the 'Mundus Novus', and the 'Quattro Viaggi' (an account of his four voyages in the form of a letter to the Florentine Piero Soderini). As Luciano Formisano usefully suggests, the most productive approach to the authorship of these texts may be to regard them as 'para-Vespuccian', an oeuvre that may be partially, but not entirely written by Vespucci, and whose organization may owe much to other hands.[20] What cannot be denied is the popularity of the 'public' Vespuccian texts: the *Mundus Novus*, a Latin translation of the lost Italian original, first appeared in print in Florence in 1502–3, was reprinted rapidly, and was swiftly translated into German (1505) and Flemish (1506); both it and the letter to Soderini were within a few years anthologized in influential collections of travel writings such as the *Paesi novamente ritrovati* of Fracanzio da Montalbaddo, and Giovanni Battista Ramusio's *Navigationi et viaggi*. The evidence of the number and frequency of editions of Vespuccian texts, when compared to the dissemination of Columbus' writings, indicates that, however briefly, during the first part of the sixteenth century Vespucci's fame displaced that of Columbus. Perhaps this success can be ascribed to Columbus' reluctance to embrace the concept of a New World, since for the most part he continued to insist that the lands he had discovered represented the easternmost extension of Asia, rather than a new continent.[21] Vespucci, on the other hand, not only embraced the concept of a New World, as the title of his work proclaims, but also provided his audience with a description of stretches of land not traversed by Columbus, particularly the coast of modern-day South America. By outlining the fact of the new continent, and the possibility of its extent far into the southern hemisphere, Vespucci expanded the accounts of Columbus, supplemented them, and at least in the minds of some of his audience, overrode them in terms of their value in constructing descriptions, verbal and visual, of 'terre nove'.[22] The ultimate success of this supplementation/supplantation was the decision of the 'Gymansium vosagense' to name the new continent after Vespucci (Amerigo: America): this was not, of course, Vespucci's own idea, but rather an act in the reception of his narratives of exploration by the scholars of Saint-Dié, in accordance with their elevation of Vespucci to the extent that, on Waldseemüller's map of 1507, Amerigo's portrait appears alongside Ptolemy, presiding over the new world now named after him.[23]

According to the 'Quattro Viaggi', Vespucci made four voyages to the 'New World' between 1497 and 1504, the first two under Spanish auspices, the latter two for the Portuguese. Of these the most spectacular voyage was the third, which departed Lisbon in May, 1501, and returned to that city in July, 1502; this is the voyage described in the *Mundus Novus*. After landing on terra firma in the region between Venezuela and Brazil on 7 August 1501, the expedition headed south along the coast, crossed the 'torrid zone' and reached a distance of some 50 degrees south, before deciding to return to Portugal. Two notable features of the description of this voyage in the *Mundus Novus* and the 'Quattro Viaggi' are Vespucci's interest in and record of celestial configurations of the southern hemisphere, and the considerable portion he devotes to ethnographic accounts of the peoples encountered in the New World. These discoveries made possible the claim in the *Mundus Novus* that Vespucci had 'discovered a continent in those southern regions that is inhabited by more numerous peoples and animals than in our Europe, or Asia or Africa' (cum in partibus illis meridianis continentem invenerim frequentioribus populis et animalibus habitatam quam nostram europam seu asiam vel africam).[24] This *inventio* expressly contradicts the opinion of 'our ancient authorities, since most of them assert that there is no continent south of the equator, but merely a great sea which they called the Atlantic; furthermore, if any of them did affirm that a continent was there, they ... denied that it was habitable land' (hec opinionem nostrorum antiquorum excedit, cum illorum maior pars dicat ultra lineam equinoctialem et versus meridiem non esse continentem, sed mare tantum, quod atlanticum vocavere. Et, si qui eorum continentem ibi esse affirmaverunt, eam esse terram habitabilem multis rationibus negaverunt).[25]

It is possible to find numerous such celebrations of the trumping of ancient knowledge by explorers in the years following Columbus' landfall. As early as October, 1493, Peter Martyr hailed Columbus' discovery of the 'western antipodes', and praised Spain for revealing so many 'hidden thousands of antipodeans' (latentes ... tot antipodum myriades).[26] The particular interest of Vespucci's narrative, however, lay in his claim to have personally encountered the antipodes, and some of the new-found antipodeans. In the third of the familiar letters to Lorenzo di Pierfrancesco de' Medici, Vespucci explicitly asserts that he reached the antipodes ('in conclusion, I was in the region of the antipodes, on a voyage which covered a quarter of the world');[27] in the first of these letters, describing his star-gazing beyond the equator, he quotes Dante's description of the southern skies in *Purgatorio* 1, and notes that 'what [the poet] says is true: for I noted four stars forming the figure of an almond, and moving little; and should God grant me life and health, I hope to return to that hemisphere soon, not leaving it until I have observed the pole' (quello che dice non salga verità: perché io notai 4 stelle figurate come una mandorla, che tenevano poco movimento; e se Dio mi dà vita e salute, spero presto tornare in quello emisperio, e non tornar sanza notare il polo).[28] Vespucci's frame of reference, comprising classical geographical authority and vernacular poetic tradition, allows him to give a providential character to his voyage to the antipodes. Disproving by his actions the opinion of Augustinian authority, he enjoys a sight which, in Dante's words, was 'not seen before except by the first people' (non viste mai fuor ch'a la prima gente). Vespucci is here aligned with both poet and primal humanity, given access to a privileged view, but his role acquires added significance from the final couplet he quotes: 'o northern hemisphere, because you were/ denied that sight, you are a widower!' (*Purgatorio* 1.26–7: oh settentrïonal vedovo sito,/ poi che privato se' di mirar quelle!). Vespucci's observations in some sense restore his hemisphere's sight, and the completion of this vision will occur with the observation of the south pole. As he crosses the torrid zone, and coasts along a new world towards the Antarctic, Vespucci overturns ancient authority, recalls lyric glimpses of the unseen, and, most dramatically, encounters antipodal inhabitants. Crucially, these encounters take place in the service of European sovereign power.

In the *Mundus Novus* encounter is given graphic form. Vespucci describes the customs and appearance of the people he saw – their dress, physique, fabrics, method of government, warfare, cannibalism, sexual and social practices, extraordinary longevity, and the absence of European religious, legal, and commercial practices. Early sixteenth-century printers played up this element of the text by taking the opportunity to insert striking images of 'natives' on its title pages. Beneath the title 'Epistola Albericii. De nouo mundo' the *Mundus Novus* printed in Rostock and tentatively dated 1505 presents a woodcut showing a naked man, bearded with cascading locks, holding a bow and arrows and regarding a startled looking naked woman, only marginally less hirsute, before a rocky landscape (fig. 29). But it would be wrong to perceive Vespucci's text as purely or fundamentally ethnographic: for in it there is no formal distinction between ethnography and other modes of description. After discussion and sketch of astral configurations ('in that hemisphere I saw things not in agreement with the calculations of the philosophers'),[29] Vespucci attempts to express the relation of the Old World to the new geometrically (fig. 30):

> Perpendicularis linea, que, dum recti stamus, a puncto celi imminente vertici nostro dependet in caput nostrum, illis dependet in latus et in costas. Quo fit, ut nos simus in linea recta, ipsi vero in linea transversa et species fiat trianguli orthogoni, cuius vicem linee tenemus cathete, ipsi autem basis; et hypotenusa a nostro ad illorum protenditur verticem, ut in figura patet.[30]

> A perpendicular line, which while we stand erect descends from the point of heaven shining over our heads, to them descends in the side and onto the back. By this reasoning if we are on a straight line, they are on the transverse. And if a kind of right-angled triangle is made, of which we occupy the perpendicular line, they occupy the base, and the hypotenuse stretches from our to their vertex: as it appears in the figure.

Vespucci's figure reveals the interconnection between spatial representation and the question of habitation. The triangle acts as a map, implicitly of land, but explicitly of human relationship; land is not named, but it is conceived of and represented in terms of human experience. 'We' (Nos) and 'They' (Illi) in some sense act as toponyms, since the question of how 'we' and 'they' stand in relation to each other cannot be divorced from the question of *where* both parties stand on the spherical earth – indeed this diagram could be transferred onto a (geographical) map. However, what is given primacy is the physical position of 'we' and 'they' (our head, their sides). Vespucci's diagram is the product of dramatic access to 'their' perspective, and the effect of it is to demand, however briefly, an alteration of 'our' perspective, to require reading in a different direction. The captions of the diagram head off in several directions: across the page, down the page, diagonally, bottom to top, upside down and back-to-front. The reader is encouraged literally to view from different angles. Nevertheless, Vespucci (and the reader) remain rooted in the world of 'nos', however much he desires a return to 'their' world: shift in perspective does not entail a shift in identity, much less a willingness to expand the notion of 'we' to include 'they'. Vespucci writes to his patron as a Florentine in the service of Portugal, and it is in this capacity that he experiences 'their' world. The right angle, after all, is a means of dividing 'nos' from 'illi' as well as bringing them into contact: it is the product of difference, as well as a moment of intersection.

The 'Gymnasium vosagense' coined the name 'America' in the *Cosmographiae introductio* of 1507, but Vespucci's discovery of new land had already been hailed by one of its members two years earlier, in Matthias Ringmann's edition of the *Mundus Novus*, printed in Strasbourg. Ringmann re-titled the *Mundus Novus* 'De ora antarctica per regem Portugallie pridem inuenta' (On the Antarctic coast discovered lately by the king of Portugal), and prefaced it with a poem in hexameter and a dedication to his friend, the law student Jacques Brun. Two woodcut scenes appear on the title page, representing the inhabitants of the newly discovered land, as well as boats bearing natives and European explorers. The dedication to Brun invokes Virgil's *Aeneid*

Fig. 29. Amerigo Vespucci, *Mundus Novus* (Rostock, 1505), title page. London, BL C.20.e.18.

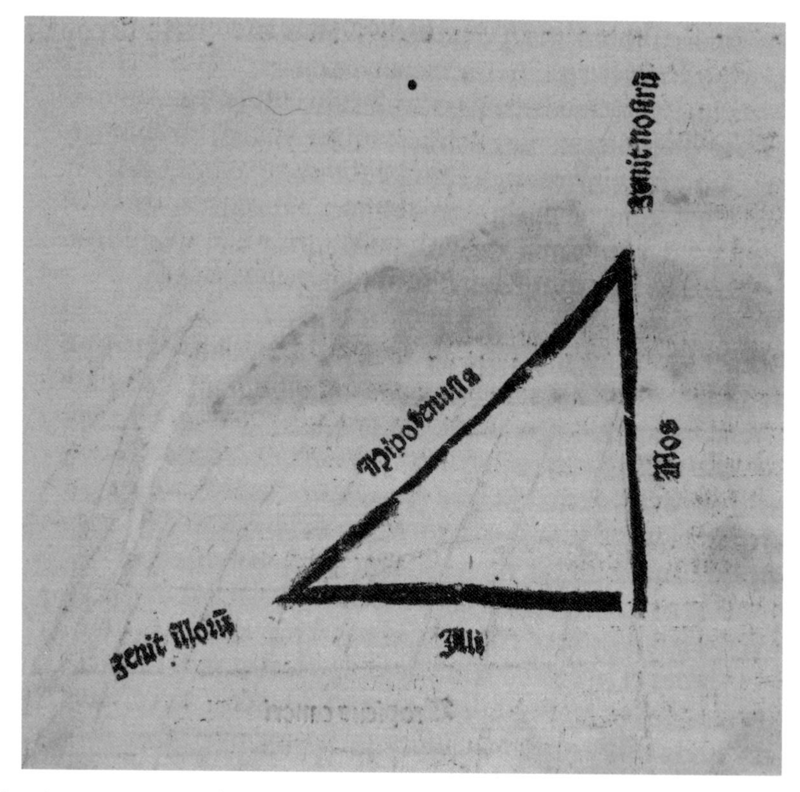

Fig. 30. Amerigo Vespucci, *Mundus Novus* (Rostock, 1505). London, BL, C.20.e.18. Vespucci's geometrical diagram shows the relationship between inhabitants of the Old World ('nos') and those of the New ('illi') as a right-angled triangle. According to Vespucci, 'we occupy the perpendicular line, they occupy the base, and the hypotenuse ('hipotenussa') stretches from our to their vertex' ('zenit nostrum', 'zenit illorum').

6.794–7 ('there is a land that lies beyond the stars, beyond the paths of the year and the sun, where Atlas the heaven bearer turns on his shoulder the firmament studded with blazing stars'), before explaining that Vespucci has related news of a 'people of the south dwelling almost beneath the Antarctic pole' (de populo Austrum versus sub Antarctico quasi polo degente).[31] The poem, meanwhile, celebrates the discovery of land not noted by Ptolemy, and especially mentions that:

> Ethiopes extra terra est Bassamque marinam
> Non nota e tabulis o ptolomee tuis
> Cornigeri Zenith cui fertur tropicus hirci
> Huic multe comes est eiaculator aque
> At procul Antarcto tellus sub cardine quedam est
> Tellus quam recolit nuda caterua virum
> Hanc quem clara tenet nunc Portugallia regem
> Inuenit missa per vada classe maris
> Et quid plura situm gentis moresque reperte
> Ille hic perparua mole libellus habet[32]

Beyond Ethiopia and Bassa in the sea there is a land not recorded on your maps, O Ptolemy, whose zenith is said to be the tropic of Capricorn, companion of Aquarius. And far away beneath the Antarctic pole, a naked band of men cultivates this land; he discovered it whom illustrious Portugal now holds as king, by a fleet sent through the shallows of the sea. And what more? This little book of trifling weight contains the location and customs of the discovered people.

Ringmann registered the excitement of new land, but also the interest in ethnographical details of its native population. The smooth transition from 'recolit' to 'invenit' effects an act of possession: the land 'cultivated' by the naked band has been 'discovered' – and claimed – by the king. And in a curious way the natives are implicitly and fleetingly aligned with classical geographic authority, both subject to the discoveries of the Portuguese fleet. The glancing reference to Ptolemy captures the dynamic in Vespucci's text of new knowledge set against old: knowledge now comes from beyond the southernmost extent of Ptolemaic geography.

The representation of the land discovered by Vespucci as an 'Antarctic coastline' seems to derive from passages that describe the discovery of a new, southern, land in the account of Vespucci's third voyage (his first for the king of Portugal), contained in both the *Mundus Novus* and the 'Quattro Viaggi'. In the *Mundus Novus*, Vespucci's expedition is said to have navigated along the coast of the new continent (i.e. Brazil), until, having passed the Tropic of Capricorn, the southern celestial pole was found to have risen over fifty degrees above the horizon. At this moment, Vespucci states, 'we were near the circle of the Antarctic itself to seventeen and a half degrees, and I will recount what I saw and knew there concerning the nature of those peoples, and of their customs and the tractability and fertility of the land …' (fuimusque prope ipsius antarctici circulum ad gradus xvii semis. Et quid ibi viderim et cognoverim de natura illarum gentium deque earum moribus et tractabilitate, de fertilitate terre … narrabo).[33] A fuller account of this discovery occurs in the 'Quattro Viaggi', in which Vespucci records that, on his third voyage, the expedition sailed south along the coast of what was to become South America for 750 leagues, noting a large number of brazilwood and cassia trees. The voyage being ten months old, it was decided to head south-east in search of another region; after progressing 500 leagues to the south, the ships struck stormy weather, in the midst of which, on 7 April 1502, new land was sighted.[34] The ships coasted this new land for around twenty leagues, finding no harbour or people, an absence Vespucci attributed to the intense cold (it was, he explains, winter in that land, with the night of 7 April lasting fifteen hours).[35] At this stage the mariners decided not to explore further, and turned back to Portugal, heading north-northeast until they reached Sierra Leone.

The accounts in the 'Quattro Viaggi' and the *Mundus Novus* evidently differ on the question of habitation of the 'Antarctic' land. The rather garbled account of the *Mundus Novus* suggests not only that the land 'near the Antarctic circle' was inhabited, but that Vespucci's ethnographic description applied to its inhabitants, rather than to the south American peoples he encountered, since the statement that he 'will recount what I saw and knew there concerning … those peoples' is followed directly by the long ethnographic description. Ringmann's reading of this truncated Vespuccian text led him, therefore, to assume not only the existence of an 'Antarctic' region, but also its habitation by a 'band of naked men'. The misapprehension seems to have been clarified by 1507: the *Cosmographiae introductio* of that year contained the longer account of the 'Quattro Viaggi', in which no inhabitants of this Antarctic land are sighted, and no landing is made; in the same year, Waldseemüller's world map showed the free-standing continent of 'America' but no Antarctic continent. The precise level of *De ora antarctica*'s influence on cartographic representation is hard to estimate. Nevertheless, Ringmann at least contributed to what appears to have been a burgeoning interest in the representation of 'Antarctic' lands. In particular, the 'Antarctic coast' appears to be the ancestor of later sixteenth-century representations of a landmass to the south of America, Africa, and Asia, at least a part of which was often described as having been discovered by the Portuguese in 1499–1500.

Terra ultra incognita: unknown land and the New World

Amerigo Vespucci's diagram in the *Mundus Novus* deserves to be recognized as one of the earliest European attempts to configure the New World alongside the old. The maps that followed Vespucci's in the first three decades of the sixteenth century, and that sought to present a more detailed image than his sketch, confronted the problem of representing the emerging space of the New World at a time of uncertainty about the precise nature of European discoveries. Land, until recently unknown, had been discovered; beyond it, undeniably, lay more *terrae incognitae.* In what manner should these further unknown lands be represented? What degree of licence was to be allowed a mapmaker at a time when the world image was in flux, and when reliable accounts of exploration were difficult to discern from unreliable ones? Three aspects characterize the changing role of unknown land on early sixteenth–century world maps. First, the function of maps as a means of registering political demarcation demanded that emergent spaces be marked in terms of European power, usually the division between Spanish and Portuguese dominion. Unsurprisingly, as well as lines of demarcation, narratives of discovery and colonization appear on maps to explain and recount recent European contact with the New World. Such political statements occurred at the very edge of knowledge, in a space of extreme fluidity, where coastlines and toponyms were highly unstable; as a consequence, considerable artistic flexibility was required to negotiate the juxtaposition of sovereignty with uncertainty and ignorance. Second, in a departure from standard practice prior to the fifteenth century, early sixteenth-century maps increasingly recorded the identity of the mapmaker, sometimes in the space of unknown land. The trend was, to some extent, the result of the increasingly commercial nature of mapmaking in the age of the printing press, in which assertion of authorship might serve as a form of advertisement. At the same time, the location of the mapmaker's name in *terra incognita* was a continuation of the meta-cartographic tradition associated with the space of the antipodes in the Middle Ages, in which unknown land functioned as a site for statements about the map as a whole. The third distinctive aspect of the representation of *terra incognita* in the early sixteenth century was, as I have already suggested, the rapid evolution of a southern land. This southern land was not contiguous with the known world in the manner of Ptolemaic *terra incognita*; nevertheless, it was defined, in part, by the confusion, or slippage, of elements from the New World. Thus place names, coastlines, and eventually ethnographic sketches found their way from the emerging space of the Americas to the entirely unknown – and entirely fictional – southern land. *Terra Australis Incognita*, spatially inchoate, was also temporally uncertain: it was land 'not yet fully known', a space to be uncovered, residing in future time, yet present on the map, and susceptible to creative intrusions.

The combination of clear political demarcation, narrative of discovery, and self-inscription in unknown land is evident in the world map produced by the Spanish pilot and navigator Juan de la Cosa in 1500 (Plate 5). De la Cosa's map, probably drawn after the return of the voyage of Alonzo de Hojeda in 1499–1500 with a crew that included both the cartographer and Vespucci, is explicitly concerned with the ownership of lands. De la Cosa shows Spanish ships and flags off northern Brazil, and notes alongside the coast 'this cape was discovered in 1499 for Castile; the discoverer is Vicente Yáñez' (este cauo se descubrio en año de mil y CCCC XCJX por castilla syendo descubridor vicentianS).[36] Discovery is dated and sovereignty assigned in a manner that accords with de la Cosa's 'liña meridional', the line of demarcation between Spanish and Portuguese territory. These signs of exploration and territorial demarcation co-exist with an evident uncertainty about the extent of the lands discovered. De la Cosa does not commit himself on the issue of whether these lands formed part of a continent or were insular. It seems likely, moreover, that on part of the coast of what was to become North America guesswork and speculation have taken the lead over observation. The area unmarked but coloured green on the map signifies space where even imagination cannot go: it is clearly *terra incognita*. However, in a continuation of the iconographic tradition of marine charts, in which paratextual elements were often located at the neck of the chart, two diverse and striking elements are located within the inchoate green area: a portrait of St Christopher bearing the Christ Child, and the inscription of de la Cosa's authorship. The St Christopher figure appears to represent the passage of Christianity to these pagan lands, as well as being an oblique reference to Columbus.[37] The inscription – 'Juan de la Cosa la fizo en el Puerto de S: ma [Santa Maria] en año de 1500' –[38] includes a different kind of topographical signification, namely the identification of the place of the map's production. The space of *terra incognita* is used here, then, for two kinds of statement. One is ideological, expressing the motive and justification for the acts of territorial demarcation represented – the conversion of newly encountered peoples to Christianity. The other is both personal and meta-cartographic, recording the circumstances of the map's making, and in so doing, drawing attention to the Old World basis of this New World cartography.

Subsequent representations of the incomplete state of knowledge of the New World, such as the Venetian Giovanni Matteo Contarini's world map of 1506, or the (closely related) Ruysch planisphere of 1507, make clever use of scrolls to cover those coastlines and interiors that remain unknown (fig. 31).[39] Such scrolls perform a number of functions. In one, Contarini addresses his readers in verse, boasting of his skill 'in the Ptolemaean art', while encouraging them to look on a hemisphere unknown to the Alexandrian: 'behold new nations and a new-found world' (nouas specta gentes orbemque recentem).[40] On the Ruysch map the scrolls and inscriptions on the 'Terra Sancte Crucis' contain a good deal of Vespuccian information, including a record of his voyage to 50 degrees south, and mention of the cannibalistic habits of the New World natives; they also, like the de la Cosa map, record European discovery and ownership of the land.

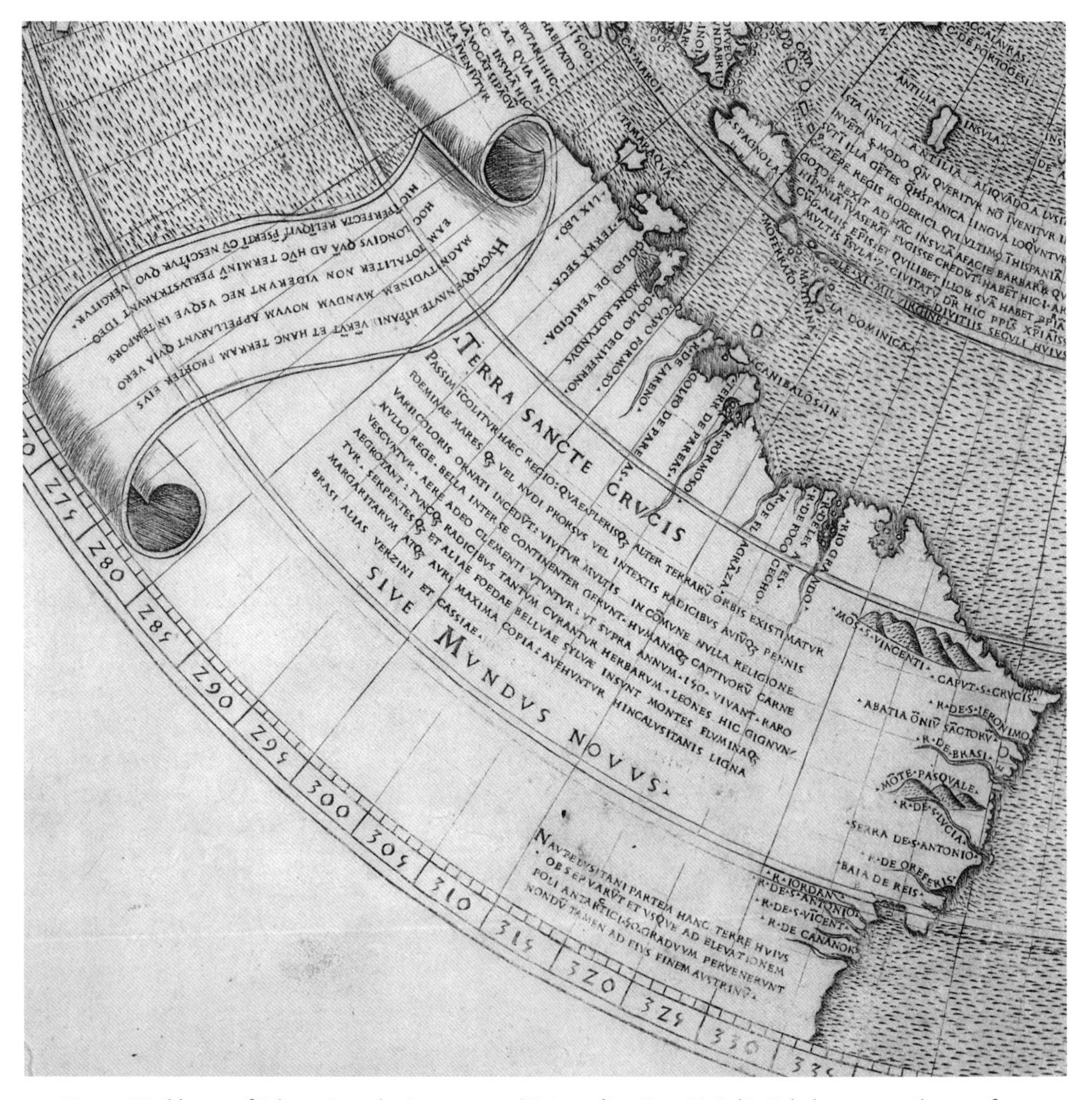

Fig. 31. World map of Johann Ruysch, Rome, 1507: 'Vniversalior Cogniti Orbis Tabvla ex recentibus confecta obervationibus'. London, BL, Maps C.1.d.6. As its title suggests, Ruysch included information derived from recent Portuguese explorations in Asia on this map, but he also made artful use of scrolls to cover areas of unknown or uncertain land in the New World. In a large scroll on the west coast of the 'Terra Sancte Crucis sive Mvndvs Novvs' Ruysch notes that the Spanish termed this land a 'New World' because of its size, and that its dimensions remain unknown.

Indeed, what gave Martin Waldseemüller's world map of 1507 its distinctive status was that it dispensed with these decorous scrolls. America was proclaimed and named on the map; it was given the form of a continent, and thus stood unambiguously apart from the known world. Waldseemüller did not hesitate to acknowledge unknown land: the area beyond the east coast of America was marked 'terra ultra incognita', and 'terra ulterius incognita' – land beyond the known. The insertion of the word 'ultra' or 'ulterius' into the familiar formulation 'terra incognita' identifies a particular moment in the process of exploration. The 'ultra' refers back to the adjacent land, which, until recently unknown, is now named and subject to discovery and dominion. However, 'ultra incognita' does not simply acknowledge the existence of more land beyond the cartographer's knowledge; it also promises a future when this further unknown land will be explored and charted.

On the 1508 world map of the Florentine mapmaker Francesco Rosselli there is a tangential relationship between the newly discovered land of the Americas and a landmass that appears in the Antarctic region (Plate 6). Rosselli's innovatory map – it is the first known use of an oval projection, and shows 360 degrees of longitude and 180 degrees of latitude[41] – unequivocally proclaimed the New World: the 'Terra S. Crucis sive Mundus Novus' (Land of the Holy Cross, or New World) occupies the western Atlantic beneath an extended Asian continent. In addition, in accordance with Columbus' conviction that he had reached the east coast of Asia, Rosselli dutifully located some of the place names recorded on Columbus' fourth voyage (e.g. 'Beragua', 'belporto', 'p. de bastimento') in the far east rather than on the 'Terra S. Crucis'.[42] The southern landmass, possibly inserted in response to the account of Vespucci's third voyage in the 'Quattro Viaggi', renews the promise of *terra incognita* by figuring the existence of yet more new worlds. The toponym 'Antarticus', referring to the Antarctic circle, and not to the landmass itself, nevertheless suggests the hypothesis (later elaborated by Gerard Mercator) of the necessity or likelihood of an Antarctic continent. Yet the southern land on Rosselli's map contains another, quite different element: his signature, located at the south pole. The specificity of its own toponymic reference – 'F. Rosello *Florentino* Fecit' – draws the reader inevitably back to the Old World. At the very point at which geographical reference points vanish, in unknown and unnamed territory beneath the Antarctic circle, the map proclaims its origins in the form of the name of the mapmaker and his city.

The southern land similarly functions as a bridge between 'old' and 'new' knowledge on the remarkable fragment that survives from the 1513 chart of the Turkish admiral Piri Reis (fig. 32). The fragment of Piri's map that survives shows the Iberian peninsula, part of the west coast of Africa, the Atlantic, Caribbean islands, and the eastern littoral of Central and South America. However, a unique feature of the map is the apparent contiguity of this portion of the New World with an austral continent: as it descends to the south, the coastline turns to the east and seems to continue, giving the impression of a landmass that extends well beyond the southern edge of the map. Precisely how the coastline of this continent continued is impossible to say in the absence of the rest of the map, although a legend in the Atlantic stating that 'coasts encircle this sea, which has taken the form of a lake'[43] suggests this austral landmass may have continued at least as far as Africa. The formation certainly bears comparison with the 1519 world map of the Portuguese cosmographer Lopo Homem, a circular image in which the coastline of a southern land, marked with numerous bays, inlets and rivers, and designated 'Mundus novus', extends from the south of Brazil beneath Africa and connects with south Asia.[44] Clearly the idea of a southern-hemisphere connection between the New World and the southern extremities of the Old had currency in the first two decades of the sixteenth century, and could be represented in a variety of ways. Piri inscribed this section of his New World continent with the words: 'this

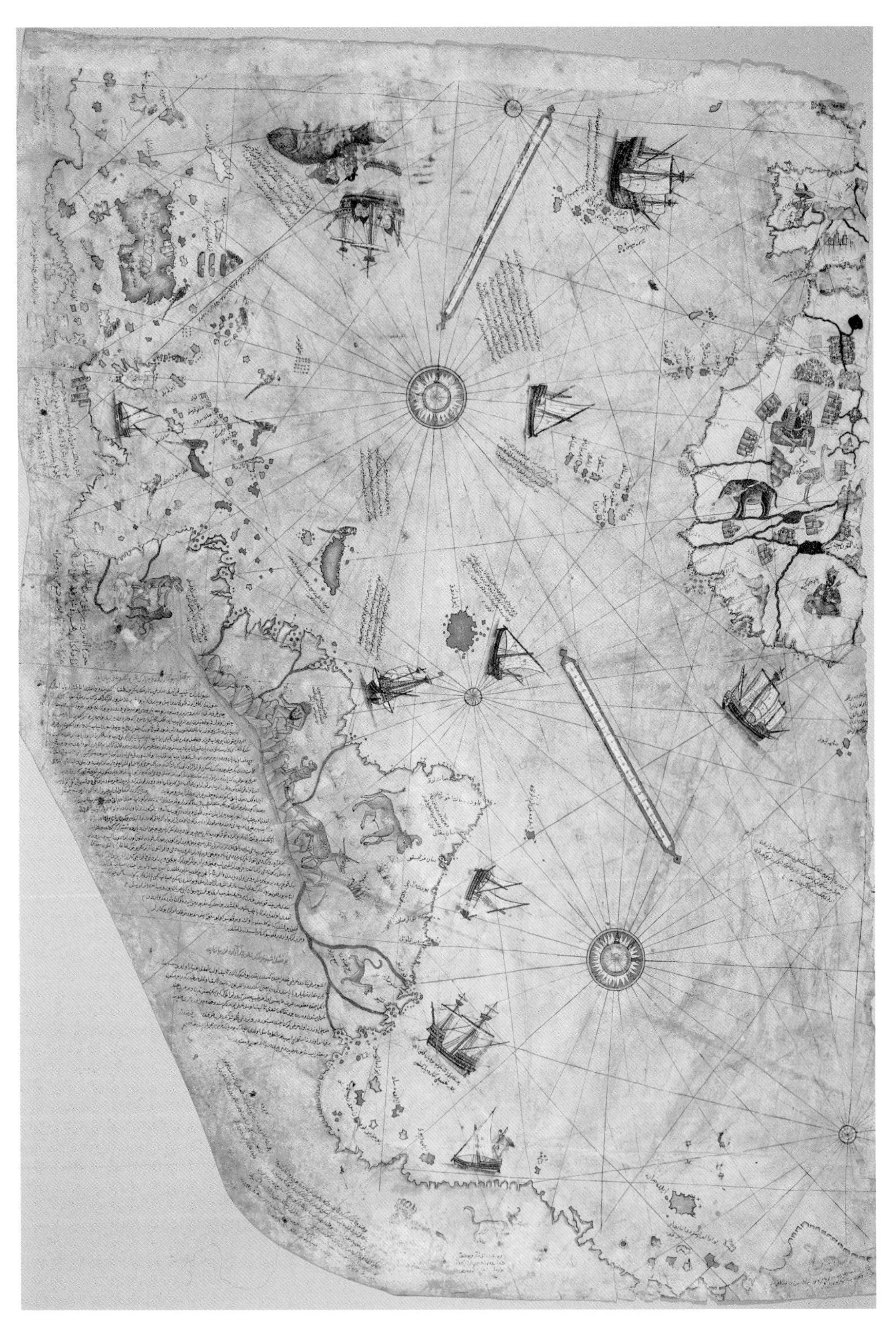

Fig. 32. World map of Piri Reis, 1513. Istanbul, Topkapi Palace Museum, MS Revan 1633m. This fragment of Piri's chart shows the eastern littoral of Central and South America apparently connected to an austral continent. A long inscription in South America relates the history of Columbus' voyages and asserts that Piri used a chart formerly owned by Columbus as a source for the information about the New World displayed on his map. Other inscriptions note the extreme heat and large snakes in the southern part of the New World, and record Piri's identity and the date of the composition of the map.

country is barren. Everything is in ruin and it is said that large snakes are found here. For this reason the Portuguese infidels did not land on these shores and these are also said to be very hot'.[45] The variety of sources that form the basis of the inscription is characteristic of Piri's work: it is a mixture of contemporary travel narratives and medieval European and Arabic geographic traditions.[46] Representation of the southern land as contiguous with the New World seems most likely to derive from an early sixteenth-century European map,[47] one perhaps influenced by Vespucci's third voyage in the 'Quattro Viaggi', in which it is stated that the Portuguese ships sailed along the coast of South America for 750 leagues, then decided to head south-east in search of another region.[48] On the other hand, Vespucci notes that the Portuguese did not land due to the intense cold of this southern region; the references in the inscription to the snakes, and the heat of the land therefore suggest the tradition of representations of the torrid zone and southern Africa on earlier *mappaemundi*.[49] This tradition also seems to have contributed to the interior of the 'new' lands; for instance, the map 'exports' to South America the iconography of monstrous races such as Blemmyae (faces in chests) and Cynocephali (dog-headed men), familiar to the *mappaemundi* tradition and usually located in southern Africa.

A number of inscriptions located on and around the newly-discovered lands on this map express the complementary discourses of supplementation and colonization. In the longest of the inscriptions, on the unknown interior of South America, Piri gives a lively account of Columbus' attempts to garner the support of European powers for his voyage of exploration. He goes on to describe the voyages, with particular attention to European encounters with natives, on the basis of the testimony of a Spaniard who had served with Columbus, but had since become a slave of Piri's uncle, Kemal. The inscription concludes with the comment that 'by now these parts have been conquered and explored in full'.[50] Moreover, the coasts and islands of the map, along with their place-names, have, we are informed, been copied from a chart owned by Columbus: this is a map, then, whose authority is located in the original acts of exploration and discovery. Just above this inscription, alongside the Cynocephali, Piri recorded his identity as the mapmaker: he states that the map was drawn in Gallipoli in the month of Muharram in the year 919 (9 March – 7 April, 1513). In a second inscription, this time just beneath that concerning Columbus, Piri continues this theme by discussing his sources: these include eight Arab world maps, an Arab map of India, four Portuguese maps of India and China, and 'a map drawn by Columbus in the West'.[51] *Terra incognita* evidently provided Piri with the opportunity for discussion of the process of mapmaking alongside the process of exploration: beyond the land tentatively mapped by Columbus and the Portuguese, Piri inscribes his own identity, locates and dates his work, and asserts its authority. Unknown land is thus the site for an account of European expansion, and for expansiveness: its meta-cartographic function, as a space where the origins and purpose of the map may be identified, persists.

'The southern land, recently discovered but not yet fully known'

The connection between the use of maps to display political demarcations and their representation of unknown land was given particular clarity of expression on fifteenth- and sixteenth-century globes. The traditions of globes and two-dimensional maps were by no means separate, but certain distinct features of each may be discerned. It is evident from their association with royal and noble patrons, as well as civic authorities, that in addition to their practical utility globes held a particular symbolic status. The orb in medieval iconography had always exuded imperial and Christological connotations.[52] To possess a globe was thus to possess a symbol of worldly power, emblematic not only of knowledge of the form of the world, but also of the ability to reduce the world to comprehensible dimensions. In this fashion the globe could serve the Ciceronian function of displaying the world while encouraging both the viewer's awareness of his or her own position on the globe, and the presence of 'those who stand differently to you' in other more distant parts. Evocative of both worldliness and humility, the globe nevertheless encouraged a consciousness of the dimensions of dominion, and the path of trade. Yet while the globe was undoubtedly the statesman's symbol – as its appearance (in both terrestrial and celestial forms) in Hans Holbein's famous painting *The Ambassadors* (1533) testifies – globes also possessed a pedagogic function, since they enabled the shape of the world to be appreciated by those with little or no learning. Here an important aspect of globes was their capacity to display antipodal regions, as the title of a 1509 publication attributed to the 'Gymnasium vosagense' bears out: 'Globe of the world. Declaration or description of the world and entire earth as a little round globe fashioned as a solid sphere. By which to whomsoever even moderately learned it is permitted to see that there are antipodeans, whose feet are opposite ours' (Globus mundi Declaratio siue descriptio mundi et totius orbis terrarum globulo rotundo comparati vt spera solida. Qua cuivis etiam mediocriter docto ad oculum videre licet antipodes esse, quorum pedes nostris oppositi sunt).[53]

It is on globes from the first quarter of the sixteenth century that an extended 'Antarctic region' first appears. The earliest surviving terrestrial globe, that of Martin Behaim, dates from 1492, and does not attempt to represent any Antarctic landmass *per se*. However, the *absence* of mappable land in the Antarctic region offered the opportunity to make a quite different kind of inscription there. Behaim's globe contains 111 miniatures, most of which depict kings and national flags, but which also include vessels, coats of arms, missionaries, and saints. The largest miniature is reserved for the space of the Antarctic circle, however: this shows the arms and eagle of the imperial city of Nuremberg (Behaim's home town, and sponsor of the globe) with the virgin's head, and the arms of the three chief 'captains' (hauptleute) of the city at the time. This image is surrounded by an inscription which notes that the desire of these men to see the production of the globe ('erdapfel') has been realized thanks to the skill ('kunst') of Behaim, an expert cosmographer who had circumnavigated one-third of the world.[54] In addition to Behaim's personal experience, the inscription credits the books of Ptolemy, Strabo, and Marco Polo (whose influence, along with Mandeville's, is certainly visible in the Asian sections of the globe) with providing information as to the shape and form of the earth's seas and lands. It concludes by recording the date of the globe's construction, and states that Behaim has left the globe behind him 'for the honour and enjoyment of the commonality of the city of Nuremberg in order that he may be kindly thought of for all time, when he shall have gone back to his wife, who lives 700 miles away, where he keeps house, in order to end his days in his island where is his home' ([s]olche kunst und apfel ist gepracticirt und gemacht worden. Nach cristi gepurt 1492 der dan durch den gedachten Herrn Martin Pehaim gemainer stadt Nürnberg zu ehren und letz hinter ihme gelassen hat Sein zu alle zeiten in gut zu gedencken nachdem Er von hinen wieder heim wendet zu seinem Gemahl, das dann ob 700 mail von hinen ist da er hauss hält und sein tag in seiner jnsel zu beschliessen da er haheimen ist).[55] This rather bathetic coda implicates the globe in an emerging colonialist discourse, since it refers to Behaim's activities as a colonist on the island of Faial in the Azores, where his father-in-law was governor: the globe is left in imperial Nuremburg for its civic authorities and citizens to inspect the world, while the globemaker and cosmographer returns to 'his' island, one marked on the globe both by his own arms and those of Nuremburg. However, it is on the neutral space of the south pole that the identity of the patrons, producer, and audience of the globe are all expressed.[56] At the ends of the earth, the underside of the apple, a portion of the globe unlikely to be easily visible, a mixture of civic and individual assertion takes the place of geographical representation.

If the unmapped space of *terra incognita* offered possibilities for self-inscription, it could also lend itself to various forms of cartographic fiction. A creative response to *terra incognita* is particularly evident on the austral and Antarctic landmasses that began to appear on globes in the first quarter of the sixteenth century, which often display a kind of pseudo-topography. The earliest surviving representation of a large-scale Antarctic landmass on a European globe may be that of the 'Globe vert', so named because of the colour of its seas. On this globe, a ring of land almost entirely surrounds the South Pole and an Antarctic sea; although wholly fictive, the landmass nevertheless contains a surprising amount of topographical representation (fig. 33). Numerous mountain ranges are depicted, along with two inland bodies of water, marked 'Lacus inter montanas' and 'palus' (marsh), which are connected to each other by a large river ('flumo').[57] Fairly narrow straits separate the 'Antarctic ring' from the continents of America, Africa, and Asia. This landmass can only partially be explained by the misreading of Vespucci's account of his South American voyage evident in Ringmann's *De ora antarctica*, since the Vespuccian narrative would not have justified the vast extent of the ring. Another influence may have been the long-standing Macrobian tradition of representing an austral landmass in the southern frigid and temperate zones, balancing the landmasses of the north. The pseudo-topography of the 'Antarctic ring' does bear strong similarities with the (equally fictive) Arctic ring that surrounds the north pole. As Monique Pelletier has pointed out, globe makers seem to have felt considerable licence to invent and speculate in the polar regions, while taking a more cautious approach to parts of the world that were known to be inhabited.[58]

Until recently, the 'Globe vert' was associated with the Nuremberg mathematician Johann Schöner, largely on the basis of the resemblance of its 'Antarctic ring' to similar landmasses on a series of globes produced by Schöner from 1515 onwards: the shape of their coastal outlines is essentially the same, and their interior contains a pseudo-topography in most respects identical to each other.[59] It has been argued, however, on the basis of similarities between the 'Globe vert' and the Waldseemüller world map of 1507, that the former was in fact the work of the 'Gymanasium vosagense'.[60] These similarities include the globe's representation of the New World as a continent, named 'America', and the appearance of the words 'terra ulterius

Fig. 33. Globe vert. Paris, BNF, Cartes et Plans, Rés. Ge A 335. This early sixteenth-century globe, traditionally attributed to the Nuremburg globe maker Johann Schöner but more recently assigned to the 'Gymnasium vosagense', is notable for its 'Antarctic ring', a landmass that almost entirely surrounds the South Pole. Numerous mountain ranges are depicted on the Antarctic land, along with two inland bodies of water, marked 'Lacus inter montanas' and 'palus' (marsh), connected by a large river ('flumo').

incognita' in North America and 'terra ultra incognita' in South America on both globe and Waldseemüller's map. Further, the appearance of the 'Globe vert's' austral ring has undergone certain modifications on Schöner's globe of 1515: Schöner has 'roughed up' the coast-line, giving it a more jagged, and therefore more realistic appearance; it is bulkier in some parts, so that, as Pelletier has noted, it is less of a 'ring', and more of a relatively dense austral continent.[61] The critical difference between the two landmasses, however, is that on the 1515 globe the toponym 'Brasilie Regio' appears in the austral land (fig. 34).

In his pioneering study of Schöner's globes, Franz von Wieser identified the source of Schöner's 'Brasilie Regio' as a broadsheet entitled 'Copia der Newen Zeytung auß Presillg

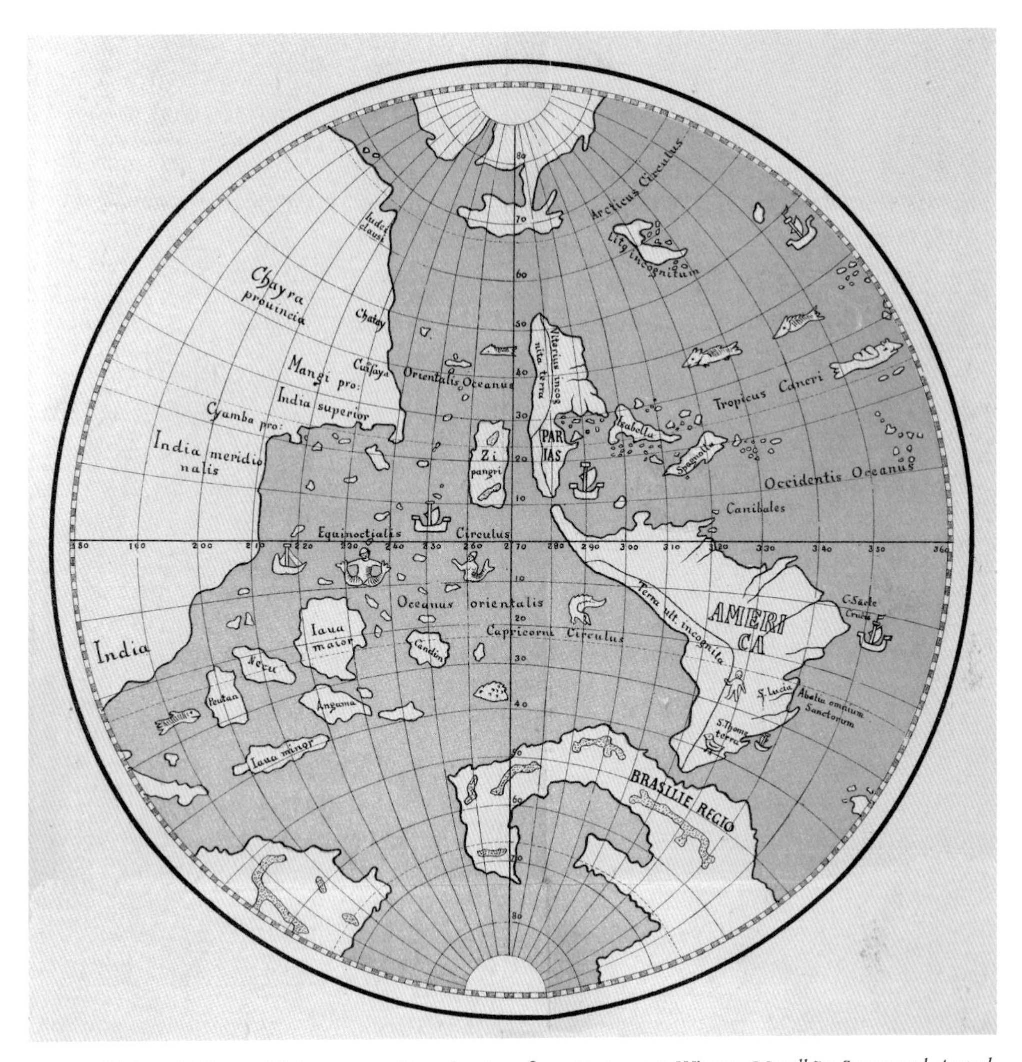

Fig. 34. Globe of Johann Schöner, 1515. Line drawing from Franz von Wieser, *Magalhães-Strasse und Austral-Continent auf den Globen des Johannes Schöner* (Innsbruck: Verlag der Wagner'schen Universitäts-Buchhandlung, 1881). Schöner's 1515 globe is preserved in Weimar, Herzogin Anna Amalia Bibliothek, Stiftung Weimarer Klassik. It contains a ring of Antarctic land similar in many ways to that of the 'Globe vert'. However, on this version of the globe Schöner has inserted the toponym 'Brasilie Regio' in the southern land. The basis for this inscription appears to have been a garbled travel narrative contained in the broadsheet entitled 'Copia der Newen Zeytung auß Presillg Landt'.

Landt', which was printed in Augsburg by Erhart Öglin, probably in 1508, and certainly by 1514.[62] The 'Newe Zeytung' reported the voyage of two ships of the trading house of Haro, which had been given permission by the king of Portugal to describe or explore (zubeschreiben oder zu erfaren) 'the land of "Brazil"' (das Presilg landt).[63] The report states that the ships, having rounded the Cape of Good Hope, discovered 'Presill' 40 degrees south.[64] The passage between the southernmost extent of Africa and 'Presill' is compared to the strait of Gibraltar; the two landmasses are thus described in terms of southern Europe and the north coast of Africa. The text then notes that the distance to Malacca from the 'start of Presill' (anfangk des Presill landt) is 600 miles;[65] it goes on to describe the inhabitants of 'Presill', drawing a distinction between the cannibals of 'lower Presill' (in dem vndtern Presill landt), and the more moral natives of upper 'Presill', who even show a knowledge of St Thomas, believing him to be a lesser god.[66] In both upper and lower 'Presill' there are no kings, but elders are followed; the people can live up to 140 years.[67] The 'Zeytung' also gives an account of 'Presill's' mountainous terrain, where snow does not melt, the land's produce, and the wildlife of the region, including the presence of lions and leopards.

It is difficult to reconstruct the actual voyage that lies behind such a garbled description. It is plausible that the 'Presill' referred to is not the Brazil of South America, but rather Sumatra, the 'Brasil' described by Marco Polo.[68] That, however, was not Schöner's interpretation of the broadsheet. In the *Luculentissima descriptio* that he composed to accompany his 1515 globe, Schöner included a description of 'Brasiliae regio' within his chapter 'On America the fourth part of the world with other new adjacent islands' (de America quarta orbis parte cum aliis novis insulis appositis). This description of 'Brasil' is essentially an edited translation of the 'Newe Zeytung': Schöner records that 'Brasil' is not far from the Cape of Good Hope, that the Portuguese circumnavigated the region and learned that it lay laterally, east to west, conforming to Europe ('which we inhabit'). He includes the 'Zeytung's' comparison with the strait of Gibraltar, and notes that this 'is shown on our globe towards the Antarctic pole' ('ut ostendet Globus noster versus polum Antarcticum').[69] He further translates parts of the 'Zeytung's' passages on the proximity of Brasil to Malacca, the mountainous terrain of the region, its animals, fruits and spices, native knowledge of metal work, the abundance of gold in certain areas, and the extraordinary longevity of the inhabitants.

Schöner's incorporation of the 'Zeytung' into his *descriptio* was a translation in more ways than one. The 'Zeytung', one of the earliest German broadsheets of its kind, was clearly written for a relatively general, geographically uninformed, audience: its language is simple, and technical terms are painstakingly glossed – for example, the term 'pilot' is gloseed 'Der Piloto, das ist der schiffuerer oder Schiflayter'.[70] Schöner's Latin text, by contrast, seems designed for an elite audience, and one, moreover, with access to his globe. This is not to say that glossing is absent – he notes, for instance, the short distance of Brasil 'from the Cape of Good Hope, which the Italians call "Capo de bona speranza"' (a capite bonae spei (quod Itali Capo debona speranza vocitant)).[71] Rather, Schöner's procedure was that of the 'Gymnasium vosagense': the translation of vernacular accounts of exploration into Latin was a means of ensuring an international dissemination within educated circles. And, as with the 'Gymnasium', the procedure involved the transition from verbal *descriptio* to visual representation. The confusing, ambiguous nature of the 'Zeytung' was smoothed out by Schöner's judicious extraction of passages from it, while the 'region of Brasil' was given an undeniable presence on his globe.

It is significant that Schöner, on the (unacknowledged) basis of the 'Zeytung', first assigned a name to the austral landmass of his globe, and then included the description of this region in the chapter on America in his accompanying geographical treatise. America in the *Luculentissima descriptio* is still the supplement to the known world: it is, Schöner maintains, the 'fourth' part of the world, distinguished from the other three by its insular nature. But the supplement now has its own supplement: Brasil, a region that lies 'as does Europe, which we inhabit', and which offers Europe attractive possibilities by means of its proximity to Malacca, and by its own alleged riches. Two points are particularly worthy of note with regard to Schöner's 'Brasilie regio': first, slippage of toponyms occurs not only between two recently explored, and recently claimed, parts of the world (the American Brazil and the Asian Brasil), but also extends to a slippage from the New World to the conjectural landmass of the austral ring. Secondly, and as a consequence, this austral landmass is given a status approaching that of America: its naming has begun, it has been tentatively identified, but remains predominantly unknown. It has, finally and most importantly, been incorporated – placed – within a scheme of colonial knowledge, having been discovered and claimed in the name of the king of Portugal. The conflation of place performed by the 'Zeytung' is, in short, given visual and verbal clarity by Schöner; the absurd upshot of this series of misunderstandings is the commencement of the mapping of the non-existent space of the austral ring.

The supplement, having spread, continued to evolve: Schöner's globe of 1520, larger[72] and more geographically detailed than the 1515 model, modified 'Brasilie regio' to give it even greater conformity to the description of the 'Zeytung'. The 1520 globe entitled the austral land 'Brasilia Inferior' (associating it with the 'Zeytung's' undter Pressill), and decorated it with trees and stylised mountain ranges. Its interior bears notable similarities to the portion of the new world entitled 'America vel Brasilia sive Papagalli Terra' (America or Brasilia or Land of Parrots), the westernmost part of which, having been designated 'terra ult. incognita' on the 1515 globe, is now marked 'Vltra Incognita Permansit' (still unknown beyond); this area is also marked with a mountain range and a lone pine tree.[73] By the early 1520s the voyage of Ferdinand Magellan – publicized by Schöner in 1523 in the small tract *De Moluccas insulis* – had given a new credibility to the representation of the southern land because of Magellan's sighting of land beyond the furthest extent of south America (Tierra del Fuego). In a manner identical to the 1531 map of the mathematician Oronce Fine,[74] Schöner's globe of 1533 (and possibly also his lost globe of 1523) extended the southern land almost as far north as the Tropic of Capricorn, and marked it with the inscription 'Terra Australis recenter inventa at nondum plene cognita' (Southern Land recently discovered but not yet fully known). No longer in the form of an 'Antarctic ring', on the 1533 globe it has two regions, since Schöner emended 'Brasilie Regio' to 'Brasielie Regio', and added the Plinian 'Regio Patalis', and the 'Psitacorum Terra' (land of parrots), another example of the slippage of information and nomenclature from America to austral land.[75] By contrast, the supplement had been subsumed into the ecumene: America is on this globe depicted as part of the continent of Asia. In the text that accompanies his 1533 globe, the *Opusculum Geographicum*, Schöner records this revision of opinion about the insular nature of America, and states that, since the voyage of Magellan, it has been shown to be connected to the 'continent of upper India, which is part of Asia'.[76]

Fine's influential 1531 bi-cordiform map of the world seems to mark a widespread acceptance of the idea of the southern continent, and the reform of its image. No longer either a 'ring', or a continent extending between the New World and southern Asia as drawn in Lopo Homem's 1519 world map, the southern land now appears as a separate yet monumental landmass (fig. 35). On this map, a revised version of an earlier world map first produced by Fine in 1519, and updated in 1534,[77] Fine marks the austral land 'Terra Avstralis recenter inuenta, sed nondum plene cognita' (southern land recently discovered but not yet fully known); he also includes Schöner's 'Brasielie Regio' and 'Regio Patalis', as well as a few hypothetical mountain ranges. It is no coincidence that by virtue of the great southern land, Fine, from 1531 Professor of Mathematics at the Collège royal in Paris, represented roughly equal proportions of land to sea in the southern and northern hemispheres. Whatever its shape, in the hands of the most sophisticated sixteenth-century mathematician-cartographers (Schöner, Fine, Mercator) the southern continent was the product of precise calculations as well as speculation; it was the necessary component of a world image that, in accord with ancient ideals, expressed theories of balance, symmetry and perfection.

During the decades that followed Fine's publications, the form of 'Terra Australis' was consolidated according to model of his maps of 1531 and 1534: it became massive; it acquired a reasonably consistent outline; it maintained a few, frequently repeated, regional toponyms. However, up to the point of consolidation, the image of 'Terra Australis' remained fluid, and susceptible to extraordinary levels of cartographic fiction, as a manuscript map in the Vatican Library testifies (fig. 36). Urbinas Latinus 274 is a late fifteenth-century manuscript of Ptolemy's *Geographia*, which contains a dedicatory letter to Paul II by Nicolaus Germanus, Jacopo d'Angelo's translation, and thirty maps. An oval planisphere, which appears to date from around 1530, has been painted on two blank leaves of the manuscript between the conclusion of the text of the *Geographia*, and the first of Ptolemy's maps.[78] This map has been characterized as 'a learned composition, which combines Ptolemaic, medieval and modern elements';[79] it is notable for its elongated representation of the Old World, and its separation of north from south America. However, its most unusual feature is without doubt its depiction of an 'Antarctic ring'. While this austral landmass resembles those of Schöner, and others such as the 'Globe vert', it differs from these representations in three principal respects: first, it circles the globe fully, enclosing an Antarctic sea; second, it displays a far more detailed coastline, comprising numerous gulfs and peninsulas, including one particularly striking example which nearly joins an elongated Asiatic peninsula. Finally, the Vatican map is marked with a detailed topography, and even more unusual, numerous toponyms, in both Latin and a vernacular that indicates Italian, and possibly Venetian authorship.[80] Most of these names identify the rivers, gulfs, and capes depicted on the map; there are four particularly prominent toponyms: 'Terra inchognita avstrale' along the southern coast of the ring, 'Terra australe' and 'Terra incognita' in the land's north, and 'Regno Patalis' in its north-eastern extremity. In addition, one legend explains that 'the race that lives in these provinces is red and idolators' (gens que habitat in has prouintia[s] sunt Rubri e[t] idolatri);[81] another, noting the length of the Strait of Magellan ('stretto de mengalianos longo lege 100 m. 300'), helps to date the map by identifying Magellan's voyage as a *terminus post quem*.

The origins of this striking representation of austral land have yet to be elucidated. As

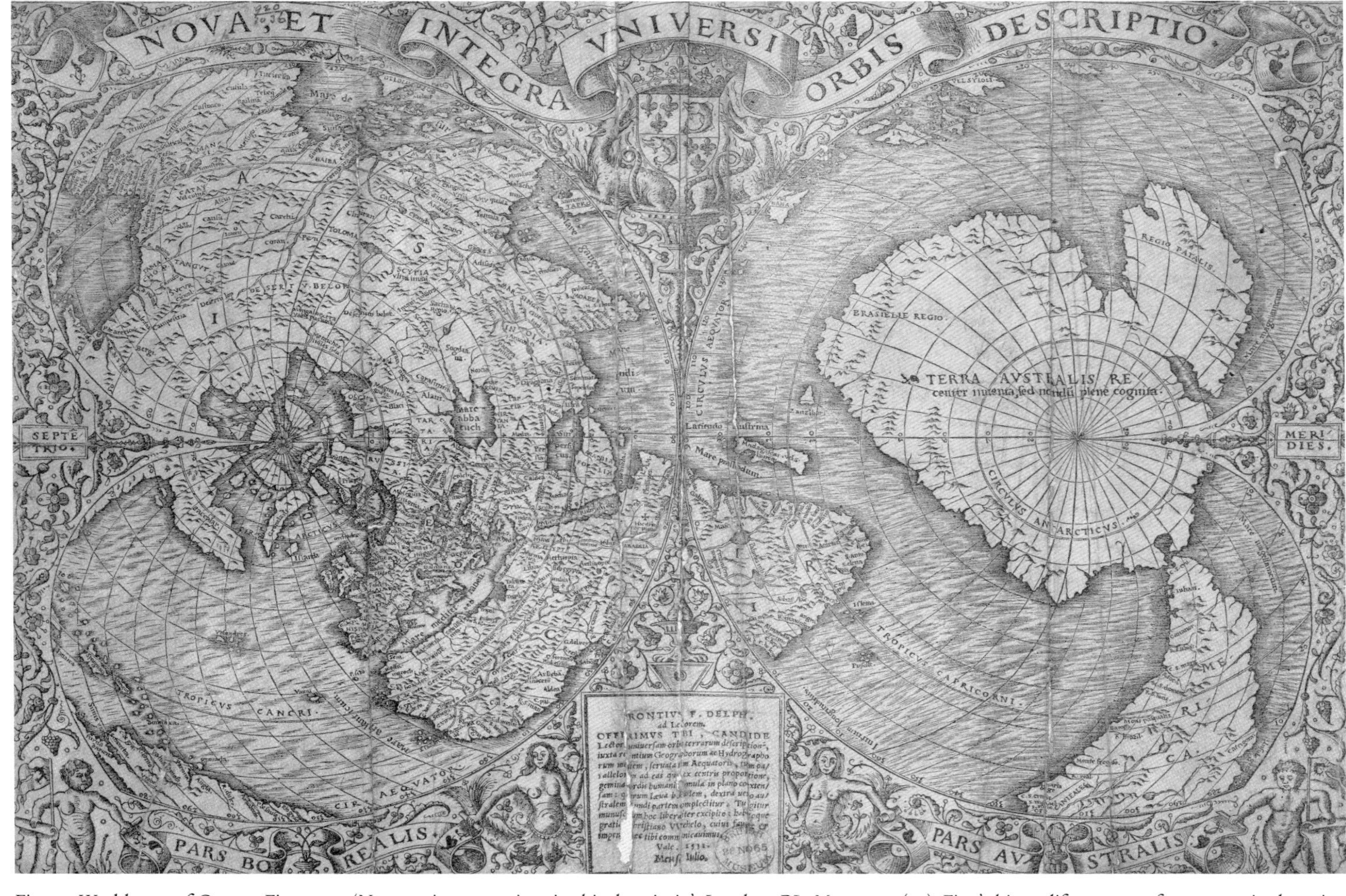

Fig. 35. World map of Oronce Fine, 1531: 'Nova, et integra universi orbis descriptio'. London, BL, Maps 920 (39). Fine's bi-cordiform map of 1531, a revised version of his 1519 map of the world, marked an important point in the transformation of the southern continent into a separate and monumental landmass. The continent is described as 'Terra Avstralis recenter inuenta, sed nondum plene cognita' (southern land recently discovered but not yet fully known); its topography includes Schöner's 'Brasielie Regio' and the Plinian 'Regio Patalis', as well as hypothetical mountain ranges.

Fig. 36. World map in Vatican City, BAV, MS Urb. lat. 274, ff. 73v–74r. This anonymous oval planisphere was painted into a fifteenth-century edition of Ptolemy's *Geographia*. It contains one of the more remarkable sixteenth-century elaborations of the southern continent. The mapmaker has supplied 'Terra inchognita avstrale' with an immensely detailed topography, including an impressively realistic northern coastline. Although many of its toponyms are generic, the southern land appears to include some places recorded on Columbus' fourth voyage.

Roberto Almagià noted, as well as possible Schönerian influence, the map's depiction of Asia aligns it with the work of the late fifteenth-century German cartographer Henricus Martellus.[82] This resemblance could equally work as evidence of the influence of Rosselli, whose work has long been linked with that of Martellus;[83] certain other similarities between the Vatican map and the work of Rosselli may be noted, such as the style and hand of the place names, and the use of small crimson marks to represent habitation.[84] The series of toponyms are for the most part relatively generic, and therefore suggestive of New World mapping practices: 'fl. d. s. gregorio' (Saint Gregory's river), 'C. d. S. Andrea' (Cape of St. Andrew), 'rio d. S. lorenzo' (Saint Lawrence river), 'rio nouo', 'rio grando' (new river, large river), 'c. d. tres puntas' (an inscription that also appears off the west coast of Africa). The level of mapping on the 'Antarctic ring' is overall no less detailed than that of the 'Mundus Novus' of north and south America, and in some parts it is more detailed. Conversely, the mapmaker has registered uncertainty about the west coast of the New World and the south coast of 'Terra australe' by sketching in these places a coastline devoid of toponyms or topographical information. Despite the mysteries surrounding its origins and purpose, two points can be made about this unusual map. First, the pseudo-topography seen on many globes around this time has reached its ultimate expression: no other surviving world map contains such a topographically- and toponymically-detailed representation of the unknown southern land. Second, and consequently, this southern land is not differentiated in any respect from those of the New World: both are partially mapped, with a well-charted coastline appearing alongside areas of *terra incognita. Terra Australis* is on this map part of the same order of exploration and colonization as the 'Mundus Novus'. It is not a site for speculation, signature, or any other form of a-cartographic representation: wholly realistic, its remarkable fictions epitomize the principle of slippage, the transference of mapping procedure from one place to another. In this case, the slippage is from the *terra inventa* of the New World, to the utterly invented *terra* of the unknown southern continent.

The possibilities of 'not yet'

Part of the intellectual attraction of the idea of the southern land for cartographers such as Fine, and the anonymous author of the Vatican map, was its position within the category of the 'not yet'. Very obviously, land 'not yet' discovered was an imaginative construct that invited mental as well as physical exploration. 'Nondum cognita' signified a state between certainty and speculation, between new and old, ecumene and *antoecumene*. As 'terra incognita' had always been, land 'not yet known' was proximate to, yet outside of, schemes of knowledge; unlike the *terra incognita* of the classical and medieval antipodes, however, *terra nondum cognita* foretold its own assimilation. The political significance of such a space can be understood in a specific and a broad sense: in terms of abstracted theories of polities, and in terms of the particular ambitions attributed to actual states and their rulers. Two texts from the period 1515–30 – Thomas More's *Utopia*, and the Flemish geographer Franciscus Monachus' *De orbis situ ac descriptione* – illustrate the ways in which spatial representation was unavoidably, and compellingly, aligned with the political, in both its broad and narrow sense, in the articulation of a New World geography that extended beyond the Americas to *Terra Australis.*

More's *Utopia* grasped perhaps better than any other work of this period the opportunities for creative disruption of old modes of spatial representation offered by the idea of passage to new worlds. *Utopia*, first printed in 1516, begins with the promise made by Peter Giles to his friend More of an account of 'unknown peoples and unexplored lands; and I know that you're always greedy for such information' (hominum terrarumque incognitarum ... historiam. Quarum rerum audiendarum scio avidissimum esse te).[85] This narrative is to be recounted by the traveller Raphael Hythloday, said to have 'joined Amerigo Vespucci [and been] Vespucci's constant companion on the last three of his four voyages, accounts of which are now common reading everywhere, but on the last voyage he did not return home with him' (Americo Vespucio se adiunxit, atque in tribus posterioribus illarum quattuor navigationum quae passim iam leguntur, perpetuus eius comes fuit, nisi quod in ultima cum eo non rediit).[86] The reference to Vespucci's 'Quattro Viaggi' is clear, as is the positioning of Hythloday's account as a supplement to Vespucci's narratives. By the end of the book, the precise location of Utopia remains unknown, despite the full report given by Hythloday of the place, its people and its form of government. This, of course, is the point: Utopia is a no-place, as well as a good place, nowhere. Yet the problem, the bait given by More to his readers, is that the island *could* be located – tentatively, vaguely – on the map. Hythloday and his companions are left by Vespucci on the coast of south America at the furthest point of the Florentine's fourth and last voyage. They travel beneath the equator through a hot, uncultivated region, then pass into a mild and cultivated zone where trade and commerce flourish; finally they reach a region where they find ships identical to European vessels.[87] The manipulation of the model of the zonal division of the earth (moving from northern to southern temperate zones through the torrid zone), and the motif of the world mirrored (identical climate produces identical naval technology) is evident. More's construction of the island as a place known only through the oral narrative of an itinerant navigator meant that, rather than *terra incognita*, or *terra inventa*, the island was for *Utopia*'s readers inscribed within the category of *terra nondum cognita.* As Guillaume Budé remarked in a letter that prefaced the 1518 printings of the text, while Utopia lies outside the bounds of the known world, Hythloday 'has *not yet* told exactly where it is to be found' (Hythlodaeus *nondum* situm eius finibus certis tradidit).[88] The simultaneous observance and breach of the traditional zonal division meant that symmetry, and inverted identities, between Utopia and the known world are continually suggested in More's text, but just as frequently frustrated. Above all, the notion of passage and commerce, of contact extending back as far as the beginnings of Utopian society itself, disturbs the notion of parallel, balancing, and separate temperate zones: the Romans and Egyptians were shipwrecked on Utopia 'twelve hundred years ago'; Hythloday introduces a bag of books (mostly works of classical Greek science and erudition) to the island; the Europeans explain the concept of paper and the printing press; 'no small number' of the Utopians convert to Christianity.[89] The antipodes remain for More hypothetical, as they were for classical writers, but unlike them, he can imagine interaction – the introduction of books and printing, evangelization, disruption. In other words, the shadow of a Vespuccian narrative.

Utopia is a political text in the sense that its purpose was to contribute to a contemporary debate with classical ancestry 'on the best state of a commonwealth' (de optimo reipublicae statu). Unlike two of its models, the *Republics* of Plato and Cicero, it uses terrestrial rather than celestial voyage to make possible a vision that acts as a critique of the known world. That the critique comes from the southern hemisphere – the old southern temperate zone – suggests More reformulated the idea of the antipodes as *recessus*, using it, in an elaboration of a New World travel narrative, as a space in which profound reflections on the organization and regulation of the *res publica* could be situated. Why this space? Again, the fundamental attraction of antipodal space appears to have been its apposition to the known world; not simply the world turned upside down, the antipodes allowed for the imagination of the known world reformed as well as reflected. And for More, who was careful to construct Utopia's position as a supplement to the New World, the attraction of the antipodes of the south lay in its new-found location within the 'not yet' of European colonization.

To understand further the ambiguities of the situation of the hypothetical *Terra Australis* on the brink of exploration, evangelization, and colonization it is necessary to turn to a political tract of a more narrowly propagandistic kind, Franciscus Monachus' *De orbis situ ac descriptione*, printed in Antwerp c. 1527–30.[90] The title page of *De orbis situ* is illustrated by a rough woodcut world map in two hemispheres, divided so as to show the demarcation of the globe between Spain and Portugal according to the Treaty of Tordesillas (fig. 37). A discrete southern landmass appears on both hemispheres, with the legend: 'this part of the world revealed to us by voyages is not yet manifest' (hec pars orbis nobis navigationibvs detecta nundum existit). The work itself originally accompanied a globe made in the atelier of Gaspard van der Heyden for Franciscus' patron Joannes Carondelet, the Archbishop of Palermo and counsellor to Maria of Hungary, who was resident in Mechelen.[91] *De orbis situ* is a mixture of geographical treatise and imperialist propaganda in support of the Holy Roman Emperor and King of Spain, Charles V. Its ostensible task is to explain the differences ('dissonantia') between Franciscus' image of the world and the texts and tradition of previous geographers. He therefore explains that the description of Ptolemy has been corrected in 'our region' and in the Indian sea by virtue of Portuguese exploration, and that 'whatever remains of the world in the east or the south either needed to be corrected or supplemented' ([q]uicquid restat terrarum orbisque in ortum, aut austrum, vel castigandum fuit, vel supplendum).[92] Franciscus is also at pains to explain the overturning of the practice of centring maps on Jerusalem, contending that this practice was metaphorical and symbolic.[93] At the same time, the treatise presents contemporary geographic debate as fluid and ongoing, arguing that America is in fact connected to Asia, a formation represented on the accompanying map, and attempting to reconcile the reports of the far east by authors such as Marco Polo and Mandeville with the accounts of western explorations provided by Hernán Cortes and Peter Martyr.[94] Towards the end of the work its imperial agenda becomes manifest: Franciscus launches into a paean to Charles, the essence of which is that, in conquering the antipodes, the Emperor has excelled Alexander the Great, Pepin, Charlemagne, and a host of other classical, biblical, and medieval rulers; this achievement, Franciscus suggests, was prophetically announced. Of particular importance here is Franciscus' insistence that the

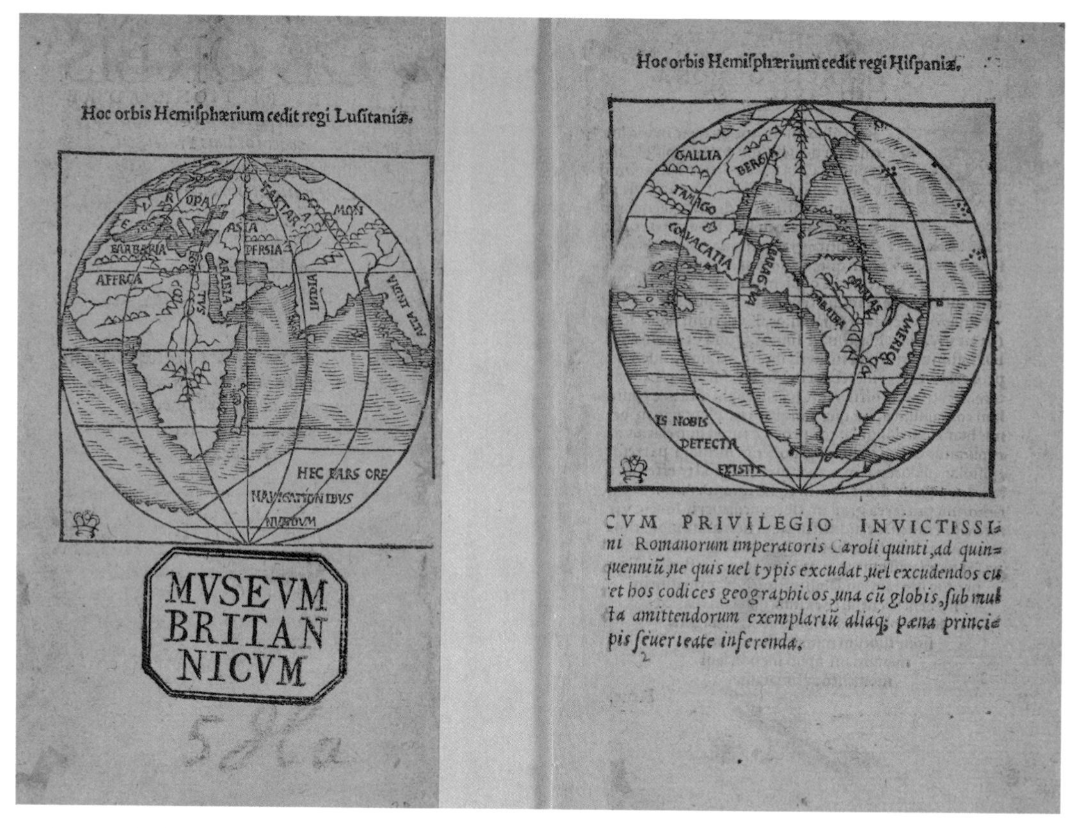

Fig. 37. World map in Franciscus Monachus, *De orbis situ* (Antwerp, c. 1527–30). London, BL C.107.bb.17 (1). Franciscus Monachus' rough woodcut map appears on the title page of his propagandistic treatise. The map shows the world in two hemispheres, divided so as to indicate the demarcation of the globe between Spain and Portugal stipulated in the Treaty of Tordesillas. A southern landmass appears on both hemispheres, bearing the legend: 'hec pars orbis nobis navigationibvs detecta nundum existit' (this part of the world revealed to us by voyages is not yet manifest).

Moluccas, subject of intense conflict between Spanish and Portuguese, fall on the Castilian side of the line of demarcation, a belief represented on his map, and presumably also on the accompanying globe.[95]

The issue of demarcation was of at least theoretical importance for that part of the world as yet unexplored. Thus Franciscus notes the emerging contours of the southern land, and laments the lack of knowledge of this region:

> Reliqua australis ore etiamnum in obscuro latent, mihi tamen admodum sit verisimile eam orbis partem non integi atque obduci pelago. Quin coniecturae sunt et argumenta vastas illic et patentes insulas, regionesque iacere, sed propter locorum intercapedinem, solique infrugiferam indolem, minus celebratas.[96]

> the remainder of the southern coast lies even now in darkness, however to me it seems very likely that that part of the earth is not covered and enveloped by ocean. Indeed conjectures and arguments hold that vast and open islands and regions lie there, but less renowned on account of the distance between places and the infertile nature of the soil.

The implication of the imperial interest in this territory has been well established by the passages that precede it in the work, which extol the greatness of the emperor in terms of geographical achievement and territorial conquest. Charles' power is destined to stretch to the ends of the antipodes, to achieve, in other words, the ultimate imperial ambition:

> An vnquam fando a condito mundo prius auditum, classem vniuersum orbis ambitum circuisse? at hoc Caroli Caesaris auspiciis dedere Superi. Cardinem antarcticum, ignotas terras, maria, populos trans mundi medium limitem videre contigit, quae essent nec ne, vix satis olim constans erat opinio. At Caesaris nostri exploratoribus multo ingens orbis pars sese aperuit, nudauit, retexit.[97]

> has it ever been heard said since the creation of the earth that a fleet circumnavigated the entire world? But God permitted this to occur under the guidance of Emperor Charles. He reached so far as to see the Antarctic pole, unknown lands, seas, people across the middle limit of the world, the very existence of which was not long ago a matter for continual conjecture. But by the explorations of our Caesar a huge part of the world opened far and wide, laid bare, revealed itself.

The key word in the description of the southern continent becomes, then, one absent from the above passage, but unambiguously present on Franciscus' world map: 'nondum'. The word signals delay, an element of future time within the representation of space. Mediating between the blankness of *terra incognita* and the seemingly inexorable progress of European exploration, *nondum* announces the provisional, incomplete status of the New World map. The imperial advance is underway, lands are invitingly baring themselves to the Emperor's gaze, but he and it have yet to reach out and incorporate the southern continent.

Just how long a land could remain 'not yet mapped' was to become a subject of some sensitivity in the second half of the sixteenth century. Yet far from disappearing as a result of the absence of secure reports of its existence, the fiction of the unknown southern land flourished. Representations of *Terra Australis* after 1530 increasingly seized upon the opportunities for signature, digression, and explanation offered in the a-cartographic space of unknown land. They also increasingly used unknown land as a site for opposition to the very practices of European expansion expressed on and by the world map.

Notes

1 *Itinerarium Sacrae Scripturae. Das ist: Ein Reisebuch* (Helmstadt, 1581).

2 Shirley, *Mapping of the World*, p. 164 (no. 142). See further Henk A. M. van der Heijden, 'Heinrich Büntings Itinerarium Sacrae Scripturae, 1581: Ein Kapitel der biblischen Geographie', *Cartographica Helvetica* 23 (2001), 5–14; Kenneth Nebenzahl, *Maps of the Holy Land: Images of* Terra Sancta *through Two Millennia* (New York: Abbeville Press, 1986), p. 88. The full title of the map makes the heraldic point clear: 'The entire world in a clover leaf, which is the arms of the city of Hanover, my beloved fatherland' (Die gantze Welt in einem Kleberblat/ Welches ist der Stadt Hannover/ meines lieben Vaterlandes Wapen).

3 My understanding of the supplement derives from Derrida's discussion of the two significations of the supplement in the writing of Rousseau. 'The supplement adds itself, it is a surplus, a plenitude enriching another plenitude, the *fullest measure* of presence ... But the supplement supplements. It adds only to replace. It intervenes or insinuates itself *in-the-place-of*; ... Compensatory [*suppléant*] and vicarious, the supplement is an adjunct, a subaltern instance which *takes-(the)-place* [*tient-lieu*]': Jacques Derrida, *Of Grammatology* (De la grammatologie), trans. Gayatri Chakravorty Spivak (Baltimore: Johns Hopkins University Press, 1976), pp. 144–5. Now, according to Derrida's analysis, mapping itself – not merely the addition of the New World to the old *imago mundi* – would be the act of supplementation. As writing is the supplement to speech (the disappearance of natural presence) so cartography supplies in place of the real – the map adds itself to the real by inscribing it, but also exists in its place. However, for the purposes of the following discussion I prefer to understand 'supplement' according to the terms in which the concept was used by early sixteenth-century cartographers when they talked of 'supplementing' Ptolemy.

4 *Itinerarium Sacrae Scripturae*, p. 7 (I refer here to the pagination of the Magdeburg, 1585 edition).

5 Amongst many examples see Lambert of Saint-Omer's 'Spera Geometrica', the Beatus maps, and the thirteenth-century Psalter map. For recent discussion of the cartographic and literary positioning of the British Isles see Kathy Lavezzo, *Angels on the Edge of the World: Geography, Literature, and English Community, 1000–1534* (Ithaca: Cornell University Press, 2006).

6 On the 'Gymnasium vosagense' see A. Ronsin, 'L'Amérique du Gymnase vosgien de Saint-Dié-des-Vosges: invention et postérité', in *La France-Amérique (XVIe–XVIIIe siècles)*, ed. F. Lestringant (Paris: Champion, 1998), pp. 37–64; Monique Pelletier, 'Le globe vert et l'œuvre cosmographique du gymnase vosgien', *Bulletin du Comité français de cartographie* 163 (2000), 17–31; and R. A. Skelton's 'Bibliographical Note' to the facsimile of the 1513 Ptolemy: *Geographię opus nouissima traductione e Gręcorum archetypis castigatissime pressum* (Strasbourg, 1513, facsimile ed. Amsterdam: Theatrum Orbis Terrarum, 1966), pp. v–xiv. (Hereafter *Geographia* 1513). Important earlier accounts include M. d'Avezac, *Martin Hylacomylus Waltzemüller: ses ouvrages et ses collaborateurs* (Paris: Challamel aîné, 1867, repr. Amsterdam: Meridian, 1963) and L. Gallois, *Les géographes allemands de la Renaissance* (Paris: Leroux, 1890, repr. Amsterdam: Meridian, 1963). Pelletier dates the formation of the Gymnasium to 1502, when Lud obtained a printer's licence, although the first book was not printed at Saint-Dié until 1507: 'Le globe vert', p. 19.

7 Franz Laubenberger, 'The Naming of America', *Sixteenth Century Journal* 13 (1982), 91–113; Skelton, 'Bibliographical Note', p. viii; Robert W. Karrow, *Mapmakers of the Sixteenth Century and their Maps: Bio-bibliographies of the Cartographers of Abraham Ortelius, 1570* (Chicago: University of Chicago Press, 1993), p. 570.

8 The *Mundus Novus* had already been translated *into* Latin from an Italian original some time before 1505 by a certain 'iocundus interpres', possibly one Giuliano del Giocondo, known to have been resident in Lisbon. See Alexander von Humboldt, *Kritische Untersuchungen über die historische Entwickelung der geographischen Kenntnisse von der neuen Welt*, 3 vols (Berlin: Nicolai'schen Buchhandlung, 1852), vol. 2, p. 400; and Ilaria Luzzana Caraci, *Amerigo Vespucci*, 2 vols (Rome: Nuova Raccolta Colombiana, 1996–9), vol. 2, p. 76. Another candidate is the Veronese humanist fra' Giovanni del Giocondo. This long-held hypothesis was vigorously questioned by Alberto Magnaghi, *Amerigo Vespucci*, 2 vols (Rome: Istituto Cristoforo Colombo, 1924), vol. 1, pp. 59–68.

9 *Cosmographiae introductio* (Saint-Dié, 1507), sig. A.iir–v, my emphasis. In a separate issue of the first edition of the *Cosmographiae introductio*, Waldseemüller's dedication was changed to credit the collective rather than the individual: the 'Gymnasium Vosagense' announces that 'we (who recently established a printing press in the Vosges of Lorraine in a town called Saint-Dié) were comparing the books of Ptolemy after a Greek manuscript ...'. The second edition of the *Cosmographiae introductio*, of 29 August 1507, similarly appeared in two states, one with an individualized and one with a collectivized dedication. Ringmann travelled in Italy on more than one occasion in

search of Greek manuscripts of Ptolemy, visiting Lilio Gregorio Giraldi and Gianfrancesco Pico de la Mirandola, who mentioned him in a letter of c. 1520 to Lefèvre: *The Prefatory Epistles of Jacques Lefèvre d'Etaples and related texts*, ed. Eugene F. Rice (New York: Columbia University Press, 1972), epistle no. 129, pp. 417–18.

10 *Geographia* 1513, 'In Claudii Ptolemei Supplementum', title page verso.

11 Skelton, 'Bibliographical Note', pp. v–xx: p. xvi.

12 'Hec terra cum adiacentibus insulis inuenta est per Columbum Ianuensem ex mandato Regis Castelle'.

13 *Geographia* 1513, 'In Claudii Ptolemei Supplementum', title page.

14 Claudius Ptolemaeus, *Geographia* (Venice, 1511; facsimile ed. Amsterdam: Theatrum Orbis Terrarum, 1969); on Sylvanus see Karrow, *Mapmakers of the Sixteenth Century*, pp. 520–4.

15 Skelton, 'Bibliographical Note', in *Geographia* (Venice, 1511; facsimile ed. 1969), p. vii.

16 Letter to don Diego Colón of 5/2/1505, in Christopher Columbus, *Lettere e scritti (1495–1506)*, ed. Paolo Taviani and Consuelo Varela, 2 vols (Rome: Nuova Raccolta Colombiana, 1993), vol. 2, p. 408.

17 The number of voyages completed by Vespucci (two, three, or four) remains one of the many 'Vespuccian questions': see the helpful summary of twentieth-century Vespucci criticism in Luzzana Caraci, *Amerigo Vespucci*, vol. 2, pp. 25–34; Luzzana Caraci argues that the second, third, and fourth voyages can be attested, but that the first (1497–8) remains uncertain: *Amerigo Vespucci*, vol. 2, pp. 119–26; 245–56.

18 For an overview see the discussion in *Letters from a New World: Amerigo Vespucci's Discovery of America*, ed. Luciano Formisano (New York: Masilio, 1992), pp. xxx–xxxv. Amongst the abundant literature on this subject important interventions were made by Magnaghi, *Amerigo Vespucci*, and Giuseppe Caraci, *Questioni e polemiche vespucciane*, 2 vols (Rome: Istituto di scienze geografiche e cartografiche, 1955–6). See more recently Luzzana Caraci, *Amerigo Vespucci*, vol. 2, pp. 35–99. Wallisch attempts to rehabilitate the *Mundus Novus* as authentic: *Der* Mundus Novus *des Amerigo Vespucci*, ed. Robert Wallisch (Vienna: Österreichischen Akademie der Wissenschaften, 2002), pp. 104–13.

19 See the analysis in Luzzana Caraci, *Amerigo Vespucci*, vol. 2. pp. 41–55.

20 *Letters from a New World*, p. xxxv.

21 On this matter see Ilaria Luzzana Caraci, *The Puzzling Hero: Studies on Christopher Columbus and the Culture of his Age*, trans. Mayta Munson (Rome: Carocci, 2002), pp. 167–201.

22 Luzzana Caraci argues plausibly that the rapid dissemination of Vespucci's work was the result of a network of Italian and German mercantile interests: *Amerigo Vespucci*, vol. 2, pp. 72–7.

23 See Laubenberger, 'The Naming of America', pp. 91–113; Harold Jantz, 'Images of America in the German Renaissance', in *First Images of America: The Impact of the New World on the Old*, ed. Fredi Chiappelli *et al.*, 2 vols (Berkeley: University of California Press, 1976), vol. 1, pp. 91–106: pp. 98–100.

24 *Der* Mundus Novus *des Amerigo Vespucci*, ed. Wallisch, p. 12

25 *Der* Mundus Novus *des Amerigo Vespucci*, ed. Wallisch, p. 12.

26 Peter Martyr to the Archbishop of Braga, 1.10.1493, *Opus epistolarum* 136: 'Colonus quidam occiduos adnauigavit – ad littus usque indicum (ut ipse credit) – antipodes'; Peter Martyr, *De orbe novo*, 1.10.4, in *Selections from Peter Martyr*, ed. and trans. Geoffrey Eatough, Repertorium Columbianum vol. 5 (Turnhout: Brepols, 1998), pp. 498, 111, 197.

27 Vespucci to Lorenzo di Pierfrancesco de' Medici, Lisbon, 1502, in Luzzana Caraci, *Amerigo Vespucci*, vol. 1, p. 290: 'fui alla parte delli antipoti, che per mia navicazione fu una ¼ parte del mondo'; *Letters from a New World*, trans. Formisano, p. 30.

28 Vespucci to Lorenzo di Pierfrancesco de' Medici, Seville, 1500: Luzzana Caraci, *Amerigo Vespucci*, vol. 1, p. 271; *Letters from a New World*, trans. Formisano, p. 7.

29 *Der* Mundus Novus *des Amerigo Vespucci*, ed. Wallisch, p. 26: 'in illo hemisperio vidi res philosophorum rationibus non consentientes'.

30 *Der* Mundus Novus *des Amerigo Vespucci*, ed. Wallisch, p. 28. On the translation of 'in costas' as 'onto the back' rather

than the more literal 'into the ribs' see Wallisch's comments: *Der* Mundus Novus *des Amerigo Vespucci*, pp. 95–6. The imaginary observers in the old and new worlds are all facing south. Cf. Vespucci's third familiar letter: 'the highest point of my zenith in those regions made a spherical right angle with the inhabitants of the northern hemisphere, who are at a latitude of forty degrees' (el punto del mio zenih più alto in quelle parte faceva uno angolo retto sperale colli abitanti di questo settantrione, che sono nella latitudine di 40 gradi): Luzzana Caraci, *Amerigo Vespucci*, 1. p. 290; *Letters from a New World*, trans. Formisano, p. 30.

31 Matthias Ringmann, *De ora antarctica* (Strasbourg, 1505), sig. A.i (verso). On Jacques Brun see Charles Schmidt, *Histoire littéraire de l'Alsace*, 2 vols (Paris: Sandoz and Fischbacher, 1879), vol. 2, p. 94 *n*24.

32 Ringmann, *De ora antarctica*, sig. A.ii (recto).

33 *Der* Mundus Novus *des Amerigo Vespucci*, ed. Wallisch, p. 18.

34 'Lettera al Soderini', in Luzzana Caraci, *Amerigo Vespucci*, vol. 1, pp. 374–5.

35 'Lettera al Soderini', vol. 1, pp. 374–5.

36 *Mapas españoles de América: siglos XV–XVII*, ed. A. Altolaguirre y Duvale *et al.* (Madrid: Real Academia de la Historia, 1951), p. 16.

37 For the suggestion that the St Christopher figure is a coded reference to a putative strait providing access to the Indian sea and the island of Taprobana, see Ricardo Cerezo Martínez, 'La carta de Juan de la Cosa (I)', *Revista de Historia Naval* 10 (1992), 31–48, esp. pp. 46–7.

38 *Mapas españoles de América*, p. 11.

39 Shirley, *Mapping of the World*, pp. 23–7, nos 24 and 25.

40 *A Map of the World Designed by Giovanni Matteo Contarini, Engraved by Francesco Roselli 1506*, 2nd edn (London: British Museum, 1926), p. 8.

41 Shirley, *Mapping of the World*, p. 32, no. 28. A similar austral landmass is represented on another map produced by Rosselli around the same time, this one a copper engraving rather than a woodcut: Shirley, *Mapping of the World*, p. 33, no. 29. On the emergence of the printed map see in general David Woodward, *Maps as Prints in the Italian Renaissance: Makers, Distributors and Consumers* (London: British Library, 1995).

42 Roberto Almagià, 'On the Cartographic Work of Francesco Rosselli', *Imago Mundi* 8 (1951), 27–34, p. 33.

43 Svat Soucek, *Piri Reis and Turkish Mapmaking after Columbus: The Khalili Portolan Atlas* (Oxford: Oxford University Press, 1996), p. 60.

44 *Portugaliae Monumenta Cartographica*, ed. Armando Cortesão and Avelino Teixeira da Mota, 6 vols (Lisbon: Commissão Executiva das Comemorações do V Centenário da Morte do Infante D. Henrique, 1960), vol. 1, pp. 55–61, pl. 16; for the argument that Homem's map relocates the Ptolemaic terrestrial continuum southwards in the still unknown austral regions as a manifestation of ancient 'terrestrial theory' (i.e. that ocean is surrounded by land and not vice-versa), see Francesc Relaño, 'Cartography and Discoveries: the Re-Definition of the Ptolemaic Model in the First Quarter of the Sixteenth Century', in *La cartografia europea tra primo rinascimento e fine dell'Illuminismo*, pp. 49–61.

45 Yusuf Akçura, *Piri Reis haritasi* (Istanbul: Devlet Basimevi, 1935), p. 15.

46 Piri acknowledged his cartographic sources in his later work, the *Kitab-i Bahriye* (Book of maritime matters), as a mixture of medieval *mappaemundi*, Arabic maps, and recent charts such as the 'one drawn by Columbus': Soucek, *Piri Reis and Turkish Mapmaking*, pp. 48–79.

47 Soucek, *Piri Reis and Turkish Mapmaking*, p. 73; Gregory C. McIntosh argues for Portuguese cartographic sources: *The Piri Reis Map of 1513* (Athens: University of Georgia Press, 2000), pp. 50–2.

48 McIntosh, *The Piri Reis Map of 1513*, p. 49 notes also the *Suma de geographia* of 1519 as a possible source for this statement.

49 See further Soucek, *Piri Reis and Turkish Mapmaking*, pp. 64–72.

50 Translation and discussion in Soucek, *Piri Reis and Turkish Mapmaking*, pp. 58–60.

51 Soucek, *Piri Reis and Turkish Mapmaking*, p. 50.

52 See the overview in Denis Cosgrove, *Apollo's Eye: A Cartographic Genealogy of the Earth in the Western Imagination* (Baltimore: Johns Hopkins University Press, 2001), pp. 55–8. See also Elly Dekker, 'Globes in Renaissance Europe', in *The History of Cartography*, vol. 3, pp. 135–59.

53 Strasbourg, 1509. This pamphlet was accompanied by a world map, which forms the basis for the attribution: see Karrow, *Mapmakers of the Sixteenth Century*, pp. 573–4.

54 E. G. Ravenstein, *Martin Behaim: His life and his globe* (London: Philip, 1908), p. 71 (Ravenstein's translation). Behaim's claim has been called into question: see Ravenstein, *Martin Behaim*, pp. 20–30; and the collection *Focus Behaim Globus*, ed. Gerhard Bott, 2 vols (Nuremberg: Verlag des Germanischen Nationalmuseums, 1991), vol. 1, pp. 173–308, especially Johannes Willers, 'Leben und Werk des Martin Behaim', pp. 173–88, which offers revisions to Ravenstein's account; see also Hans Werner Nachrodt, 'Martin Behaim und sein "Erdapfel": Bemerkungen zur Lebens- und Wirkungsgeschichte', *Jahrbuch für Frankische Landesforschung* 55 (1995), 45–64; and, on the commercial impetus of the globe, Jerry Brotton, *Trading Territories: Mapping the Early Modern World* (London: Reaktion, 1997), pp. 67–71.

55 Ravenstein, *Martin Behaim*, p. 71. For the argument that this inscription was added not by Behaim but by the Nuremberg council after the completion of the globe (possibly in 1494, possibly as late as Behaim's death in 1507) see Willers, 'Leben und Werk des Martin Behaim', and 'Die Geschichte des Behaim-Globus', in *Focus Behaim Globus*, pp. 184, 209.

56 This practice was relatively common: the coat of arms of Cardinal Ferrero, Bishop of Bologna, is painted near the south pole on a celestial globe possibly possessed by Pope Julius II: Edward Luther Stevenson, *Terrestrial and Celestial Globes: Their history and construction including a consideration of their value as aids in the study of geography and astronomy*, 2 vols (New Haven: Yale University Press, 1921), vol. 1, p. 63.

57 Paris, BNF, Cartes et Plans, Rés. Ge A 335. Chet Van Duzer points out that the river strongly resembles the mountain ranges on the ring and suggests that it was labelled 'flumo' to rectify a mistake made by the cartographer, who was copying from another source: 'The Cartography, Geography, and Hydrography of the Southern Ring Continent, 1515–1760', *Orbis Terrarum* 8 (2002), 115–58: p. 127.

58 Pelletier, 'Le globe vert', p. 24.

59 There are clear similarities of toponymy: for comparison see Van Duzer, 'Southern Ring Continent', pp. 126–7.

60 Pelletier, 'Le globe vert', pp. 17–31. Note also the similarities between the 'Globe vert' and the early sixteenth-century Acton globe: they possess similar inscriptions, as well as an Antarctic continent: Stevenson, *Terrestrial and Celestial Globes*, vol. 1, pp. 79–80.

61 Pelletier, 'Le globe vert', p. 19.

62 Franz von Wieser, *Magalhães-Strasse und Austral-Continent auf den Globen des Johannes Schöner* (Innsbruck: Verlag der Wagner'schen Universitäts-Buchhandlung, 1881), pp. 28–99. Von Wieser argued also that the 'Zeytung' came to Germany as a result of the Augsburg-based Welser merchant house (pp. 96–9).

63 'Copia der Newen Zeytung auss Presillg Landt' (Augsburg), p. 2: facsimile printed in H. H. Bockwitz, 'Die "Copia der Newen Zeytung auss Presillg Landt"', *Zeitschrift des Deutschen Vereins für Buchwesen und Schrifttum* 3 (1920), 27–35.

64 'Copia der Newen Zeytung auss Presillg Landt', p. 2.

65 'Copia der Newen Zeytung auss Presillg Landt', p. 2.

66 'Copia der Newen Zeytung auss Presillg Landt', p. 3.

67 'Copia der Newen Zeytung auss Presillg Landt', pp. 3, 5.

68 J. Enterline, 'The Southern Continent and the False Strait of Magellan', *Imago Mundi* 26 (1972), 48–58.

69 Johann Schöner, *Luculentissima descriptio* (Nuremburg, 1515), sig. L.ii recto–verso.

70 'Copia der Newen Zeytung auss Presillg Landt', p. 2.

71 *Luculentissima descriptio*, sig. L.ii, recto.

72 The 1520 globe is just under 90 centimetres in diameter as against the 26.8 centimeter diameter of the 1515 globe: Oswald Muris and Gert Saarmann, *Der Globus im Wandel der Zeiten* (Berlin: Columbus Verlag Paul Oestergaard, 1961), pp. 80–1.

73 Muris and Saarmann, *Der Globus im Wandel der Zeiten*, p. 82.

74 Karrow, *Mapmakers of the Sixteenth Century*, p. 179; von Wieser argued that the influence could only have gone one way, i.e. from Schöner to Fine and not vice-versa: *Magalhães-Strasse und Austral-Continent*, pp. 78–81; see the arguments to the contrary of Gallois, *Les géographes allemands*, pp. 92–6.

75 See Richardson, 'Mercator's Southern Continent', p. 82. A number of globes from the late 1520s and early 1530s testify to a strengthened interest in the austral continent, probably as a result of the influence of Schöner's globes: these include the 'Gilt globe' (c. 1528) and the 'Nancy globe' (c. 1530): Stevenson, *Terrestrial and Celestial Globes*, vol. 1, pp. 98, 106.

76 Johann Schöner, *Opusculum geographicum* (Nuremberg, 1533), chapter 20 ('De regionibus extra Ptolemaeum'): 'eam terram inuenerunt esse continentem superioris Indiae, quae pars est Asiae'.

77 See Karrow, *Mapmakers of the Sixteenth Century*, p. 171, and Shirley, *Mapping of the World*, pp. 72–3, 77 (nos 66 and 69).

78 Roberto Almagià, *Monumenta Cartographica Vaticana*, 4 vols (1944–55), vol. 1: *Planisferi, carte nautiche e affini dal secolo xiv al xvii esistenti nella Biblioteca Apostolica Vaticana* (Vatican City: Biblioteca Apostolica Vaticana, 1944), p. 59

79 Almagià, *Monumenta Cartographica Vaticana*, vol. 1, p. 58. More recently the map has been discussed by Chet Van Duzer, 'Cartographic Invention: The Southern Continent on Vatican MS Urb. Lat. 274, fols. 73v–74r (c. 1530)', *Imago Mundi* 59 (2007), 193–222. Van Duzer argues that some of the toponyms on the map's southern continent derive from Columbus' fourth voyage.

80 Almagià, *Monumenta Cartographica Vaticana*, vol. 1, p. 59.

81 Almagià read 'iudei' for 'Rubri': *Monumenta Cartographica Vaticana*, vol. 1, p. 59.

82 Almagià, *Monumenta Cartographica Vaticana*, vol. 1, p. 58.

83 S. Crinò, 'I mappamondi di Francesco Rosselli', *La Bibliofilia* 41 (1939), 381–405, p. 394.

84 Almagià remarks on the 'notevoli analogie' between the Vatican map and Rosselli's planispheres: *Monumenta Cartographica Vaticana*, vol. 1, p. 59.

85 Thomas More, *Utopia*, ed. and trans. George M. Logan, Robert M. Adams, and Clarence H. Miller (Cambridge: Cambridge University Press, 1995), pp. 42–3. All translations are from this edition.

86 *Utopia*, pp. 44–5.

87 *Utopia*, pp. 44–8.

88 *Utopia*, pp. 14–15; my emphasis.

89 *Utopia*, pp. 180–4; 218–21.

90 The first edition of this work is undated, but a reference to land discovered at 52 degrees south in 1526, the date of the discovery of South Georgia, establishes the work's *terminus post quem*. A further edition appeared in 1565, lacking the map of the title page: Karrow, *Mapmakers of the Sixteenth Century*, pp. 407–8; Shirley, *Mapping of the World*, p. 61 (no. 57). Van der Krogt argues for a date of 1526–7 for the text and the terrestrial globe it accompanied: Peter van der Krogt, *Globi Neerlandici: The production of globes in the Low Countries* (Utrecht: HES, 1993), pp. 43–4.

91 Karrow, *Mapmakers of the Sixteenth Century*, pp. 407–9; van der Krogt, *Globi Neerlandici*, pp. 42–8. Franciscus Monachus was born in Mechelen, and was a student at the University of Louvain.

92 *De orbis situ* (Antwerp, 1565), sig. F4r.

93 *De orbis situ*, sig. Gv–G2r.

94 *De orbis situ*, sig. F4v. De Smet argues for the influence of Franciscus' globe on Fine's world maps of 1531 and 1534, based on the representation of Asia and America as a single landmass, and the similarity of inscription on the southern continent: 'L'orfevre et graveur Gaspar Vander Heyden et la construction des globes à Louvain dans le premier tiers du XVIe siècle', *Der Globusfreund* 13 (1964), 38–48, pp. 43–4.

95 The Treaty of Saragossa of 1529, in which the King of Castile gave his right to the Moluccas and other neighbouring lands to the King of Portugal for 350,000 ducats, established a line of demarcation in the Pacific 'from pole to pole' by reference to two 'padrones' (navigational charts): *European Treaties*, pp. 174/188.

96 *De orbis situ* , sig. H2r.

97 *De orbis situ*, sig. G8v.

Chapter Eight

Still Unexplored:
Terra Australis Incognita *1531–1610*

The historian of geography Numa Broc has written that in the sixteenth century the representation of unknown lands such as *Terra Australis* was 'characteristic of the intellectual fermentation of the Renaissance', the product of a thirst for knowledge but also of a 'disorderly' curiosity:

> Assailed by new information but incapable of distinguishing true from false, avid for 'curiosities' and for the marvellous, geographers anticipated the reality and filled the lacunae of knowledge with the imagination. Thus can be explained that 'horror of the void' which provides at once the charm and the weakness of the cartographic production of the Renaissance. Without doubt, at the origin of geographic myths we find nearly always some grain of truth, but that often modest truth is embellished, exaggerated to give birth to grandiose constructions of the mind.[1]

These lines exemplify many easy – and initially compelling – assumptions about the significance of unknown lands in sixteenth-century cartography. At first glance the proposition that geographers were unable to distinguish false from true information, and consequently represented the fictional as well as the real on their maps, seems self-evident. But closer examination of cartographic practices during this period suggests several problems with Broc's account. To begin with, the opposition between true and false risks misunderstanding and misrepresenting sixteenth-century geography. I do not mean that mapmakers of the era did not distinguish between truth and falsehood, or that they were uninterested in true representations, or unconcerned to eliminate false information from their maps; on the contrary. However, they operated primarily on axes slightly different to that of truth-falsehood: sixteenth-century geographers, like their medieval counterparts, dealt in certainties and uncertainties, distinguishing between probabilities and improbabilities, the attested and the unattested. Far from being avid for 'curiosities', many mapmakers were notably resistant to the marvellous – at least in the sense of the fantastic or monstrous. If they placed a scene of cannibalism on a New World interior, it was often because they had read a reliable report of such practices, which they then interpreted in visual form. If they sought information about newly documented peoples it was not out of mere curiosity, but frequently from a sense of Christian universalism. Cartographers supplied the lacunae of knowledge with calculation as well as imagination. The unknown southern land might have been a fiction, but its shape and features were not random: they were the result of careful consideration and interpretation, however speculative. Further, it is a mistake – although a common one – to assign to 'horror vacui' the role of driving force of cartographic production. A better expression would be the manipulation, rather than abhorrence, of the void. The void, the unknown, was within the scheme of representation – hardly evidence of dread. Finally, Broc's notion that geographic myths were the products of the embellishment or exaggeration of truths is also initially appealing. But what was more true for sixteenth-century geographers: the belief in a significant austral continent, based on theories of climate and land-to-sea ratio, the product of a long-standing, if now modified, tradition of antipodal lands – or the accounts of explorers who had discovered (or seemed to have discovered) land at various points on this austral landmass? The answer is that they were equally true. *Terra Australis* was not an exaggeration of the truth of exploration, and nor were the accounts of exploration exaggerations of *Terra Australis*; the process was not that of the formation of grandiose embellishment from a kernel of truth, but rather of the conjunction of theory with narrative.

The shape of New World regions subject to a process of exploration and/or colonization required fairly constant updating, and the need to redraw maps was of course a boon to those who were in the business of producing a seemingly infinite succession of 'new' and 'exact' descriptions of the world. However, the new-old world of the southern continent – the familiar conjecture of classical and medieval geography, the space beyond – was destined to be explored by the European imagination even as it was sought out by its navigators. The means of filling the southern land became increasingly varied and flamboyant over the course of the century. Images of animals, plants, and people vied with words for the indeterminate space. The meta-cartographic function of antipodal land reached its zenith in the magnificent wall maps of Gerard Mercator and his imitators, as great panels of explanatory text filled *Terra Australis*. Other cartographers, such as the Italians Giacomo Gastaldi and Paolo Forlani, resorted to zoological and botanical display, occupying the southern land with a variety of plants and animals real and imaginary, ancient and modern.[2] Still others followed the principle of slippage from the New World, using the space of the antipodes for ethnographic drawings of South American and south-east Asian natives. By the last decade of the sixteenth century, and the first decade of the seventeenth, cartographers and writers had begun to use the southern land as the locus of allegory and satire, directed, in a return to Menippean traditions, at the follies of the known world. The elusive status of *nondum cognita* had not been resolved by discovery and exploration, so the blank space of *terra incognita* invited an efflorescence of representation: a flourishing of images, but also the flourishing of the idea of the unknown southern land, with all its possibilities, mirror of the Old World, mimic of the New.

The triumph of meta-cartography: Mercator's Third World

The key figure in the sixteenth-century consolidation of the southern continent was Gerard Mercator. From his first world map of 1538, to his masterpiece, the wall map of 1569, Mercator developed and promoted the idea and the image of the southern continent. He was well versed in the theory of the antipodes, which was a consistent feature of the textbooks produced by his teacher at Louvain, Gemma Frisius, and by Peter Apian and others from the 1530s onwards.[3] Mercator collaborated on Gemma's terrestrial globe of 1536, on which an incomplete southern land appears, with coastlines marked only beneath Africa and to the south of the Strait of Magellan.[4] During the same period, Mercator seems to have developed the mathematical theories deployed by Johann Schöner and Oronce Fine on their maps and globes. Mercator's 1538 map of the world is clearly indebted to Fine's map of 1531: both are double cordiform (i.e. consisting of two heart-shaped hemispheres), and apart from the fact that Mercator, unlike Fine, represents America as a continent separate from Asia, both show similar geographical features, including a significant Antarctic continent. Fine's *Terra Australis*, replete with a pseudo-topography consisting of mountain ranges and rivers, extended well beyond the Antarctic circle, but did not cross the Tropic of Capricorn, as more extensive southern continents were to do. Mercator removed the mountains and rivers from an essentially identical outline, and added the inscription 'it is certain that there are lands here, but how great their extent, and with what borders is uncertain' (Terras hic esse certum est sed quantas quibusque limitibus finitas incertum).

Three years later, when he produced his terrestrial globe of 1541 (fig. 38), Mercator had made some significant alterations to the image of the southern continent.[5] This globe, designed specifically for navigators, shows a southern landmass basically similar to that of the 1538 map, but with two significant additions: a 'Region of Parrots' (Psitacorum regio) to the south of Africa,[6] and a promontory in south-east Asia that extends above the Tropic of Capricorn to a point just south of 'Java Major' (whose inhabitants, a caption notes, are cannibals).[7] On this promontory are located the province of 'Beach' and the kingdom of 'Maletur'. Both these toponyms, as Mercator explains in a caption underneath them, were derived from the *Divisament dou monde* of Marco Polo and confirmed by the early sixteenth-century itinerary of Lodovico di Varthema: 'he who reads book 3, chapters 11 and 12 of the Venetian M. Polo, and compares it with book 6, chapter 27 of Lodovicus Patricius of Rome will easily believe that there are most vast regions here' (Vastissimas hic esse regiones facile credet qui 11 & 12 caput lib: 3 M: Pauli Veneti legerit collato simul 27 capite libri 6 Lud: Rom: Patricij). Mercator was referring specifically to the versions of Polo and di Varthema contained in the 1532 edition of the Basel theologian Simon Grynaeus' *Novus Orbis Regionum ac Insularum Veteribus Incognitarum*.[8] In book 3, chapter 11 of the *Divisament* in Grynaeus' edition, Polo describes 'Boeach' as a very large and well-endowed region south of Java, with its own king and language, a fierce and idolatrous people, bears, elephants and gold. In chapter 12 he mentions the kingdom of Maletur, 'where there is a great wealth of spices, and the inhabitants have their own language' (ubi maxima est copia aromatum: & habent incolae eius propriam linguam).[9] For Mercator it was significant that the version of Polo he read did not refer to Beach and Maletur as islands but, by using the terms 'provincia' and 'regnum', implied that they were part of a continent, located somewhere to the south of Java.[10] Corroboration appeared to come from the *Navigationes* of di Varthema, a native of Bologna who reached India in 1505 and claimed to have travelled from there to parts of south-east Asia, including Java and the Spice Islands.[11] Di Varthema referred to certain inhabitants of

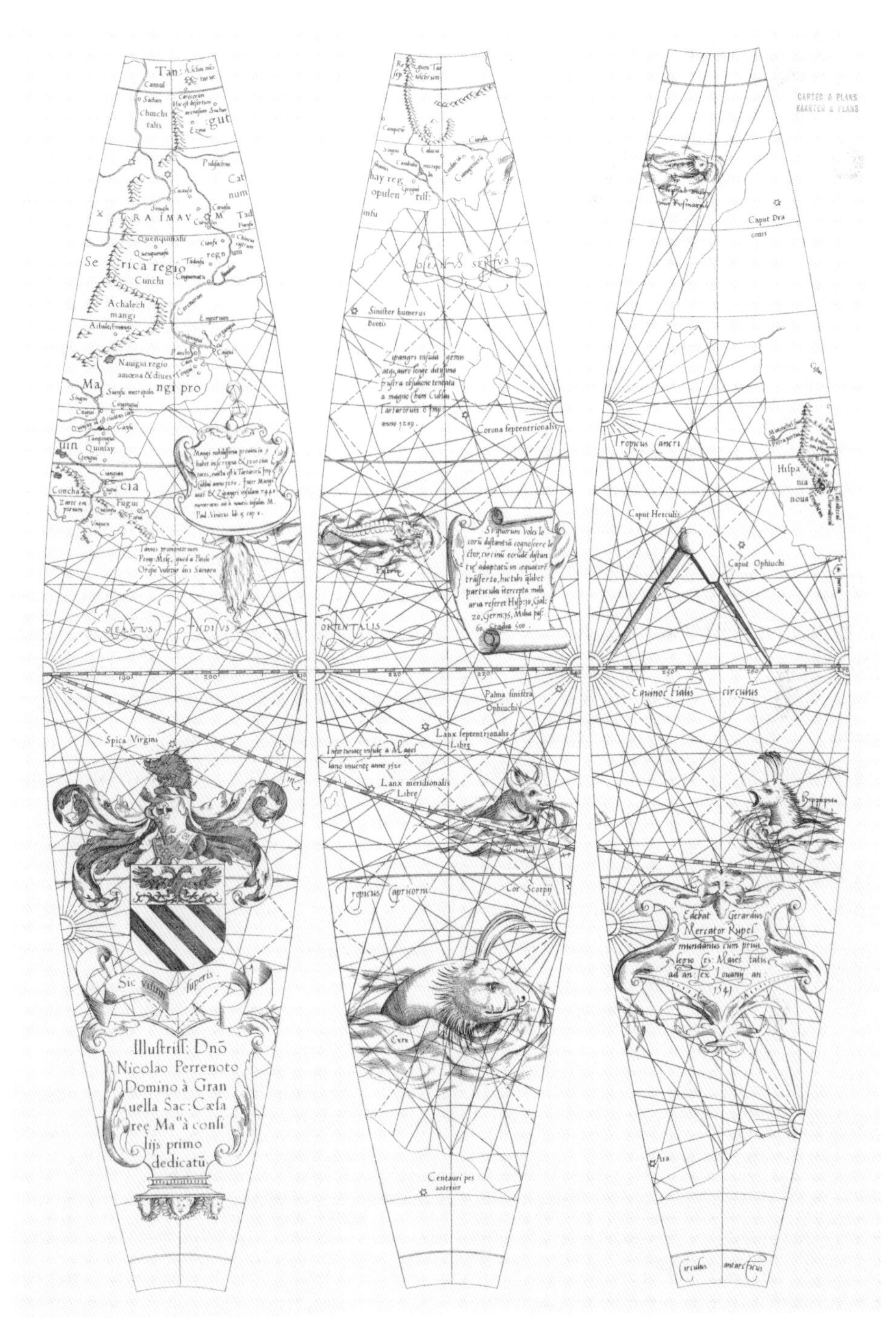

Fig. 38. Gerard Mercator, Globe gores of 1541 (detail). Brussels, Bibliothèque royale, 7D 148 (RP). Mercator's globe of 1541 contains a prominent southern continent. The innovations to the image of the southern land on this globe include the addition of a 'Region of Parrots' (Psitacorum regio) to the south of Africa, and a province of 'Beach' and kingdom of 'Maletur' on a promontory that extends above the Tropic of Capricorn. As Mercator made clear, he derived these two toponyms from the travels of Marco Polo. The southern continent is also the location for a dedicatory cartouche to Mercator's patron Nicolas Perrenot de Granvelle, the chief advisor to Emperor Charles V.

Java as 'antipodeans' in relation to north-eastern Europe, and reported the habitation of the frigid zone around the Antarctic pole. Mercator's inscription testifies to the importance of medieval and early modern narrative description in the formation of the world image. The explicit citation of sources also reveals a new use for the space of the southern continent: that of bibliography, a function reiterated further along the continent by the remark that 'our little book will indicate where and with which arguments from whose edition, reader, we described' (Vbi & quibus argumentis Lector ab ahorum desciuerimus aeditione libellus noster indicabit).[12] The southern land draws comment, and requires explanation; it becomes a discursive space in which the mapmaker addresses, instructs, and answers the imputed questions of the reader.

On Mercator's globe of 1541, *Terra Australis* also provided space for the acknowledgement of patronage. Mercator interrupted the imagined coast of the southern continent to insert a dedicatory cartouche to Nicolas Perrenot de Granvelle, the diplomat and chief advisor to Emperor Charles V.[13] In a letter of 1540 Mercator had described his project of a geographical globe to Perrenot's son, Antoine, bishop of Arras. From the letter, in which Mercator records certain of his 'inventiones' – features of the globe, particularly in Asia, that differ from contemporary representations and which were based on a critical study of Ptolemy and Strabo – at least four audiences for the globe can be discerned. Intended for navigators (it is replete with aids to navigation such as compass roses, loxodromic lines and the locations of stars), it was also designed for a mercantile audience interested in emerging markets, and for 'studiosi', learned men familiar with classical and more recent geographical authority; it was, finally, prompted by the interest and encouragement of influential patrons.[14] One of the points of conjunction of these audiences is land 'about to be known'. Mercator's cartography was one of future as well as present spaces, and those spaces were of primary importance to Europe's ministers, the navigators they employed to explore them, traders who stood to profit from them, and the geographers who represented them on maps and globes.

Mercator was convinced that the southern continent, which at this stage he termed the fifth part of the world, not only existed, but was greater than all other parts. As he put it in an inscription located on the Antarctic pole of his 1541 globe: 'this fifth and indeed, as one may conjecture, most ample part recently acceded to our globe, but in truth as yet it has been explored in only a few shores' (QVINTA haec, & quidem amplissima pars, quantum coniectare licet, nuper orbi nostro accessit, verum paucis adhuc littoribus explorata). Mercator's explanation for this belief is clearly set out in one chapter of his treatise 'De creatione et fabrica mundi', an essay on the creation of the earth that was incorporated in his *Atlas*, first published posthumously in 1595. There he explains that the equilibrium of the 'machine of the land and sea' (machina terrae et maris) depends on a balance of the two elements; hence it is necessary for landmasses of equal weight to be found in all parts of the sphere. If the ancients had known and examined these principles they would have determined the position and size of the new continent 'discovered by our age', and the southern continent as yet unexplored 'lying beneath the Antarctic pole':

> Etenim cum terrae veteribus cognitae 180. gradibus longitudinis comprehendantur, hoc est, dimidiam tantum sphaeram occupent, necessarium erat tatundem terrarum in altera medietate extare.

> Et cum Asia, Europa & Africa pro maxima parte vltra aequinoctialem, versus boream sint sitae, necesse erat tantam continentem sub antarctico poli existere, quae cum Asiae et nouae Indiae, siue Americae partibus meridianis, reliquis terris aequiponderaret.[15]

> For, since to the ancients known lands were contained within 180 degrees of longitude, that is, they occupied just half the sphere, precisely as much land had to exist in the other half. And since Asia, Europe and Africa for the most part are located beyond the equator, towards the north, it followed that an entire continent existed beneath the Antarctic pole, which along with the southern parts of Asia and New India or America equaled [the northern hemisphere] in weight.

Mercator conceived of his geographic projects not simply in terms of the provision of accurate, useful, and up-to-date images of the world, but also in terms of the reconstitution of classical knowledge. In the 1560s he planned and began work on a great cosmography; this was never realized, although a number of its constituent parts were completed in his lifetime. The work was to include both the modern *Atlas* eventually published in 1595 and the edition of Ptolemy published by Mercator in 1578, as well as 'De creatione et fabrica mundi'. According to his friend and biographer, Walter Ghim, book five of Mercator's projected cosmography was to have been devoted to geography. It would, Ghim stated, have adopted a revolutionary arrangement which no one had thought of or attempted before:

> orbem in tres aequales continentes dividere decreverat, quarum unam Asiam, Africam et Europam constituisset, alteram Indiam occidentalem cum omnibus regnis et provinciis illi contiguis, tertiam vero, etsi adhuc latentem et incognitam esse non ignoraverit, solidis tamen rationibus atque argumentis demonstrare ac evincere se posse affirmabat; illam in sua proportione geometrica, magnitudine et pondere ac gravitate ex duabus reliquis nulli cedere aut inferiorem vel minorem esse posse, alioquin mundi constitutionem in suo centro non posse consistere, haec Australis continens a scriptoribus appellatur.

> he had determined to divide the world into three equal continents, one comprising Asia, Africa, and Europe, the other West India with all its adjacent kingdoms and provinces, and a third, which he realized was unknown and still awaiting discovery, but whose existence he thought he could clearly prove by solid reasoning and argument. It could not be less in its geometric proportions, size, weight, and gravity than the other two, otherwise the world would be unable to remain balanced on its axis. Writers call this the Southern Continent.[16]

Mercator's giant world map in twenty-one sheets of 1569, which featured a vast southern continent (fig. 39), appears to confirm Ghim's account of his theories. Mercator intended the 1569 map to be used by mariners – it is a description explicitly 'ad usum navigantium' – as well as by overland travellers and scholars.[17] He was consequently at pains to make sure that the map represented the size and shape of lands and seas as exactly as possible, and that it facilitated the accurate calculation of distances between places. One of the techniques Mercator used to achieve these objectives was the representation of meridians of longitude as parallel, rather than curved, lines; to do this, however, he was obliged to increase the distance between parallels of latitude as they approached the poles, resulting in the appearance of enlarged northern and southern regions, and the absence of the poles themselves. This innovation enabled rhumb-lines (lines that intersected all meridians at the same angle), to be represented by a straight line, rather than

Fig. 39. Gerard Mercator, 'Nova et aucta orbis terrae descriptio ad usum navigantium emendata et accommodata' (Duisburg, 1569). Paris, BNF, Cartes et Plans, Rés. Ge A 1064. Mercator's world map of 1569 is one of the masterpieces of cartography. Extending across twenty-one sheets, it is most famous for introducing the Mercator projection, but one consequence of this mode of representation was the enlargement of the northern and southern regions, and the absence of the poles. Mercator filled the expanse of *Terra Australis* with an array of cartouches, devoted variously to a description of the northern polar regions and the history of their exploration, a disquisition on classical geographical knowledge of the Ganges, and technical information such as the 'Method of measuring the distance between places', and the 'Use of the Diagram of Courses'. In between these blocks of text Mercator retained the now relatively standard pseudo-topography of the southern land

curved as they were drawn on spheres.[18] How, given Mercator's belief in the 'third part of the world', was the space of the southern continent, known only in a few shores, to be filled? The vast, hypothetical southern continent of the 1569 wall map contains five cartouches dense with text, a view of the northern polar region, with its own explanatory cartouche, and a 'diagram of routes' (organum directorum). The quantity of text on the continent is immense, and overwhelmingly concerned with other parts of the world and geographical theory; from time to time a 'Pars Continentis Australis' peeps through this series of tables, but for the most part *Terra Australis* has been overlaid by a wealth of explanation, exemplification, and dedication. Two of the cartouches deal with the history of geography: the view of the north pole is accompanied by a description of the northern regions and the history of their exploration. A long rectangular panel beneath 'Psitacorum regio', meanwhile, argues that classical geographers such as Ptolemy were well aware of the correct location of the Ganges, and had mapped Asia to its easternmost point. Other cartouches provide practical instructions for using the map: these offer advice on the 'Method of measuring the distance between places' (Distantiae locorum mensurandae modus), and the 'Use of the Diagram of Courses, in brief' (Brevis Usus Organi Directorii), accompanied by an illustration. The cartouche could also be used to make statements of political and personal interest: an inscription beneath Terra del Fuego notes Alexander VI's 1493 bull of demarcation, while an oval panel beneath 'Beach' asserts the royal and imperial privileges granted to Mercator for the publication of the map.

Distinct from these cartouches are the continent's few topographical reference points: 'Terra del Fuego', 'Promontorium Terrae Australis', 'Psitacorum regio', and the regions of 'Beach', 'Lucach',[19] and 'Maletur', located on a promontory that extends to 15°S, adjacent to Java Minor. A nearby chunk of text explains the derivation of the toponyms from the accounts of Polo and di Varthema, and also cites *Mandeville's Travels*; two chunks of text further west explain the basis for 'Psitacorum regio' (represented also by two large parrots) in Portuguese navigations, and the nature of ocean currents in the area. These occasional reference points serve to remind the reader/viewer that while this space appears as a backdrop to learned essays, treatises, and panegyric, it still has a geographical signification. To read the southern continent in its entirety on Mercator's map is therefore to shift from one form of reference to another: the first cartographic, the second meta-cartographic. Mercator uses the agency of the cartouche in *Terra Australis*, North America, and the south and north seas to provide a plurality of descriptions and declarations. To read the words in these cartouches is to be drawn away from the text of the map, to texts literally and figuratively *on and about* the map. It is to move from a reading of topography, of the shape and names of land and sea, to a reading of text that refers not only to the map, but to the map's construction and history, and that draws the reader's attention away from the southern land, redirecting it towards places as far flung as the north pole and the Ganges.

It seems likely that Mercator's 1569 map was designed to be read in two forms: as a wall map, and divided up into atlas form. The latter is the form of one of the few surviving copies of the map;[20] moreover, atlas form would presumably have been more practical 'ad usum navigantium' than a multiple-sheet wall map. The map was conceived, then, in both monumental and codex form, as an item of display and as a text for private study. These different forms necessitate, of course, different modes of interaction with the text. In atlas form some of the panels – the address to the reader; a poem to Mercator's patron, the Duke of Juliers, Cleves, and Berg; the notice of imperial privilege; the discursus on the Ganges – have been removed from the map and placed on the title-page and following pages. This proves the point that the map incorporates the paratext of the book in its display; what it also suggests is that the obvious location of paratextual display is *terra incognita*. In its form as a wall map, actually to read the words in the southern continent the reader must be up close to the map; moreover, the proximity demanded by the act of reading the cartouches is such that it is difficult for more than one reader to see this text at the same time. The spaces of the southern continent and the other areas of *terra incognita* filled with panels of text – the address to the reader in the interior of North America, and a disquisition on Prester John in North Asia, to cite just two examples – invite the gaze of a lone reader, unlike the space of the whole map which may be read by a number of people at the same time. Yet such is the illegibility of the words in these panels when seen from any distance that one must conclude that the multiplicity of text on *terra incognita* – and especially *Terra Australis* – is designed to be seen primarily as a whole, and not read. Or rather, that instead of being a display of text intended to be read, it is intended to be read as a display of text. The southern continent, once again, is a space taken up, written over with erudition and theory whose practice may be seen in the representation of the world above.

Mercator's 1569 world map formed the basis for the 'Typus Orbis Terrarum', the world map that appeared the following year in the *Theatrum Orbis Terrarum* of Abraham Ortelius (fig. 1). The success of Ortelius' *Theatrum*, the first systematically-compiled printed atlas, is indicated by the fact that, at his death in 1598, twenty-four editions had appeared in the space of twenty-eight years (further editions continued until 1612, by which time the number of sheets of maps had swollen from the original 53 to 166).[21] These editions included Dutch, German, French, and Spanish translations; English and Italian translations were to appear after Ortelius' death. In the preface to the 'Typus Orbis Terrarum', Ortelius acknowledged the theories behind Mercator's world map: he set forth Mercator's idea of the three continents, and outlined the distinction between modern and ancient cartography.[22] The division between the old and new is made manifest on the world map. Just as in the *Liber Floridus* of Lambert of Saint-Omer, the division of the codex separates one world from another, only here the division is between the New World (on the left-hand page) and the Old (on the right), rather than Lambert's division between the known world on the left and the unknown world on the right. However, underlying Ortelius' image of the world is precisely the sixteenth-century reinvention of the medieval 'Plaga australis': *Terra Australis Incognita*, stretching beneath both old and new worlds, 'as yet revealed in only a few shores' (paucis hactenus littoribus detectam), awaiting by implication exploration, colonization, and subsumption within the world known to Europeans.

Ortelius stripped the southern land of much of the extraneous matter that filled Mercator's continent, leaving only abridged versions of his topographical captions. These note the presence of 'most vast regions' according to the accounts of Marco Polo and Lodovico di Varthema, and the naming of the continent 'Magellanica', after Magellan. Such captions assert a tentative European experience of the southern land, in marked contrast to the medieval dependence on assumption and theorization of unknown regions. At the same time, as well as reducing and simplifying Mercator's wall map, Ortelius added an important ideological dimension to it (or, at least, made this dimension more explicit). In a caption that appears beneath the image, Ortelius invoked the classical tradition of displaying the world in order to undermine the pursuit of worldly knowledge and glory, by adducing a quotation from Cicero's *Tusculanae Disputationes* (4. 37): 'Quid ei potest videri magnum in rebus humanis, cui aeternitas omnis, totiusque mundi nota sit magnitudo' (what can seem great in human affairs to the one to whom all eternity and the magnitude of the whole world is known).[23] So Ortelius, while promoting the new geography by emphasizing the extent of European territorial and intellectual expansion after Columbus' 1492 landfall, nevertheless recurred to the Ciceronian tradition of belittling the world by displaying it.

Recent scholarship on Ortelius has attempted to connect his numerous cartographic and antiquarian projects with his religious beliefs, in particular his probable membership of an irenic fellowship known as the Family of Love.[24] The clandestine nature of the Family, founded by the merchant Hendrik Niclaes and consolidated in the Low Countries in the second half of the sixteenth century, makes it difficult to give detailed information about its doctrines and practices. Defamed from the Protestant as well as the Catholic side, the Family's impulses appear to have been fundamentally conciliatory, with a strong emphasis on interior spirituality, the extension of God's grace across the community of Christians and even beyond to Jews, Muslims, and pagans, and the possibility of a new Adamic age, in which dialogue between heaven and earth would be resumed.[25] At the same time the Family practised outward conformity to official religious and political authority, completely eschewing seditious activities. Ortelius' associations with the Antwerp printer Christophe Plantin and other members of the fellowship (possibly including Mercator) are well documented, and the presence of aspects of the Family's iconography – in particular the flaming heart, representative of Charity – in his cartographic works has been seen as evidence of the infusion of Familist beliefs throughout his scholarly and commercial output.[26] Ortelius' expressions of neo-Stoicism, in particular the position of simultaneously representing and condemning the world, have also been held to emanate from Familist thought. A particular example of his neo-Stoicism is found in the inscription on his funeral monument in the church of St Michael, Antwerp. On the memorial stone a sculpted portrait of Ortelius appears with the emblems of a geographer, including a globe with the device 'Contemno et orno, mente, manu' (literally 'I scorn and I adorn, with mind, with hand', paraphrased by Giorgio Mangani as 'I scorn the world with the mind, but at the same time I glorify it with my work').[27] Or as the bookseller Arnold Mylius put it in a well-known letter to Ortelius in 1596: 'I congratulate you on your contempt of the world and your service to it' ([g]ratulor tibi de tuo isto mundi contemptu et simul ornatu).[28] According to the interpretation of Ortelius as a Familist neo-Stoic, the motto 'contemno et orno' is neither paradoxical, nor contradictory, but consistent with the neo-Stoic posture of both enduring and enriching the world: steering clear of its violent conflicts and impious preoccupations and maintaining a sense of its transience – laughing along with Democritus and crying with Heraclitus – all the while working in and for the world and developing a 'third way', that of interior spirituality.[29]

On the face of it, the representation of unknown land seems to have occupied no place within such affiliations. Certainly the surviving correspondence of Ortelius reveals no burning debate about the southern continent. Just one letter amongst a collection of over three hundred preserved by Ortelius' relatives after his death contains a request for 'anything written about that southern land as yet insufficiently known, which is beyond the strait of Magellan' (aliquid scriptum de terra illa australi hactenus parum cognita quae est vltra fretum Magellanicum).[30] However, the neo-Stoic posture, whether or not it is an expression of specifically Familist sympathies, does seem to raise implications for *terra incognita*. First, is *terra incognita* part of the scorned world? It might be said to be an awkward presence, disrupting both the action of scorning (can the unknown be treated with contempt?) and that of adorning, since the unknown can only adorn in so far as it is used as a blank space to be filled. Second, how significant was the New World context for neo-Stoicism? The world to be adorned had changed in form; but more than that, new peoples had been revealed as well as new lands, and a new political disposition had emerged, with the result that the prospects of a universal monarchy were not as far-fetched as they had previously seemed – and as they would come to seem in later centuries. At the same time, the venality of European actions in the New World were, by the second half of the sixteenth century, widely known, augmenting the atrocities of religious and political conflict committed daily in the Old World, several of which we know Ortelius and his colleagues to have witnessed.[31] The unknown had, it seems, to be accommodated within the thought of universalists, and in the maps of those, such as Mercator and Ortelius, who wished to represent the entire world. According to universalizing Christian thought, the world in its entirety had to be evangelized; equally, those desiring universal monarchy sought to include new and unknown lands within European imperial delineations. But precisely because of its inscription within the category of the 'not yet', *terra incognita* in the sixteenth century might be said to hover disconcertingly outside scorn and glorification – beyond family and perhaps also beyond love.

Some of the difficulties caused for sixteenth-century fantasies of universality by the liminal status of the southern land are evident in the writings of another of Ortelius' correspondents, the French polymath and orientalist, Guillaume Postel. Postel revived the old function of the antipodes as incitement to expansion as part of a radical, although biblically-inspired, reconceptualization of world geography. He argued for the renaming of all continents based on their relationship to the sons of Noah: Europe should be called 'Iapetia' after Japhet, Asia 'Semia' after Shem, and Africa 'Chamia' or 'Chamesia' after Cham. America, according to Postel, should be known as 'Atlantis', since it was so identified by Plato. Finally, the fifth continent, admittedly only known in part, should be named 'Chasdia', since reported sightings of black inhabitants had convinced Postel that, like Africa, it was populated by the sons of Cham:

> De illa uero parte quam Chasdiam debere dici supra notaui, nil adhuc potest referri: nisi quis, ex eo quod Mauros homines et nigerrimos habeat, instar Chamesiae (in qua ex Chuso filio Chamesis, alioqui albi ex coniuge alba procreato, & non loco, sed scelere patris sic infecto, sunt nati Æthiopes) dicat esse orientalem Aethiopiam, cuius ex Homero meminit Strabo. Nam in ea littoris parte quae est reperta, sunt uisi summa nigredine homines, quales in sola Chamesia sunt.[32]

> About that continent which, as I noted above, should be called Chasdia, nothing can as yet be related: save only that because it has Moors and very black men one might think it were east Ethiopia (which Strabo mentions, following Homer), the likeness of Chamesia [i.e. Africa]. (In Chamesia Ethiopians were born from Chusus the son of Cham; originally white, born of a white woman, he was darkened not by the properties of place but by the sin of the father). For in that part of the coastline that has been discovered, men were seen of great blackness, of the sort that are found only in Chamesia.

Postel, who like Mercator was convinced of the vastness of the southern continent, expressed his theories about continental nomenclature in a polar planisphere of 1578, in which an austral continent is labelled 'Chasdia'.[33] It is clear from the activities of both Postel and Mercator that a key element of their cosmological studies was the reconciliation of global geographical knowledge with biblical history. For Postel, the map demanded explanation in terms familiar from Augustine's discussion of the antipodes: how far had Christianity spread, how far could it spread, and what were the origins of the peoples of the world? The connection of the continent with Cham, and thus, according to Postel's logic, with Africa is significant, since, for Postel, Cham and his descendents were associated with rebellion against the law of the father: it was for this reason that they were black, and it was the ultimate mission of the King of France, as the heir of Japhet, to lead an age of restitution in which such rebelliousness would be subdued and peoples and religions reunified under the kingdom of God.[34]

The idea of French colonization of the southern land was advanced with even greater urgency, though on far more pragmatic grounds, by Postel's contemporary, the Calvinist historian and Huguenot officer in the Wars of Religion, Henri Lancelot Voisin de la Popelinière. In his lengthy treatise 'Les trois mondes', printed in 1582,[35] La Popelinière (c. 1541–1608) argues that French interest will be served by colonization of the unknown southern land for two reasons: first, it will enrich and enhance the state; but second (and arguably more significantly) it will divert 'les passions des plus mutins' which currently endanger the state.[36] This is precisely why Spain sent 'the worst rascals of its realm' (le plus mauvais garnemens de son royaume) to the Indies, especially those who did not want to return to their vocations after the wars against the Moors, and it also explains why the Indies have been racked by sedition and nearly completely ruined. The direction of unhealthy elements outwards is, La Popelinière points out, the action of a doctor who purges or bleeds his patient to restore health, and the unknown southern land is the ideal receptacle:[37]

> Voila un monde qui ne peut estre remply que de toutes sortes de biens & choses tres-excellentes. Il ne faut que le descouvrir. Il servira du moins cy apres pour recevoir la purgation de ce royaume: les autres nations nous ont frayé un si beau chemin. Sans doubte, si elles estoient si fournies d'hommes que la France, elles n'eussent tant esté à le peupler et cultiver. Car il ne peut estre qu'aussi beau et autant riche que l'Amerique.[38]

> Behold a world which can be filled by all manner of good and most excellent things. All that is required is to discover it. At the least it will be of use after discovery to receive the purgation of this realm: the other nations have cut a very fine path for us. Without doubt, if they were as furnished with people as France is, they would not make so much delay in peopling and cultivating it. For it must be as beautiful and equally rich as America.

La Popelinière's words encapsulate his critique of the French state. His dissatisfaction is two-fold. On the one hand, there is the path not taken: France's failure to colonize the Indies when it had the chance has enabled the growth of Spain's military and political might within Europe. La Popelinière's other point is less pragmatic: in line with a morality that was increasingly finding its way onto world maps, he draws attention to the unfortunate behaviour not only of individuals, but of 'the states themselves' – in particular their use of violent means to achieve transitory (and highly destructive) ends. Against this background, a 'third' world potentially filled with various riches offers the prospect of redirection, of destructive energies channeled outwards. Moreover, it will enable the realm, purged of its inner strifes, to redeem itself for its earlier folly in not pursuing an expansionist path.[39]

At the beginning of book one of 'Les trois mondes' appears 'La carte des trois mondes', a world map based on Ortelius' 'Typus Orbis Terrarum' of 1570 or a subsequent copy.[40] The presence of this map suggests not only the influence of Mercator's theories of a vast southern continent, retailed by Ortelius, but also the power of the map itself to express in visual form the new continent ripe for exploration. In the mentality of La Popelinière, as well, certain ambiguities that characterize the political philosophy of Mercator and Ortelius are evident. States are, on the one hand, castigated for their brutality and short-sightedness – their pursuit of worldly vanity. Yet on the other hand, they are encouraged to expand, to push their frontiers into new worlds, despite the disastrous consequences of such expansion evident to La Popelinière in places such as the Indies. For La Popelinière the promise of the third world ranges from, at best, great riches, to, at the least, a 'purgation' of evil influences and impulses. Absent, then, is the mystical vision of Postel centred on world unification under a single ruler and the Christian religion. What is left is a vision at once patriotic, idealistic, and cynical, in which the failings of the 'first' world can be written on the unknown land of the southern continent.

Terra Australis, the space endorsed and given iconic form by Mercator, evidently acquired different meanings in the visions of his contemporaries. What for the great cartographer was the backdrop to a celebration and summation of geography was, for his friend Ortelius, a blank space linked to the Ciceronian tradition of contempt for the world. For their correspondent Postel it was an ethnographic extension of Africa, and hence of the sin of rebellion; for La Popelinière, it was the means to bleed France of her ills. In the maps of the sixteenth and early seventeenth centuries this diversity of interest in the practical and ideological meanings of the southern continent found generous expression. Although *Terra Australis Incognita* was not discovered, explored, or settled by Europeans before 1600, the possibilities for its representation were fully colonized. Its imagined peoples, landscapes and resources were all expressed in visual form – with the result that the space never lost its capacity to reflect, and to expose, Old World ambition and pretension.

Ethnographic dislocation: the Dieppe school

The appearance of ethnographic drawings in the space of the unknown southern land on sixteenth-century maps was a significant departure from earlier representational strategies.

During the Middle Ages, lack of knowledge and the theologically problematic nature of the idea prevented mapmakers from depicting people in unknown parts of the southern hemisphere. Yet when they did appear in the sixteenth century the peoples drawn were not, in fact, natives of any part of *Terra Australis*; instead, the ethnographic images in the southern land were in most cases the product of European encounters with natives of various parts of the New World, principally South America. Such images were frequently not the result of precise observations: engravings of native peoples in works such as Theodore De Bry's *America* were constituted by what has been termed 'iconographic bricolage', since they consisted of a melange of costume and ritual taken from distinct and geographically separate cultures (Aztec and Tupinamban, for example) and combined with some European elements to form a homogenous image of the American 'savage'.[41] When they appeared on maps, ethnographic sketches of the kind popularized by De Bry were usually located in the interiors of New World regions, often intended to intimate the types of peoples and practices seen by Europeans along the coasts. Even in these instances, then, there is an element of displacement of the native image from coast to interior. But when placed in the unknown southern continent, the image of Brazilian, Floridian, or even Sumatran native certainly does not signify the presence of such people in that or any nearby location. In one sense, the native in unknown land can be accounted for as a decorative element, a means, along with many others (animals, plants, fictive topographies, paratextual material), of filling otherwise empty space and entertaining the viewer of the map. More particularly, images of natives in *Terra Australis* on world maps must be understood in terms of the interrelation of ethnographic images. All sketches of New World natives and native life on sixteenth-century world maps signified the anticipation of contact with non-European peoples as well as its historical occurrence. The native American, whether in the unexplored interior of Brazil or in the unknown southern land, represented the imagined presence of further, as yet unknown, peoples in the New World. The particular function of the ethnographic drawing in *Terra Australis* is to amplify – to reiterate and to expand – a discourse of colonial encounter by means of dislocation. The New World native is repeated, but the image is also transferred, signifying no longer an identity (a particular tribe, a particular place, a particular incident) or actual presence but a process by which such peoples, places, and incidents were recorded and then disseminated in iconographic form. Abstracted, emblem of a process, the native in *Terra Australis* is no longer native, but a sign of indigeneity.

Ethnographic illustrations on sixteenth-century maps differed in one significant regard from the drawings of beasts and marvellous races that are found on the encyclopedic world maps of the thirteenth and fourteenth centuries, such as the Hereford *Mappamundi*, the Ebstorf, and the Aslake world maps, and in early printed books such as Hartmann Schedel's *Nuremberg Chronicle* of 1493: the sixteenth-century images make an implicit claim to representational accuracy based upon experience, if not of the mapmaker himself, then of those whose reports form the basis of the information represented on the chart.[42] The assertion of empiricism enhanced the value of ethnographic representations within the map's summation of knowledge, testifying not only to new places but to the peoples therein. Such information was, in the formulations of several cartographers of the sixteenth century, the special concern of sovereign power. In particular, the work of the 'Dieppe school', a group of cartographers based in or nearby the French port city in the 1540s and 1550s, made explicit the connection between the sovereign gaze and the lavish representations of native inhabitants, architecture, animal and plant life that appeared on their maps.[43]

At the beginning of his *Boke of Idrography* (1542) Jean Rotz, a member of the Dieppe school until his defection to the service of Henry VIII of England, addressed his new master. 'Vous pourrez', he wrote:

> pour la recreation de vostre noble esprit apprez lauoyr trauaille pour les affaires et bien public de vostre tres exellent Royaulme et peuple, voyr et congnoistre quelles terres des costes de la marine ioygnent et Regardent les vgnes aulx aultres et par quel nombre de lieues et aussy par quel nombre de degrez de latitude elles sont Auec les facons et manieres tant des maisons, des habitz et coulleurs de corps, que des armes et aultres vsages des habittantz en chascune desdictes costes qui nous sont les moins congnues. Et ce au plus certain et vray quil ma este possible de faire, tant par mon experience propre que par la certaine experience de mes amys et compagnons nauigateurs.
>
> after labouring for the affairs and public good of your most excellent Kingdom and people, you may for the recreation of your noble mind observe and learn which coastal lands adjoin or face one another, how many leagues apart they are and in what latitude, together with the style and manner of houses, clothes and skin-colour, as well as arms and other features of the inhabitants of all those coasts which are least known to us. All this I have set down as exactly and truly as possible, drawing as much from my own experience as from the certain experience of my friends and fellow navigators.[44]

Coastal geography is complemented by interior ethnography. And one of the key features of this new ethnography, which (at least in Rotz's *Boke*) represents the inhabitants of all non-European continents, not just those of the New World, is the representation of the 'scene'. The scene, of course, carries its own particular form of authority in that it represents what people do, not just what they look like, and therefore attests to observation of their customs and behaviour. The scene depicted on these maps is usually one of ritual life – a procession, a dance, punishment, hunting, religious practices, sleeping arrangements, the layout of a village. Where specific, rather than generic, incidents are shown these may involve contact between Europeans and indigenes, including contact of a violent nature (such as the Rotz atlas' representation of the murder of French mariners in Madagascar [ff. 13v–14r]). The scene, then, represents the interior practice of the land, but also the intrusion of the external elements of exploration, including cartography itself.

The connection between sovereign power, universal cosmography, and the unknown – already explicit in Rotz's *Boke* – was spelled out in a work that may have influenced the Dieppe school's representation of the southern land: the *Cosmographie* of the experienced Norman pilot, Jean Alfonse. Completed in 1545,[45] Alfonse's work includes confident assertions of the extent of 'La terre australle' from the Strait of Magellan, five or six hundred leagues beneath the Cape of Good Hope, to 'Jave la Grande', and as far south as the Antarctic pole.[46] The proof of this vast land, Alfonse admitted, was lacking:

> J'estime que ceste coste de la mer Occéane qu'est dicte coste Australle se va rendre en Oriant, à la Jave, du cousté d'occident de ladicte Jave. Toutesfoys jusques à présent n'est point découvert

parceque l'on n'y ose aller à cause des froidures et tormentes du polle antar, et ne sçait l'on que les gens de la terre Australle croyent.[47]

I reckon that this coast of the Ocean sea which is the foresaid southern coast extends itself in an eastward direction as far as Java, starting from the west coast of this same Java. However up to the present day it has not been discovered because no-one has dared go there due to the cold and the tempests of the Antarctic pole, and it is not known what the people of the southern land believe.

It is significant that Alfonse included a description of the southern land (with one or two passes at the appearance of its inhabitants and topography) in a work designed to provide the French king with knowledge of 'all the provinces of the entire world presently seen, discovered and known by Europeans' (toutes les provinces de l'universel monde, lesquelles jusques à présent ont esté veues, descouvertes et congneues par ceulx de nostre Europe), and specifically with knowledge of the non-Christian areas of the world to which he might, in time, extend his rule.[48] Here, as in the cosmographic schemes of Postel and Ortelius, the inclusion of the southern land is demanded by the pretence to a universal cosmography.

Alfonse's work helps to explain why the maps of the Dieppe school show a large landmass in the South Pacific to the south of Java, called Java-la-grande, La Jave Grande, or simply 'The Londe of Jaua', the presence of which has given rise to considerable speculation that it represents the results of early sixteenth-century Portuguese (or possibly French) exploration of the north-eastern coastline of Australia.[49] The land was represented variously: on the Rotz atlas it extends from a promontory at around 8°S to an inchoate continental coastline traced as far as 60°S;[50] on the 'Harleian' world map (BL, MS Additional 5413) it appears as a mainland, separated from the island of 'Iave' by a narrow river marked 'R. grande' (fig. 40);[51] on Pierre Desceliers' map of 1550 it is a landmass that resembles the great southern continent of printed maps,[52] with a coastline extending between Tierra del Fuego and Java in a series of generic capes, rivers, gulfs and bays, marked simply 'R', 'Cap', or 'Plaine', broken up by the occasional 'Terre non du tout descouuerte'.

Many of the ethnographic scenes that appear in the interior of Java-la-grande on Dieppe school maps were transferred from south-east Asia, but close inspection reveals the eclectic nature of the assemblages in the southern land. The Vallard Atlas of 1547 filled Java-la-grande's interior with a quite stunning series of images derived from the Parmentier expedition to Sumatra in 1529–30.[53] These images display a dense landscape rich in detail of costume, weaponry, and architecture, with vignettes of hunting and gathering, and social and political life. On his 1550 chart (BL, MS Additional 24065) (Plate 7), Pierre Desceliers located eight text boxes on 'Iaua', just one of which concerned it: the others relate to islands of the Indian Ocean, including 'Seilan' and 'Samatra'. The box dedicated to Iaua in fact describes the south-east Asian island, rather than the extensive southern land which it appears to designate; the text describes Iaua as a large island with an idolatrous population ruled by five kings, gives an account of its commodities, the various forms of idolatry practiced, and cites Lodovico di Varthema's account of the sale of the elderly to cannibals. The iconography of 'Iaua' has been supplemented by Desceliers: a group of idolators worships the sun, while above them a pair of agricultural labourers toils on a landscape dotted with various dwellings (huts, clusters of round shelters),

Fig. 40. 'Harleian world map'. London, BL, MS Additional 5413 (detail). This product of the 'Dieppe school' of mapmakers shows 'Iave la grande' as a mainland, separated from the island of 'Iave' by a narrow river marked 'R. grande'. As in the atlas of Jean Rotz, 'Iave la grande' extends south from around 5°S and presumably continues beyond the southernmost extent of the map at 64°S; its toponyms range from the anonymous 'R de' in its far west through 'Coste blanche', and 'Baye bresille', to the east coast's 'Coste Dangereuse', 'Baye perdue', 'R de beaucoup disles', and 'Coste des herbaiges', before reaching the 'C: de fremose', several more unnamed rivers and a 'Gouffre'. The interior of 'Iave la grande' is essentially divided into three landscapes, decorated with hills, trees, animals, inhabitants apparently going about their daily lives, and numerous dwellings. In the eighteenth century this map was consulted by the hydrographer to the East India Company, Alexander Dalrymple, who saw it as evidence of the sixteenth-century European discovery of the north-east coast of Australia.

animals (guanacos, elephants), and palm trees; in Java's deep south a pair of cynocephali (dog-headed people) dismember a corpse. To the right of the cynocephali appears the Montmorency coat of arms,[54] and beneath it the place and date of the map and the name of its maker.

The effect of ethnographic dislocation is not to integrate the southern land with the rest of the world, but to use its space as a repository of information about *terra cognita.* On Desceliers' 1550 chart, only the illustrations of idolators and men working the land seem to purport to represent the southern land; the depiction of dog-heads, for example, clearly illustrates the text describing the islands of 'Angania' (the Andaman Islands). Indeed every single feature of Desceliers' southern land can be found on another portion of the map, including the cynocephali, one representative of which is located in the far north of Asia (although not engaged in butchery). Knowledge, new and old, has slipped down south, in some cases from nearby Sumatra, in others from Africa and America, in still others from the texts of classical antiquity. What has been seen elsewhere is projected onto the unfilled screen of the southern land, so that not for the last time it represents a combination of diverse world detail, an amalgam of the foreign, orchestrated by the principles of European craftsmanship and patronage so clearly enunciated in the map's bottom right corner.

In one sense it is possible to make a clear distinction between the pseudo-ethnography of the Dieppe school's Java-la-Grande and that published in the first quarter of the sixteenth century in More's *Utopia.* To construct his fiction More drew on the classical idea of the antipodes in the context of the emergent space of the New World. The Utopians were not the kind of New World natives described by Amerigo Vespucci, but an amalgam of classical ethnography, philosophical speculation, satirical observation, and (to a limited degree) more recent reports of non-European peoples.[55] Working twenty-five to fifty years later, the Dieppe school cartographers located in the space of the antipodes an amalgam of a different kind: an array of toponyms generic and transferred, and the images of peoples encountered in several different parts of the world, almost in the manner of encyclopedic overspill. Yet it is possible to see continuities with More's project in two manuscript atlases associated – although perhaps not compellingly – with the Dieppe school. These atlases contain remarkable fantasias of the southern land in which the guise of empiricism is cast aside and in which, as in *Utopia,* the function of the antipodes is to act as *recessus,* as depth of imagination and representation.

The *Cosmographie Universelle selon les Navigateurs, Tant anciens Que modernes,* produced by the Norman pilot Guillaume Le Testu in 1556, extends to fifty-six maps; twelve of them are devoted to the southern land (fig. 41). Le Testu clearly stated that in the *Cosmographie* the southern land was marked and depicted 'only by imagination [ce n'est marquée que par l'imagination] ... for there has never yet been any man who has made a certain discovery of it'.[56] However, the cosmographer goes on to declare, 'I have marked and named some promontories or capes in order to align the pieces in which I depict the area' (j'ay marqué et dénommé quelques promontoires ou caps pour radreser les pieches qui pour ce sont cy depeinctes).[57] The place names, and certain geographical features, are not, in other words, the product of mapping at all, but aids to the composition of the book. The toponyms themselves, a mixture of French and Portuguese expressions,[58] raise further questions: at least one of them, 'Cap de More', seems to

Fig. 41. Unknown southern land in Guillaume Le Testu, *Cosmographie Universelle selon les Navigateurs*, fol. 33v (1556). Vincennes, Bibliothèque du Service historique de l'armée de terre, Rés. DLZ 14. East at top. The Norman pilot Guillaume Le Testu devoted twelve of the fifty-six maps in his *Cosmographie* to the southern land. He frankly admitted that the southern land was the product of imagination and that its coastline was shaped more by the exigencies of producing a coherent series of regional maps than by observation. Le Testu's maps are remarkable, however, for their elaborate and vibrant interiors. Here a portion of the southern land is shown beneath Sumatra and the Moluccas, divided between 'Petite Iave' and 'Grande Iave'. Its inhabitants wear a mixture of European and native American costumes.

make a play, by no means the first, on the name of the author of *Utopia*.[59] Another, 'Terre de Offir', located in the region of 'Grande Jaue', invokes Ophir, the biblical region that supplied Solomon with gold, precious stones and wood, and perpetuates a sixteenth-century preoccupation with attempting to locate this region in the New World.[60] Le Testu's ornate, beautiful interiors on the southern land resemble those in his representations of the Americas: the principal recurring motifs are scenes of hunting and warfare, with elements taken from both classicizing and New World iconography.[61] But the detail is abundant, including the occasional representative of a monstrous race (men with huge ears, useful as quilts),[62] an array of mythical and real animals, landscapes, architecture. This is not, as Le Testu confesses, empiricism: we are obviously not presented with the natives of 'Terre Australe', since 'there has never yet been any man' to report on them, and since even the coastal toponyms are aids 'to align the pieces', rather than records of observation. It is not even conjecture, really. Space filling? But there are many ways of filling space: Le Testu does it here by means of iteration, by the arrangement of variations of a limited number of images. And that principle may also apply to the construction of the massive southern coastline on this and other Dieppe school maps. The coastline is the product of imagination, but imagination of a certain kind: it is a multiplication of the syntax of coastlines. Bay, inlet, promontory, offshore island, smooth, jagged; the cartographer knows how coasts iterate, and how to reiterate them when he ventures into uncharted waters. A 'universal' cosmography must show the whole world; but this explains only in part the need to lavish so much attention on the southern land. There seems to be a further statement implicit in Le Testu's 'Terre Australe': that cosmography is not yet universal; that it still of necessity acts as its own supplement.

Le Testu's *Cosmographie* appears to have influenced just one other representation of the southern land. Around 1587, the last of the surviving Dieppe school works, an anonymous, unfinished world sea atlas, located in the interior of its own 'Terre Australe' 'perhaps the most wholehearted elaborations of imaginary features known to the history of cartography'.[63] There are two significant differences between the 1587 atlas and Le Testu's in terms of their representation of the southern land. First, the 'Terre avstrale' of the 1587 atlas contains a vastly expanded number of toponyms (over 120). Second, the interiors on the southern land, as on other regions, are heavily abstracted: 'scenes' are rarely depicted; architectural structures, cities, towns (European and non-European) and animals appear much more frequently than peoples – partly no doubt as a result of the unfinished state of the maps, but also because there is minimal aspiration to depict 'actual' inhabitants of these areas. Instead, what is shown is generic, remarkable only in the way it fuses the figures of different cultures to form a melange, landscapes simultaneously of all places and of none.

The principle of ethnographic dislocation, according to which scenes of New World peoples were transferred to the southern continent, was evident in early Dieppe school maps, but in the school's later productions it changed to one of synthesis, in which a frankly creative cartography found its complement in utopian interiors. These later representations of 'Terre australe' make no claim to specificity of place or time. For this reason their iteration of coastlines, scenes, buildings, and animals acquired the quality of a generic statement about land and habitation. Yet if the unmoored and repetitive displays such as that of Le Testu's atlas marked the return of the southern land to its function of *recessus*, an interior space of the imagination, it was a recess that contained no philosophical import, and a limited political one. To see the survival of those elements of antipodal representation it is necessary to turn to the grand allegorizing and satirical gestures made by cartographers at the end of the century.

Speculum Orbis Terrarum: allegory and satire on the southern land

One function of the antipodes had always been to turn the European gaze back upon itself. In this regard, *Terra Australis Incognita* was no different to its predecessors. Its space and its emergent form encouraged cartographers to use allegory and satire either to celebrate Europe's achievements, or to take aim at its failings. One contributing factor to the reconfiguration of the allegorical and satirical traditions of the antipodes was the high degree of recognizability acquired by the coastline of *Terra Australis Incognita*, established on so many maps and copies of maps in the second half of the sixteenth century. The result was a paradox: although the great southern land signalled the incomplete nature of the mapping of the world, it had become an integral part of the world map. And what made it integral was not simply cosmographical theory that demanded the presence of a vast, balancing, landmass; it was also the possibility that *Terra Australis* allowed for non-cartographic modes of expression to be located within the frame of the map.

Allegory on *Terra Australis* served diverse political ends. On Willem Janszoon Blaeu's 'Nova Orbis Terrarum Geographica ac Hydrographica Tabula' of 1606–7 it expresses a triumphalist celebration of European mastery of people and science (fig. 42).[64] Blaeu's map depicts on the southern land a representation of Europe, with the emblems of arts, sciences, and industry at her feet, receiving homage from Asia (with loaded camel), Africa (riding a crocodile and holding a parasol), and the Americas (with attendant armadillo). A magnificent double border surrounds the map, containing twenty-eight town views, thirty groups of natives in national costume, countries ranging from Ireland to the Moluccas, and an extensive 'explanatio' of the earth's physical, political, and ethnic divisions, illustrated by ten engravings.[65] The top of the frame is dedicated to panels of 'ten images of the most powerful Princes, who rule the earth in our age' (decem potentissimorum Principum, qui toti terrarumorbi nostro hoc saeculo imperant, effigies). These potentates are European and Asian, and the verses that appear beneath them urge the Europeans to resist the incursions of the Turks and the Tartars.[66] In contrast to this warlike array, Blaeu's southern panorama offers a stately declaration of European sovereignty over the other continents in the form of a fantasy of homage. Accompanying verses attribute this supremacy to 'arms and practical skill' (marte et arte), and indeed the map seems to be making a strong case for the role of scientific mastery in Europa's triumph. Its declaration of technical sophistication encompasses the combination of chorographic and cartographic achievement evident in the town views, the ethnographic survey, and the world image itself (an explanation of the Mercator projection appears alongside the tableau on *Terra Australis*).

A rather more pointed and polemical use of allegory is evident in an image that appeared a

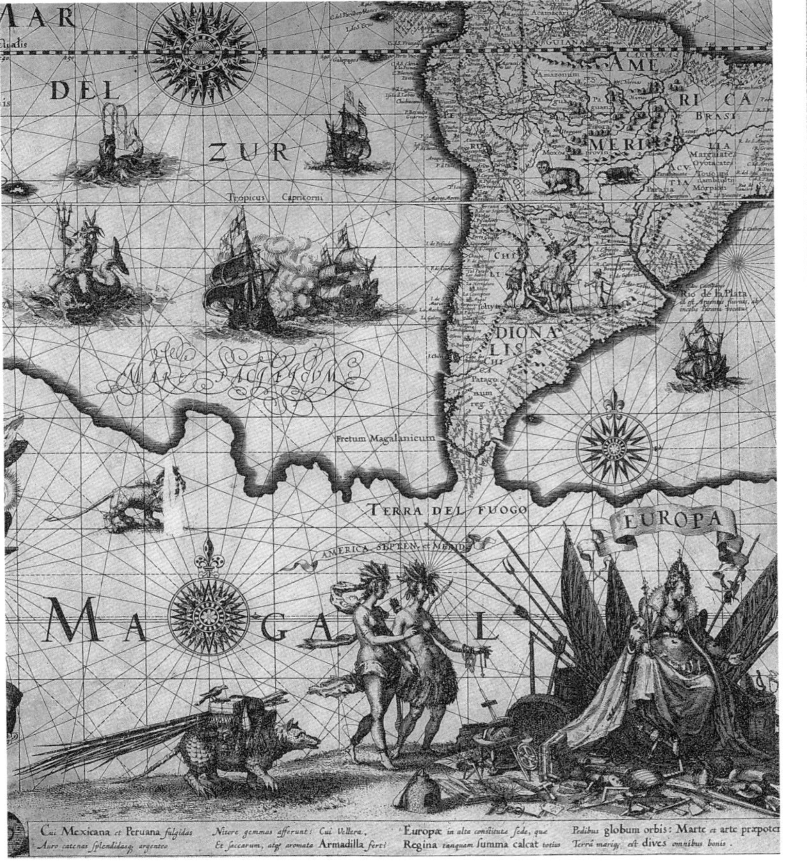

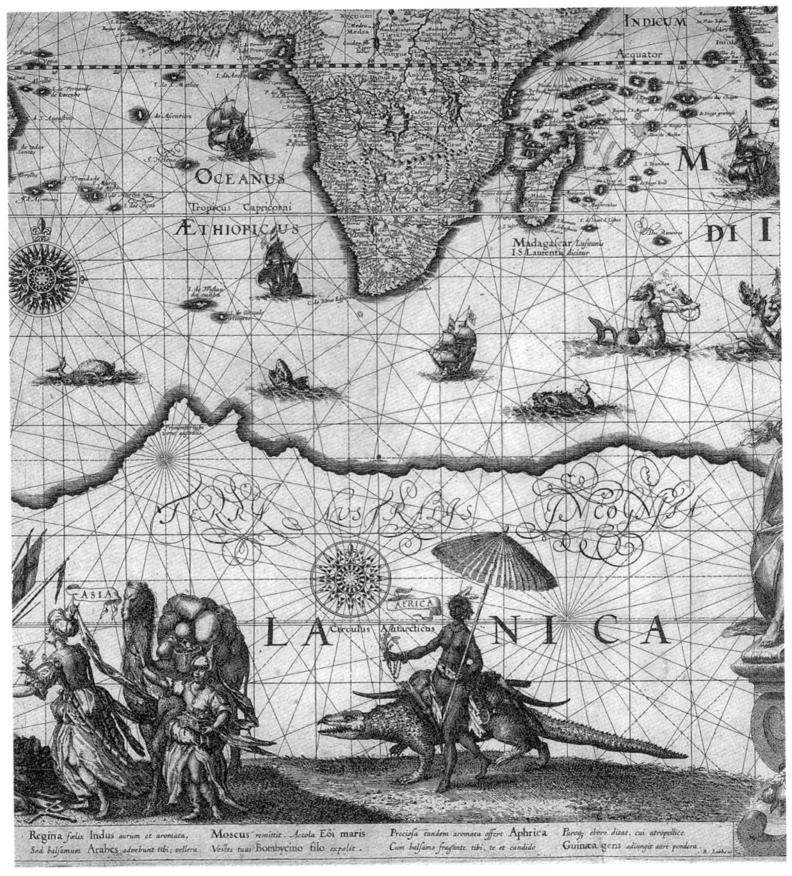

Fig. 42. Willem Janszoon Blaeu, 'Nova Orbis Terrarum Geographica ac Hydrographica Tabula' (1606–7) (detail). Bern, Stadt- und Universitätsbibliothek, Collection Johann Friedrich Ryhiner, vol. 5, nos 61–4. On Blaeu's world map the southern land is the site for allegory: Europe, with the emblems of arts, sciences, and industry at her feet, receives homage from Asia (with loaded camel), Africa (riding a crocodile and holding a parasol), and the Americas (with attendant armadillo).

few years before Blaeu's world map: Jodocus Hondius' 'Christian Knight Map of the World', or 'Image of the whole world, in which the struggle of a Christian knight on the earth (in the service of zealous piety) is represented pictorially, by Iud. Hondius, engraver' (Typus Totius Orbis Terrarum, in quo et Christiani militis certamen super terram (in pietatis studiosi gratiam) graphicè designatur. à Iud. Hondio caelatore) (fig. 43).[67] Hondius shows the central figure of Spiritus trampling Caro (Flesh), while being assaulted by the serpent of Peccatum (Sin), and threatened by the figures of Diabolus, and Mors (Death). Mundus (World), crowned with an orb, looks on. Each figure is surrounded by biblical quotations; this extends even to their implements, such as Spiritus' sword and shield.[68] A framed dedication to three English mathematicians, Richard Brewer, Henry Briggs, and Edward Wright, appears to the left of the figure of Mundus, at the far left of the southern continent; on the right a similar frame quotes Paul's Letter to the Ephesians 6.18, a text that calls for the armour of God to be used to combat the devil, principalities, powers, and 'spiritual wickedness in high places' – loins girt with truth, breastplate of righteousness, feet shod with gospel of peace, the shield of faith, the helmet of salvation, and the sword of spirit.

The iconography employed by Hondius in his tableau was by no means original. As John Gillies has pointed out, at least some of its origins lie in the morality play.[69] More immediately, the scene was derived from an engraving of c. 1580 by Hieronymus Wierix (after Maarten de Vos), which shows almost identical figures.[70] Hondius' originality lies rather in his placement of these figures on a map, on the world's stage. Hondius had emigrated to London from Ghent in 1583 to avoid religious persecution, and, according to one interpretation, in this map he altered the features of the characters to make explicit the image's Protestant agenda: the Knight resembles Henri IV of France, at the time a Protestant rallying point; he is opposed to the forces of the Papacy, represented by Mundus, whose iconography is that of certain early modern representations of the Whore of Babylon.[71] Yet Hondius' map is innovatory in strictly cartographic as well as in dramatic or imaginative terms. The projection used by Hondius was one of the first to be based on Mercator's projection as expounded by Edward Wright, to whom, along with Brewer and Briggs, the map is dedicated.[72] The dedication to these mathematicians seeks to draw attention to the originality and scientific sophistication of the map. Hondius was, moreover, concerned to represent accurately the northernmost parts of the world, noting uncertainties and probabilities about the shape of North America (and the elusive North-West passage), and leaving its northern coastline incomplete. This is not, then, an expression of neo-Stoic scorn for mapping the world; rather, the allegorical scene that is staged upon the southern continent seems to suggest an alliance between science and a kind of militant Protestantism, headed by the sovereign figure of the Knight – and orchestrated by the mapmaker, whose name, scarcely visible, appears in the far right of the image ('I. Hondius excudit'). At the same time, the figure of the world, accompanied as she is by biblical injunctions, testifies to the dangers of mapping the world for the world alone.

The role of cartography in the expansion of European knowledge and power was always at issue in allegorical displays in the southern land. In 1603 Hondius produced a spectacular world map, clearly influenced by Petrus Plancius' eighteen-sheet world map of 1592,[73] which featured

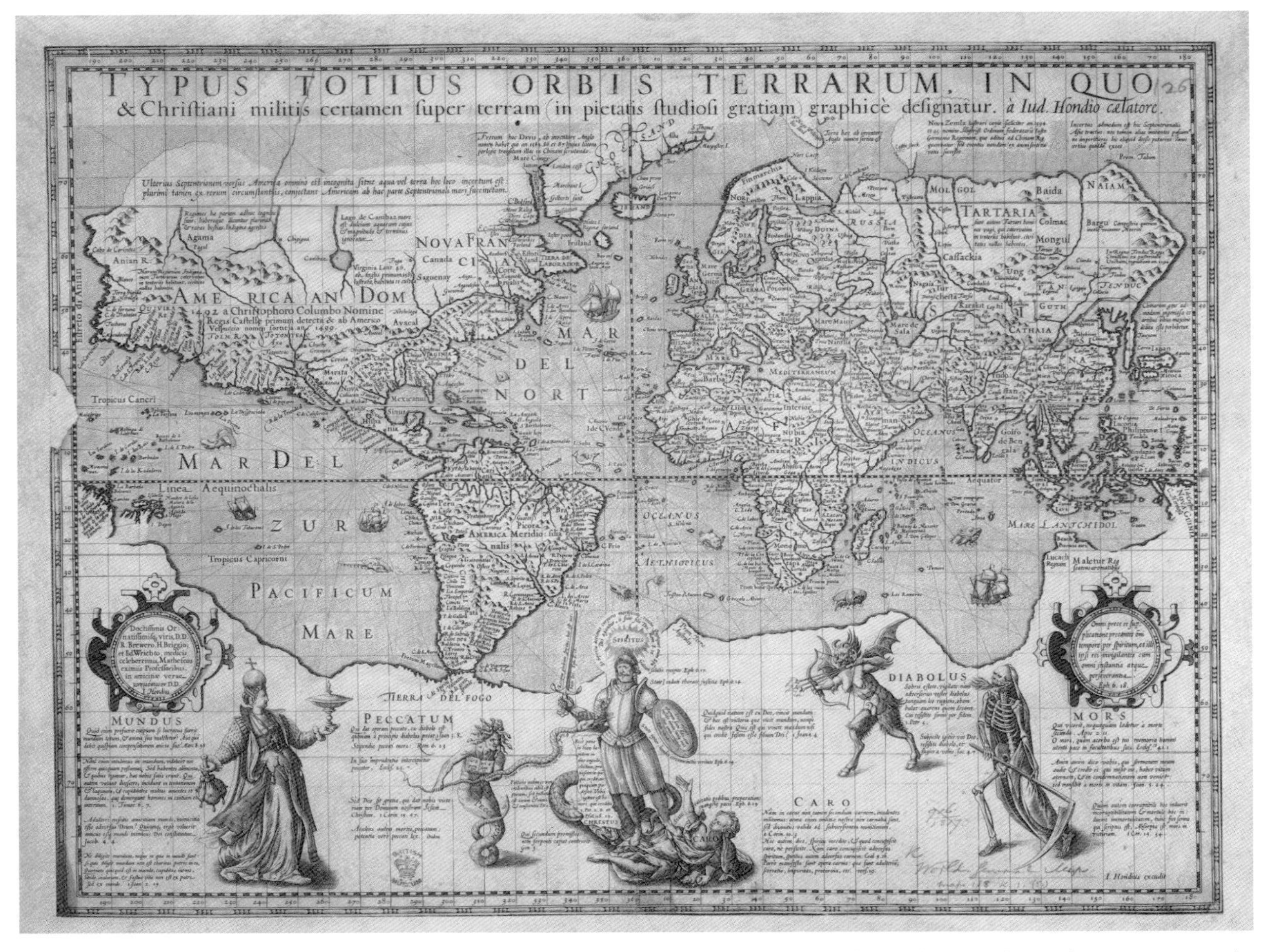

Fig. 43. Jodocus Hondius, 'Christian Knight World Map' (1598). London, BL, Maps 158.K.1 (56). Hondius uses the southern continent as the stage for a tableau: the struggle of the Christian Knight on earth. The central figure of Spiritus tramples Caro (Flesh), while being assaulted by the serpent of Peccatum (Sin), and threatened by the figures of Diabolus, and Mors (Death). Mundus (World), crowned with an orb, looks on. The Knight appears to bear a resemblance to Henry IV of France, suggesting that one function of the image was to act as Protestant propaganda.

the standard vast southern continent.[74] However, Hondius made no attempt to repeat his Protestant allegory on it. Instead, in a return to Mercator's meta-cartography, he combined portrait with cartographic image to celebrate the idea of geography (fig. 44). On the far left of the southern continent Hondius displays a portrait of Columbus above a panel entitled 'The mode of representing the globe on a plane surface, or the calculation of delineation' (Modus fabricandi globum in plano sive ratio delineandi).[75] To balance that of Columbus, in the far right of the continent a portrait of his contemporary, Vespucci, tops a table concerned with the rectification of errors caused by different systems of hydrographic description. But the attention grabber on *Terra Australis* is the series of portraits of Ptolemy, Magellan, Francis Drake, Thomas Cavendish, and Olivier de Noort atop small oval world maps. The mini-map beneath Ptolemy shows the world known to Europeans before the age of exploration; the maps beneath the four explorers indicate their respective routes through the world by means of a line.

The effect of the portraits is not simply to represent geography, but to represent the history of exploration, and its impact upon cartography. The invocation of Ptolemy demonstrates the classical heritage of geography, and also conveniently shows how much less of the world was known before the voyages of European explorers from 1492 onwards. Significantly, all the maps beneath the four explorers – all, that is, apart from Ptolemy's – show a vast southern continent. The appearance of the southern land is thereby linked with the process of exploration, as the force that has brought the shape of the unknown into clearer definition. Once more, *Terra Australis Incognita* acts as a space for summation, a space where the world can be seen in miniature – in this case a whole series of worlds, strung out as cartography's tribute to itself. But there is more to this image than self congratulation: the maps within the map are a sort of *mise-en-abîme*, in that they represent the writing, and re-writing of maps. *Terra Australis Incognita* is an abyss, a chasm where cartography acts in the *absence* of exploration.

The world map – and in particular the southern continent – offered the opportunity not only for a history of the world image but also for its satirical refashioning. In Joseph Hall's *Mundus Alter et Idem* of 1605 it is precisely the absence of exploration that encourages mock-cartographic representation of *Terra Australis*.[76] Unlike Hondius and Blaeu, Hall was not a professional cartographer, and his narrative, the unreliable account of the travels of one Mercurius Britannicus to the great southern land, is accompanied by a series of maps whose target is precisely the intellectual opulence of the kind that characterizes maps such as those produced by the two Dutchmen. On Hall's world map the southern continent is exaggeratedly large,[77] and divided into the regions encountered by the narrator on his travels to the hitherto unknown 'other world'. The map plays on actual toponyms (New Guinea, Virginia, Patagonia) and cartographic convention, including amongst its regions 'Terra Sancta (Ignota etiam adhuc)' (Holy Land even yet unknown), 'Aphrodysia Nova Gynia vel Viraginia', and 'Pamphagonia'.[78] A series of regional maps accompany the text (fig. 45), so that the reader is able to follow Mercurius' adventures in food and drink-sodden Crapulia, female-run Viraginia, overpopulated Moronia, and dangerous Lavernia (a region of rogues and thieves), and to trace his progress through such cities as Marza-pane, Mortadella, the fetid Fourmagium, Gynaecopolis (capital city of Viraginia), Orgilia, Pazzivilla, and Larcinia. The locations of these places are given with reference not only to parts

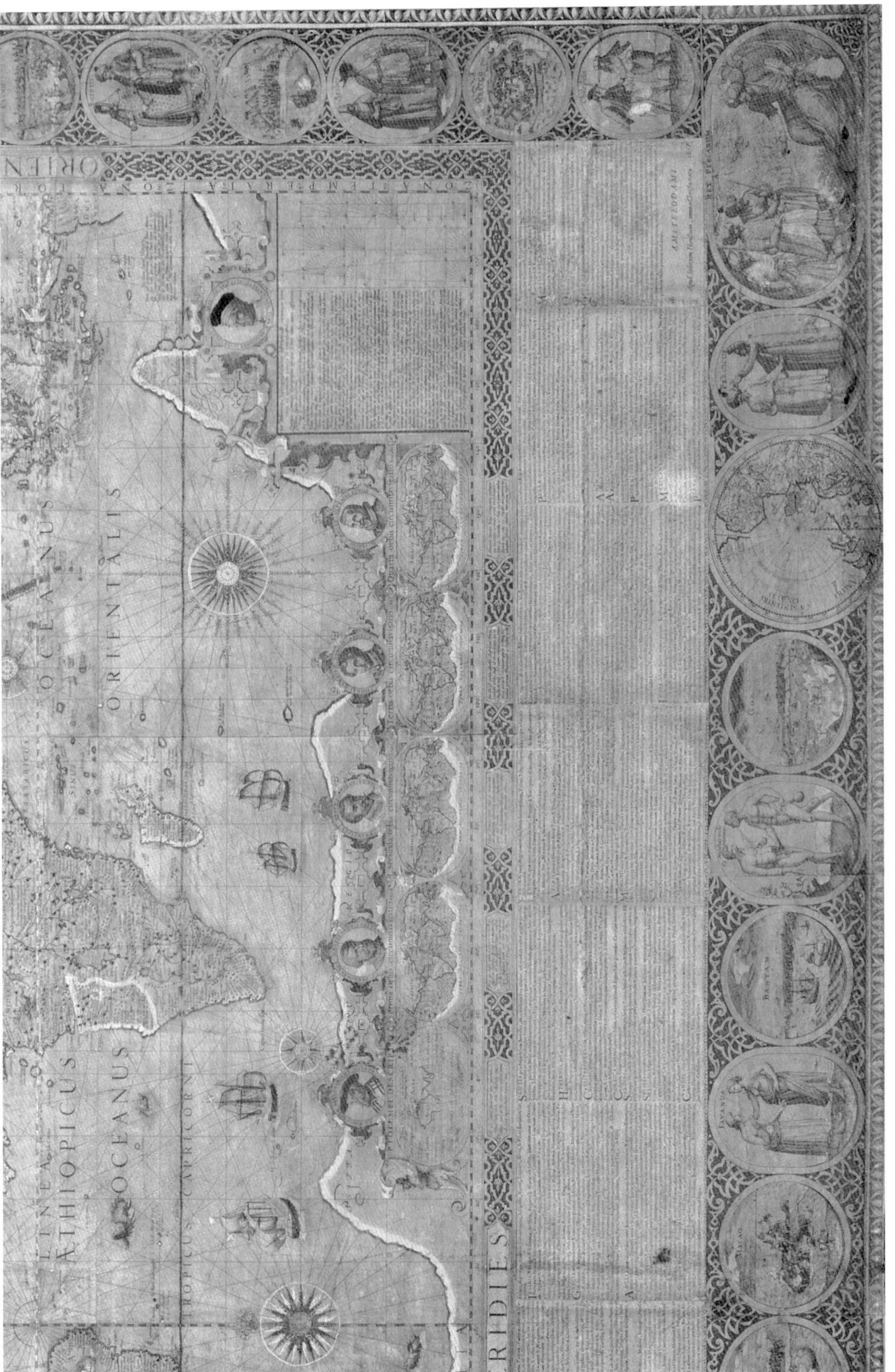

Fig. 44. World map of Jodocus Hondius, 1608 (detail). Royal Geographical Society, London. In Hondius' magnificent world map of 1603, reissued in 1608, the southern continent is the site for a display of portraiture. Hondius presents portraits of Columbus and Amerigo Vespucci respectively at the far left and right of the southern continent. Most striking of all is the series of portraits of Ptolemy, Magellan, Francis Drake, Thomas Cavendish, and Olivier de Noort placed above mini oval world maps. The mini-map beneath Ptolemy shows the world known to Europeans before the age of exploration; the maps beneath the four explorers indicate by means of a line their respective routes through the world.

Fig. 45. 'Crapulia' from Joseph Hall, *Mundus alter et idem* (1605). London, BL, 684.d.5. Hall printed maps of the regions of *Terra Australis* to augment the world map which appeared in *Mundus alter et idem.* His mixture of real and fantastic geography sees here the region of Crapulia, divided into two provinces, Yvronia (= intoxicated) and Pamphagonia ('the most voracious gluttons'), with the latter sub-divided into the 'Golosinius Tract' and the 'Lecanican Plain'. Crapulia is found to the west of Loçania and Viraginia, to the north of Moronia Felix, to the east of the Tryphonian Swamp – and to the south of the Cape of Good Hope, and south-east of America. Amongst other features the map represents the fertile fields of Offulia and Lardana, the city of Mortadella, the fetid village of Fourmagium, and the metropolis of Artocreopolis.

of the known world, but also to parts of *Terra Australis Incognita* that had become familiar to a cartographically-aware audience. Viraginia, for instance, is said to be 'located where European geographers depict the Land of Parrots' (sita est, vbi Geographi Europei Psittacorum terram depingunt).[79] The satire is specifically anti-Catholic in its orientation, and in that regard not out of sympathy with allegories such as Hondius' Christian Knight map. The section of Hall's work devoted to Moronia includes extended, and not particularly subtle, attacks on friars, pilgrimage, and the papacy; the thinly-veiled leader of the province, 'Il Buffonio Otimo Massimo', is said to be 'midway between an emperor and a priest' (medius inter Imperatorem, et sacerdotem).[80] Hall, future Anglican bishop of Exeter and Norwich, omits England from the map.

Cartographic practice itself is deeply implicated in this folly. In the guise of a fantasy of the revelation of *Terra Australis* the known world is ridiculed, and the sham newly-mapped world laughs back at the old. Hall's satire seeks to explode the idiocy at the heart of the vision of the world, of the desire to see the world, and in so doing must undermine the expansionist impulse that makes such vision possible. Contrary to its Ortelian formulation as 'the eye of history', geography appears in the maps of the *Mundus Alter et Idem* as the eye of infamy. Again, though, one notes the referentiality of this dystopian geography. The explication of the upper world – its gluttony, folly, shrewishness, and so on – is seen in the lower, and vice-versa. Any idea, then, of representing alterity is completely antithetical to Hall's design. The map of 'mundus alter' is, the joke goes, that of 'mundus idem', man cannot help but map the earth according to his folly, since to map is, anyway, to be a fool.[81] The southern land possesses, according to Hall's satiric vision, a revelatory quality, rather than that of a mirror. It does not reflect, so much as strip away pretence and leave bare the corruption of human society. It is the means by which the world can be seen for what it is: mad, lustful, greedy, stupid, and impious.

The contemporaneity of Hall's satire with Willem Blaeu's allegory of European supremacy indicates, not unexpectedly, that profound differences existed within Europe with regard to the political and moral meaning of the world map, and with regard to the significance of unknown land on it. The multiplicity of signification that *Terra Australis* could hold is striking, yet consistent with its classical and medieval traditions. Sign of mastery or sign of folly, the antipodes had always been malleable. However, never before – and never again – would the map be the form of text that expressed in such vivid detail the signifying potential of unknown land.

Postscript: The Sacred Heart

One might assume that, with the beginnings of the Dutch exploration of what was to become Australia in the first quarter of the seventeenth century, the days of the great southern continent were numbered, and that, since the formulations of Mercator and other cartographers had been shown to be erroneous, the antipodes could be reconstituted only by their removal. Certainly it is the case that some mapmakers after around 1625 ceased to show 'Terra Australis Incognita' or 'Magellanica', and chose to depict instead the emerging outline of New Holland. An example of this change (by no means the earliest) is the world map that appeared in Joan Blaeu's *Atlas Maior* of 1662 (fig. 46).[82] In this map (copied from a reduction of Blaeu's wall map of 1648) the

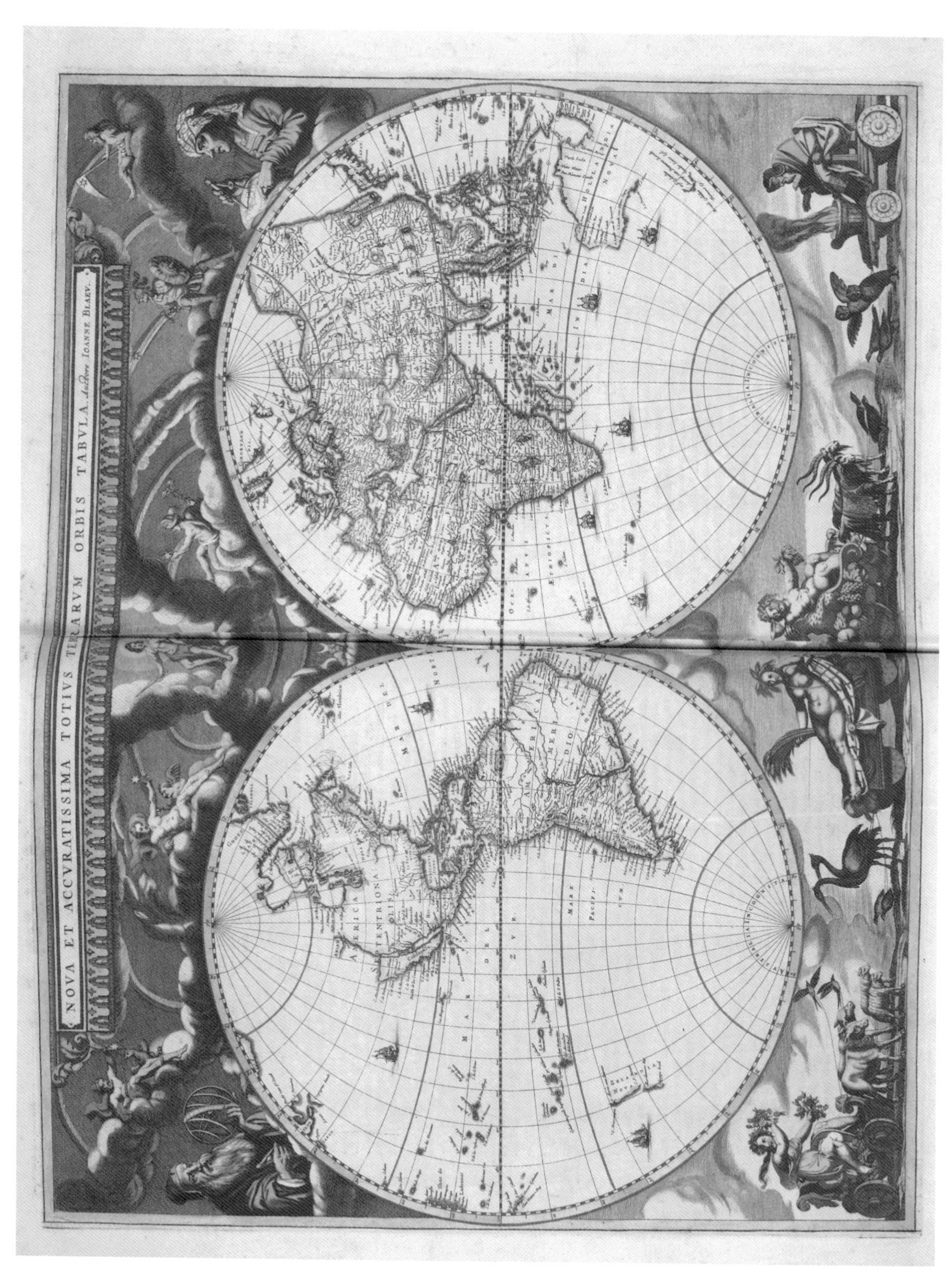

Fig. 46. World map in Joan Blaeu's *Atlas Maior* (1662). London, BL, Maps C.4.c.1. This map has abandoned *Terra Australis Incognita*, instead representing the west, north and south coasts of New Holland ('Hollandia Nova'), as well as portions of Van Diemen's Land and New Zealand, based on Abel Tasman's voyages of 1642–4.

west, north and south coasts of 'Hollandia Nova' are shown to have been charted in significant part, as well as portions of Van Diemen's Land and New Zealand, based on Abel Tasman's voyages of 1642–4; there is no trace of the great(er) southern land.

On the other hand, many cartographers were reluctant to discard the hypothesis of *Terra Australis Incognita*, and they continued to show its outlines, coexisting if necessary with New Holland. Pierre Duval's 'Carte Vniverselle du Commerce' of 1674,[83] designed to show the principal trading routes to and from the East Indies, also shows an extensively mapped 'Nouvelle Holandt'; to its south, however, Duval has retained the expanse of 'Antarctic lands called Southern and Magellanic, and unknown' (Terres Antarctiques dites Australes et Magellaniques et Inconnues). He also notes that the 'Terre des Perroquets' and the realms of Psitac, Beach, Lucac and Maletur are located by some on the continent, which now incorporates the additions of seventeenth-century exploration: 'Terre de Quir' (land discovered in 1606 by the Spanish explorer, Pedro Fernández De Quirós, initially thought to be part of the southern land) and New Zealand. In fact, it was not until the voyages of James Cook in the second half of the eighteenth century that the non-existence of *Terra Australis Incognita* was conclusively established. As much as this longevity was due to the tenacity of classical, medieval and early modern geographical theory, perhaps equally the iconographic allure of the southern continent ensured its survival.

A most remarkable image from a mid-seventeenth century Jesuit apologia (1664), which depicts the spread of the Society of Jesus throughout the world, gives an indication of why the possibilities for visual representation offered by the southern land continued to be particularly attractive well after news of its non-existence (Plate 8).[84] From the side-wound of Christ at the top of the engraving four rays of light shoot forth to the heart of the Order's founder, St Ignatius Loyola. The rays strike Loyola's heart, and (now seven) fan out over an assembled throng of the peoples of the world. American Indians, African kings, turbaned Asians; all receive the light of Christ via Loyola's heart. In amongst this assembly of nations and races prominent Jesuits, such as St Francis Xavier, lead forth their converts. These beneficiaries of the Order have been led to … a map, in the shape of a heart, before which a young member of the Society in a surplice, possibly St Aloysius Gonzaga, kneels with a small child. The child leans towards the map, arms outstretched; the young Jesuit wags the forefinger of one hand in admonition or instruction, and with the forefinger of the other he points. To what does he point? To which place on the map does he direct the keen gaze of the boy? Which land is drawn to the attention of the assembled ranks of rapturous onlookers, where is the eye of the reader encouraged to pause?

The misshapen heart at the bottom of the map has no known name, nor a known people. No missionaries can lead happy converts from its shores to stand alongside the peoples of the world, and to experience Christ's love. Not all the world contained within the sacred-heart map is truly within it. The pointing finger touches on that which is outside, that which has not been reached, that which will be reached. *Terra incognita* seems to be the necessary inversion of the sacred heart, the ignorance that accompanies dreams of world expansion, of universal reach, always receding into the distance. This unknown land underwrites knowledge, forms part of its interior space; it can be celebrated, manipulated, and mocked, but it cannot be described. It belongs, in short, to a different order of representation: a signifier without signified, it possesses no representatives.

Perhaps, then, *terra incognita* is the first and last of lands, not-here, always there, constantly renewed; emblem of ambition (where we will go next) and humility (where we have not yet gone), of breadth of vision and insignificance, its richness of non-signification relinquished after a most thorough search, and only then with the greatest of reluctance.

Notes

1 Numa Broc, *La géographie de la Renaissance 1420–1620* (Paris: Éditions du CTHS, 1980, repr. 1986), p. 172: 'Assaillis d'informations nouvelles mais incapables de distinguer le vrai du faux, avides de "curiosités" et de merveilleux, les géographes anticipent sur la réalité et suppléent par l'imagination aux lacunes du savoir. Par là s'explique cette 'horreur du vide' qui fait à la fois le charme et la faiblesse de la production cartographique de la Renaissance. Sans doute, à l'origine des mythes géographiques, trouvons-nous presque toujours quelque parcelle de vérité, mais cette vérité, souvent modeste, est embellie, amplifée pour donner naissance à de grandioses constructions de l'esprit.'

2 See particularly Gastaldi's 'Cosmographia Vniversalis', a world map extending over nine woodcut sheets compiled in or around 1561 (with lions, elephants, centaur, Cyclops, dragon-like creatures, bears, lynx, and frog on the southern land), and Forlani's world map of 1565 (lions, camels, griffins and the occasional unicorn): Shirley, *Mapping of the World*, pp. 122–5, no. 107, and 133–5, no. 115.

3 See, for example, Peter Apian, *Cosmographicus Liber* (Landshut, 1524), chapter 16: 'De Perioecis, Antoecis, Antipodibus siue Antichthonibus, Peristiis et anphistiis'; Gemma Frisius, *De principiis* (Leiden, 1544), ff. 18v–19v; Sebastian Münster, *Cosmographiae universalis* (Basel, 1550), p. 16; Hieronymo Girava, *Dos Libros de Cosmographia* (Milan, 1556), p. 41.

4 Johannes Dörflinger, 'Der Gemma Frisius-Erdglobus von 1536 in der Österreichischen Nationalbibliothek in Wien', *Der Globusfreund* 21–3 (1973), 81–99, p. 91.

5 van der Krogt, *Globi Neerlandici*, pp. 62–7, 413.

6 *Les sphères terrestre et céleste de Gérard Mercator 1541 et 1551*, ed. A. De Smet (Brussels: Editions Culture et Civilisation, 1968). Mercator's caption reads: 'Psitacorum regio a Lusitanis anno 1500 ad milia passuum bis mille praeter uectis sic appellata quod psitacos alat inauditę magnitudinis, vt qui ternos cubitos æquent' (region of parrots, so called by the Portuguese who sailed along it in the year 1500 for two thousand miles, because it has parrots of an unheard-of size, such that they equal three cubits). On the sources for Mercator's identification of this land see Richardson, 'Mercator's Southern Continent', pp. 78–86.

7 This information originally appeared in Polo's account of Java Minor, which was said to include amongst its peoples idolatrous mountain-dwellers who consumed human flesh: *Divisament dou monde*, chapter 166 (p. 543). Di Varthema also located cannibals amongst the Javans: *Navigationes*, in Simon Grynaeus, *Novus Orbis Regionum ac Insularum Veteribus Incognitarum* (Paris and Basel, 1532), book 6, chapter 29.

8 Richardson argues that Mercator is likely to have used the Paris rather than the Basel edition: 'Mercator's Southern Continent', p. 74 *n*23.

9 Simon Grynaeus, *Novus Orbis Regionvm ac insularvm veteribvs incognitarvm* (Paris, 1532), p. 351. Various suggestions have been made about the actual identity of these places: see, for example, *The Book of Ser Marco Polo the Venetian Concerning the Kingdoms and Marvels of the East*, ed. and trans. Henry Yule, rev. Henri Cordier, 2 vols (London: Murray, 1920), vol. 2, pp. 278, 280.

10 Richardson, 'Mercator's Southern Continent', p. 77.

11 Richardson, 'Mercator's Southern Continent', pp. 74–8.

12 This 'libellus', presumably produced to accompany the globe, is now lost. It may be similar, however, to the 'Declaratio insigniorum utilitatum quae sunt in globo terrestri, caelesti et anulo astronomico' that Mercator produced for Charles V in 1552: Stevenson, *Terrestrial and Celestial Globes*, vol. 1, p. 129.

13 Shirley, *Mapping of the World*, pp. 87–9; Stevenson, *Terrestrial and Celestial Globes*, vol. 1, pp. 124–34, fig. 61.

14 'Itaque, cum sperem huiusmodi nostras inventiones gratas fore studiosis adeoque vendibiles, statui iamiam globum geographicum edere, quod institutum vehementer D. secretario Morilloni, ac D. Adriano Amerocio placet, atque hoc ipsum hortantur ...': *Correspondance Mercatorienne*, ed. M. Van Durme (Antwerp: Nederlandsche Boekhandel, 1959), no. 3, p. 16. Mercator refers to Gui Morillon, secretary to Charles V, and Adrien Amerocius or Gueneville, a tutor of the sons of Nicolas Perrenot: De Smet, *Les sphères terrestre et céleste de Gérard Mercator*, p. 5. See also the discussion of audience in van der Krogt, *Globi Neerlandici*, pp. 69–70.

15 G. Mercator, *Atlas, sive Cosmographicae meditationes de fabrica mundi et fabricati figura* (Duisburg, 1595), p. 22.

16 Walter Ghim, *Vita Mercatoris*, ed. Hans-Heinrich Geske, *Duisburger Forschungen* 6 (1962), 246–72, pp. 260–1; trans. A. S. Osley, in *Mercator* (London: Faber, 1969), pp. 185–94, p. 190.

17 Ghim, *Vita Mercatoris*, pp. 252–3; trans. Osley, p. 187: 'he set out, for scholars, travellers, and seafarers to see with their own eyes, a most accurate description of the world in large format' (novum opus, scilicet universi orbis exactissimam descriptionem, in amplissima forma intuentium doctorum hominum ac peregrinantium et navigantium oculis conspectuique exhibuit atque proposuit).

18 On the initially muted impact of Mercator's innovations see Mireille Pastoureau, 'The 1569 World Map', in *The Mercator Atlas of Europe*, ed. Marcel Watelet (Pleasant Hill: Walking Tree Press, 1998), pp. 79–88.

19 'Locach regio' had been added to the toponyms of the southern continent on Abraham Ortelius' 1564 cordiform world map. In fact 'Beach' and 'Locach' were one and the same place (or non-place). See chapter 9.

20 See *Gerard Mercator's Map of the World (1569) in the Form of an Atlas in the Maritiem Museum 'Prins Hendrik' at Rotterdam*, reproduced and introduced by B. van't Hoff, *Imago Mundi supplement 6* (Rotterdam, 1961). See also the facsimile of the Basel copy of this map, one of the two surviving copies in wall map form: *Weltkarte ad usum navigantium, Duisburg 1569*, ed. Wilhelm Krücken and Joseph Milz (Duisburg: Mercator-Verlag, 1994). The evidence that the atlas form is contemporary includes the fact that the paper is sixteenth-century, and that Mercator's hand is found on some sheets which differ from the wall map form. For example, Mercator has added 'Pars Continentis Australis' to the atlas' Sheet 7, in place of the panel concerning the Ganges and Golden Chersonese, which had been relocated. See also Pastoureau, 'The 1569 World Map', p. 83.

21 Karrow, *Mapmakers of the Sixteenth Century*, p. 9.

22 Abraham Ortelius, *Theatrum Orbis Terrarum* (Antwerp, 1570; facsimile edition Amsterdam: Israel, 1964).

23 The third state of the 'Typus Orbis Terrarum' (dated 1587) added a further four quotations – two from Cicero and two from Seneca – on the theme of man's contemplation of the world, and the folly of human conflict. The central quotation from Cicero remained. For discussion see Lucia Nuti, 'The World Map as an Emblem: Abraham Ortelius and the Stoic Contemplation', *Imago Mundi* 55 (2003), 38–55.

24 See especially Giorgio Mangani, *Il 'mondo' di Abramo Ortelio: Misticismo, geografia e collezionismo nel Rinascimento dei Paesi Bassi* (Modena: Panini, 1998), and Mangani, 'Abraham Ortelius and the Hermetic Meaning of the Cordiform Projection', *Imago Mundi* 50 (1998), 59–83.

25 For a summary of the thought of Niclaes see Alastair Hamilton, *The Family of Love* (Cambridge: Clarke, 1981), pp. 34–9; for the thought of Niclaes' schismatic disciple Hendrik Jansen van Barrefelt ('Hiël') and its influence on the Antwerp humanists see Hamilton, *Family of Love*, pp. 87–102; Mangani, *Il 'mondo' di Abramo Ortelio*, pp. 89–94.

26 Mangani, *Il 'mondo' di Abramo Ortelio*, pp. 94–146, and Mangani, 'Abraham Ortelius and the Hermetic Meaning of the Cordiform Projection', pp. 68–71; but see Hamilton for the argument that '[l]ike Plantin, like Justus Lipsius … [Ortelius] was a humanist and a *politique* before he was a Familist': *Family of Love*, pp. 65–74.

27 Mangani, *Il 'mondo' di Abramo Ortelio*, pp. 15–19, 135–9; see also Marcel P. R. van den Broecke, 'Introduction to the Life and Works of Abraham Ortelius (1527–1598)', in *Abraham Ortelius and the First Atlas: Essays commemorating the Quadricentennial of his Death 1598–1998*, ed. Marcel van den Broecke, Peter van der Krogt and Peter Meurer (Utrecht: HES, 1998), pp. 29–54: p. 43.

28 *Ecclesiae Londino-Batavae Archivvm, vol. 1: Abrahami Ortelii et virorvm ervditorvm ad evndem et ad Jacobvm Colivm Ortelianvm Epistvlae*, ed. J. H. Hessels (Cambridge: University Press, 1887), no. 295, p. 698; cited also in van den Broecke, 'Introduction', p. 43.

29 Mangani, *Il 'mondo' di Abramo Ortelio*, p. 264.

30 Joannes Vindvillius to Ortelius, 21/6/1568: *Abrahami Ortelii … Epistulae*, no. 25, p. 60.

31 See for example Ortelius' letters to to Emmanuel van Meteren of 3/7/1559 and 13/12/1567; Geeraert Janssen's letter to Jacop Cool of 14/11/1576: *Abrahami Ortelii … Epistulae*, nos 8, 23, and 64, pp. 17–19, 52–5, 145–53.

32 Guillaume Postel, *Cosmographicae disciplinae compendium* (Basel, 1561), p. 14. The source of Postel's claim that the inhabitants of 'Chasdia' were black may have been André Thevet's report of a conversation with an English mariner who told Thevet he had landed on Tierra del Fuego and encountered black inhabitants. Thevet, however, doubted that people on a latitude equivalent to that of England and Scotland could have black skin: *Le Brésil d'André Thevet:*

Les Singularités de la France Antarctique (1557), ed. Frank Lestringant (Paris: Éditions Chandeigne, 1997), chapter 56, p. 217.

33 Shirley, *Mapping of the World*, pp. 167–8, no. 144. The map is on a north polar projection and extends as far as the equator. The southern hemisphere is shown in two quarter segments located in the top right and left corners of the map. See also Marcel Destombes, 'Guillaume Postel cartographe', in *Guillaume Postel, 1581–1981* (Paris: Tredaniel, 1985), pp. 361–71, esp. pp. 366–71.

34 Claude-Gilbert Dubois, 'La mythologie nationaliste de Guillaume Postel', in *Guillaume Postel, 1581–1981*, pp. 257–64: pp. 259–62; Dubois, 'La fonction du roi de France dans la fantasmagorie politique de Guillaume Postel', in *L'Image du souverain dans les lettres françaises des guerres de religion à la révocation de l'Edit de Nantes*, ed. Noémi Hepp and Madeleine Bertaud (Paris: Klincksieck, 1985), pp. 141–51. On the adaptability of Postel's political positions see also William J. Bouwsma, *Concordia Mundi: The Career and Thought of Guillaume Postel* (Cambridge: Harvard University Press, 1957), esp. pp. 216–30; Yvonee Petry, *Gender, Kabbalah and the Reformation: The Mystical Theology of Guillaume Postel* (Leiden: Brill, 2004), pp. 67–8.

35 *Les Trois Mondes de La Popelinière*, ed. Anne-Marie Beaulieu (Geneva: Droz, 1997). Beaulieu (p. 48) suggests that La Popelinière was influenced by the proposals written by the cosmographer André d'Albaigne in 1571, 'Au Roy et à Monsieurs de son conseil privé', and 'Navigation et commerce. Discours D'albaigne'.

36 *Les Trois Mondes*, p. 78.

37 *Les Trois Mondes*, p. 78.

38 *Les Trois Mondes*, p. 417.

39 On the widespread extent of this theory at the time of the Wars of Religion see Frank Lestringant, *Le Huguenot et le Sauvage: L'Amérique et la controverse coloniale, en France, au temps des Guerres de Religion (1555–1589)* (Paris: Amateurs de livres, 1990), pp. 233–4.

40 Shirley, *Mapping of the World*, p. 171, no. 148. The only significant difference between this image and Ortelius' map is that the quotation from Cicero has been translated into French.

41 Frank Lestringant, *Le Huguenot et le Sauvage*, pp. 183–202.

42 For an overview see William C. Sturtevant, 'First Visual Images of Native America', in *First Images of America*, ed. Chiappelli, vol. 1, pp. 417–54.

43 For a summation of the work of the Dieppe school see Helen Wallis, 'The Boke in the context of the Dieppe school of hydrography', in *The Maps and Text of the Boke of Idrography presented by Jean Rotz to Henry VIII*, ed. Helen Wallis (Oxford: Roxburghe Club, 1981), pp. 38–9.

44 *The Maps and Text of the Boke of Idrography presented by Jean Rotz to Henry VIII*, p. 79 (translation p. 80). The *Boke* was initially conceived as a project dedicated to François I, before Rotz switched allegiance to the English court.

45 Authorship of this work has been traditionally attributed solely to Alfonse despite the assertion made in its opening and conclusion that authorship was shared by Raulin Sécalart: see Jean Fonteneu dit Alfonse de Saintonge, *La Cosmographie avec l'espère et régime du soleil et du nord*, ed. Georges Musset (Paris: Leroux, 1904), pp. 38–9.

46 *La Cosmographie*, pp. 343, 399–400.

47 *La Cosmographie*, p. 427.

48 *La Cosmographie*, p. 59.

49 See Wallis' summary in 'Java-la-Grande: the first sight of Australia?', in *The Maps and Text of the Boke of Idrography*, pp. 58–67. The case for French discovery is made by Elizabeth Bonner, 'Did the French discover Australia? The first French scientific voyage of discovery, 1503–1505', in *Revolution, Politics, and Society: Elements in the Making of Modern France*, ed. David W. Lovell (Canberra: Australian Defence Force Academy, 1994), pp. 40–8. It seems likely that at least some of the toponyms that recur in several Dieppe school maps on the northern coasts of Java-la-grande have their basis in Portuguese mapping of south-east Asia; however, the case for an Australian basis for these names has not been, and perhaps cannot be, proven: see W. A. R. Richardson, 'Jave-la-Grande: a case study of place-name corruption', *The Globe* 22 (1984), 9–32.

50 London, BL, MS Royal 20 E.IX, ff. 29v–30r, 9v–10r, respectively.

51 On the toponyms see Richardson, 'Jave-la-Grande: a case study of place-name corruption'.

52 Guillaume Brouscon's world map of 1543 (now in the Huntington Library, San Marino) also shows 'La Jave Grande'. Brouscon, a pilot from Le Conquet, marked the southern continent around Tierra del Fuego 'MAGAILLAN'; the land to the south of La Java Grande extends towards the Antarctic and is marked 'TERRE OSTRALE'. See Michelle Mollat du Jourdin and Monique de la Roncière, *Les portulans: cartes marines du XIIIe au XVIIe siècle* (Fribourg: Office du Livre, 1984), no. 42, pp. 227–8.

53 Wallis, 'Java-la-Grande: the first sight of Australia?', p. 59.

54 A. Anthiaume, *Cartes marines*, 2 vols (Paris: Dumont, 1916), vol. 1, p. 89. Two other coats of arms appear: one of Henri II, the other of the Admiral of France, Claude d'Annebaut: see Mollat du Jourdin and de la Roncière, *Les portulans*, no. 47, pp. 231–2.

55 On this last point see the argument of Romuald Lakowsky, 'Geography and the More Circle: John Rastell, Thomas More, and the "New World"', *Renaissance Forum: An Electronic Journal of Early-Modern Literary and Historical Studies* 4 (1999).

56 Translation from Frank Lestringant, *Mapping the Renaissance World: The Geographical Imagination in the Age of Discovery*, trans. David Fausett (Berkeley: University of California Press, 1994), p. 133.

57 Lestringant, *Mapping the Renaissance World*, p. 133; A. Anthiaume, 'Un pilote et cartographe havrais au XVIe siècle, Guillaume le Testu', *Bulletin de géographie historique et descriptive* 1–2 (1911), 135–202, p. 179.

58 Anthiaume, 'Un Pilote et cartographe havrais', p. 177.

59 Frank Lestringant, 'Fictions cosmographiques à la Renaissance', in *Philosophical Fictions and the French Renaissance*, ed. Neil Kenny (London: Warburg Institute, 1991), pp. 101–25, p. 117.

60 1 Kings 9:28, 10:11. Ophir had already been located near the southern land by Alfonse: *La Cosmographie*, p. 400.

61 On the American natives depicted in Le Testu's atlas see Sturtevant, 'First Visual Images of Native America', pp. 430–3.

62 Again this may reveal the influence of Alfonse, who mentions large-eared men amongst the inhabitants of the southern land: *La Cosmographie*, p. 399.

63 Tony Campbell, 'Egerton MS 1513: A Remarkable Display of Cartographical Invention', *Imago Mundi* 48 (1996), 93–102, p. 99. As Campbell points out, the dating of the atlas, sometimes erroneously described as the Pasterot atlas, to c. 1587 is insecure but plausible: 'Egerton MS 1513', p. 93.

64 Günter Schilder, *Monumenta Cartographica Neerlandica*, vol. 3 (Alphen aan den Rijn: Canaletto, 1990), pp. 58–67; Shirley, *Mapping of the World*, pp. 270–6, nos 255 and 258. Blaeu's wall map of the world in two hemispheres of 1605 located an equally remarkable cartouche on 'Magallanica', in which two groups of mathematicians and navigators are depicted discussing the optimum 'method of measuring the distances between places' beneath the presiding figure of 'Geometria'. On the allegorical representation of the continents on maps see Günter Schilder, *Monumenta Cartographica Neerlandica*, vol. 5 (Alphen aan den Rijn: Canaletto, 1996), pp. 17–34, and John Gillies, *Shakespeare and the Geography of Difference* (Cambridge: Cambridge University Press, 1994), pp. 162–3; 179–82.

65 The source for the town views and maps was Georg Braun and Frans Hogenberg's *Civitates Orbis Terrarum*, a town atlas of the world: Günter Schilder, 'Willem Jansz. Blaeu's Wall Map of the World, on Mercator's Projection, 1606–07 and its Influence', *Imago Mundi* 31 (1979), 36–54.

66 See Günter Schilder, *Three World Maps by Francois van den Hoeye of 1661, Willem Janszoon (Blaeu) of 1607, Claes Janszoon Visscher of 1650* (Amsterdam: Israel 1981), pp. 25–7 for a translation of these verses.

67 On Hondius see Antoine De Smet, 'Jodocus Hondius, continuateur de Mercator', *Industrie* 17 (1963), 768–78; reprinted in *Album Antoine De Smet* (Brussels: Centre National d'Histoire des Sciences, 1974), pp. 305–27.

68 Compare the world map probably engraved by Hondius in Hugh Broughton's *A Concent of Scripture* (London, c. 1590). This version of Ortelius' 'Typus Orbis Terrarum' of 1570 is entitled 'A map of the earth with names (the most) from Scriptures', and contains several scriptural quotations; the southern continent is labelled simply 'The Chamberes of the South Iob 9': Shirley, *Mapping of the World*, p. 191, no. 173.

69 Gillies, *Shakespeare and the Geography of Difference*, p. 178.

70 *The New Hollstein Dutch and Flemish Etchings, Engravings and Woodcuts 1450–1700: The Wierix Family Part VIII*, ed. Jan van der Stock and Marjolein Leesberg (Rotterdam: Sound and Vision, 2004), pp. 134–5, no. 1795. Wierix's engraving in turn has a long heritage, including fifteenth-century woodcuts and later engravings, such as Albrecht Dürer's 1513 'Knight, Death, and the Devil'.

71 Peter Barber, 'The Christian Knight, the Most Christian King and the rulers of darkness', *The Map Collector* 52 (1990), 8–13.

72 Wright accused Hondius of making unauthorized use of these theories on Hondius' world maps, as well as his maps of Europe, Asia, Africa, and America: Schilder, *Monumenta Cartographica Neerlandica*, vol 5, pp. 50–5. Wright gave his account of the dispute in his *Certaine Errors in Navigation* (London, 1599; repr. Amsterdam: Theatrum Orbis Terrarum, 1974), sig. ¶¶ 4v–¶¶¶r. For the map of 1599 ascribed to Wright see Shirley, *Mapping of the World*, pp. 238–9, no. 221.

73 Plancius' monumental 'Nova et Exacta Terrarum Orbis Tabvla Geographica ac Hydrographica' bears no less than ten tablets on its version of 'Magallanica', as well as polar projections, a dedicatory coat of arms with a publisher's imprint, and a large picture of sage teachers of geography and their students: see F. C. Wieder, *Monumenta Cartographica*, 5 vols (The Hague: Nijhoff, 1925–33), vol. 2, pp. 27–36 and plates 26–38; and Günter Schilder, *Monumenta Cartographica Neerlandica*, vol. 7 (2003), pp. 105–10.

74 Hondius reissued his 1603 world map in 1608 to compete with Blaeu's twenty-sheet world map of 1605. The 1608 version survives in the Royal Geographical Society, London; the 1603 map only resurfaced relatively recently: see Paul E. Cohen and Robert T. Augustyn, 'A Newly Discovered Hondius Map', *The Magazine* 153 (1999), 214–17. For descriptions of the 1608 version of Hondius' map see Shirley, *Mapping of the World*, pp. 279–82, no. 263; and Schilder, *Monumenta Cartographica Neerlandica*, vol. 3, pp. 71–5. Blaeu's 20-sheet world map of 1605, and his 1606–7 world map are also important examples of large, elaborately decorated wall maps. My discussion here relates principally to the 1608 version of Hondius' map, the decorative frame of which differs in significant detail from the 1603 version. A facsimile of the 1608 map was published by Edward Heawood, *The Map of the World on Mercator's Projection by Jodocus Hondius* (London: Royal Geographical Society, 1927).

75 For an outline of the history of Columbus portraiture in painting see Carla Rahn Phillips, 'The Portraits of Columbus: Heavy Traffic at the Intersection of Art and Life', *Terrae Incognitae* 24 (1992), 1–18.

76 Hall's work was published under the name 'Mercurius Britannicus', apparently at Frankfurt. It spawned several imitations and outright copies, including *The Travels of Don Francisco de Quevedo Through Terra Australis Incognita. Discovering the Laws, Customs, Manners and Fashions Of The South Indians. A Novel. Originally in Spanish* (London, 1684); it was loosely translated into English as early as 1609 by John Healey. On this translation and the question of authorship see Richard A. McCabe, *Joseph Hall: A Study in Satire and Meditation* (Oxford: Clarendon Press, 1982), pp. 321–39.

77 Shirley, *Mapping of the World*, p. 267, no. 251. The map is a folded insert before book one of the text: 'Crapulia'.

78 For discussion of these names see *Another World and Yet the Same: Bishop Joseph Hall's* Mundus Alter et Idem, ed. and trans. John Millar Wands (New Haven: Yale University Press, 1981), pp. 118–26; 198–200.

79 'Mercurius Britannicus', *Mundus Alter et Idem Siue Terra Australis ante hac semper incognita longis itineribus peregrini Academici nuperrime lustrata* (Frankfurt, [1605]), p. 91; *Another World and Yet the Same*, trans. Wands, p. 57.

80 *Mundus Alter et Idem*, p. 190; *Another World and Yet the Same*, trans. Wands, p. 105. Hall's targets also included heretical sects: see McCabe, *Joseph Hall*, pp. 104–5.

81 An obvious parallel is the Fool's Cap World map of 1590, in which a world map was substituted for the face of a fool, beneath the heading 'Nosce te ipsum' (Know yourself): see Shirley, *Mapping of the World*, p. 189, no. 170, and (for an earlier version of the image) p. 157, no. 134.

82 Shirley, *Mapping of the World*, pp. 449–51, no. 428.

83 Shirley, *Mapping of the World*, p. 482, no. 465.

84 The image, engraved by Bartholomew Kilian, is one of a series of images in a work dedicated to the Archbishop of Brixen: see further Mangani, 'Abraham Ortelius and the Hermetic Meaning of the Cordiform Projection', p. 71.

Chapter Nine

Much Lost Knowledge

In 1686 a curious debate erupted about geographic falsehood. The spark was a two-page list that had appeared five years earlier as an appendix to Michel-Antoine Baudrand's *Geographia Ordine litterarum disposita.* Baudrand's *Geographia* attempted to provide a compendium of place names of the entire world, arranged in alphabetical order, with each toponym recorded in Latin alongside a vernacular translation or equivalent. To enable those who knew the vernacular but not the Latin name to use the volume, an appendix was provided in which toponyms were organized alphabetically, with vernacular name first, followed by Latin translation. Baudrand's mode of alphabetical ordering was designed to improve ease of reference to both classical and modern place names, and thereby to facilitate the comprehension of ancient and modern geography by holding the toponyms of each in parallel. The project was frankly universalist in its ambitions: Baudrand explained in the dedication of his work to Colbert, the chief minister to Louis XIV, that 'Geography, or the description of the world, is the complete sum of all things of whatever number, in which constant comparison of ancient with recent Geography is made' (Geographia est integra, sive Orbis descriptio, omnium quotquot unquam fuere maxima, in qua veteris Geographiae cum recenti perpetua fit collatio).[1] But that comparison was not purely inclusive; it had also to exclude. Quoting with approval the preface to Philipp Clüver's *Germania antiqua,* Baudrand defined the task of geography as one not merely of collection, but of the discernment of true from false:

> haec enim scientia rebus concisis, & minutissimis saepe vocabulis constat, quae in tanto auctorum numero vage ac fuse dispersa, hinc inde conquirere, in unum locum colligere, & sub unum aspectum subjicere, falsa à veris discernere, haec confirmare, illa convellere, invicem pugnantia nonnunquam conciliare, denique obscuris lucem, vetustis novitatem, novis authoritatem, ac dubiis fidem dare, plusquam Herculei est laboris.[2]

> for this science is founded on small matters, and often the most tiny words, which are widely and copiously dispersed in the greatest number of authors: to search here and there, to collect in one place and to make subject to a single inspection, to set apart false from true, to confirm these, to overthrow those, at times to reconcile inconsistencies with each other, at length to give light to the dark, freshness to the old, authority to the new, and credence to the uncertain, is more than a labour of Hercules.

At the end of his lengthy list of toponyms, Baudrand included a series of extra sections – lists of archbishoprics and bishoprics throughout the world; of universities; of free and imperial cities of Germany; of cities, dominions, and provinces sold and pledged – and a list of 'regions, towns, and places that never were, even though they have been marked on maps' (regiones, urbes, & loca quae nusquam fuerunt, licet in chartis designentur).[3] In that list appeared the toponyms 'Beach' and 'Java Minor'. Of Beach, Baudrand recorded that it was a large region, located on *Terra Australis* in many maps, 'but there is profound silence amongst authors as to where it is, and by whom it was discovered; in addition, more recent accounts of those parts have established that there is no region of that name in all of *Terra Australis*, and in those parts discovered by Europeans' (sed ubi fuerit, aut à quibus detecta, altum inest inter authores silentium, et ex recentioribus relationibus illarum partium constat, nullam esse regionem sic dictam in omni Terra Australi, et in partibus ab Europaeis detectis).[4] Baudrand therefore places Beach in the same category as such figments of New World geography as the Strait of Anian (the legendary North-West passage), El Dorado, and Frislandia, the large North Sea island said to have been discovered in the fourteenth century by the Venetian Nicolò Zeno. Polo's Beach is displaced by the toponyms derived from seventeenth-century Dutch mapping of New Holland such as 'Concordiae Regio' (*t'landt van Eendracht*), 'Diemeni Regio' (*het landt van Diemens*), Nova Hollandia and Nova Zelandia, all of which are given the stamp of authenticity and included in the main section of Baudrand's work.

There can be little coincidence in the appearance of Baudrand's *Geographia* in the same year as the publication of Jean Mabillon's seminal *De re diplomatica*. In that multi-volume work Mabillon attempted to set down the principles and also the means for an 'art of diplomatic', by which truth might be discerned from falsehood in medieval documents.[5] Baudrand's parallel project was to weed out places that were not and had never been. It was not, however, to exclude those places that had existed in antiquity but had since disappeared, or changed names, of which there were many. On the contrary. It had long been the task of the geographer to preserve, indeed promote knowledge of such places, lest the works of classical antiquity should become baffling and meaningless. The radical act performed by Baudrand's list of fictitious places was to create a category of the 'never was', of places that had existed only in the imagination, yet whose fictitiousness had not prevented them finding their way onto maps. But at what cost did one designate a place name as false?

For at least one of Baudrand's readers, the cost was the malignant rewriting of history. In his *Riflessioni geografiche circa le terre incognite* (Padua, 1686), the Benedictine prior Vitale Terrarossa unleashed a vitriolic attack on Baudrand for denying the crucial role of Venetian travellers and explorers in the history of discovery. Terrarossa was particularly incensed by Baudrand's inclusion of Beach, Java Minor, and Frislandia in his list of fictitious toponyms. Dedicating his work to three *procuratori* of San Marco, Venice, Terrarossa attempted to prove that Venetians had discovered all lands unknown to the ancients, including America and *Terra Australis*, and he argued for 'an exact and perfect concordance of ancient and modern accounts of the terrestrial globe, in honour of the nobility and the Republic of Venice'.[6] To deny the validity of the toponym 'Beach' – to deem it a falsehood, a lie – was to deny the Venetian Marco Polo the honour of discovering and naming this part of the southern land. Indeed Terrarossa called for the southern continent to be renamed 'Pola' instead of 'Magellanica', since Polo had reached its shores first and penetrated further than any other early explorer into the great landmass.[7] Similarly, the relegation of Frislandia to the nebulous realm of fictitious geography denied the role of the Venetian Zeno, an account of whose exploits had been widely disseminated in Ramusio's *Navigazioni*.[8] In Terrarossa's anguished response a geography of patriotism is vividly expressed – but one that shares with Baudrand and other geographers of the period the desire for concordance, for parallel between ancient and modern. Terrarossa noted that the regions previously called Locach and Maletur were now, following Dutch explorations, called 'Terra Concordiae' and 'Nova Hollandia' on maps.[9] These were, he insisted, merely the same places already discovered by Polo re-seen and renamed: names change, and the fact that more recent explorers had found no places with these names did not mean that they never existed. The answer to the problem was to unite past and present nomenclature. As at least one mapmaker had already done, Polo's toponyms should be displayed along with the new vocabulary (fig. 47).[10]

Terrarossa's claims may on the surface seem anachronistic, absurd, and tendentious in the extreme. They are nevertheless important because they reveal something of what was at stake in the classification of certain places as fictive or fabulous. It may be tempting to see Terrarossa's plea for the removal of such names as Beach from the realm of pseudo-geography as a counter-Enlightenment move against the scientific purging of fantastic accretions from geographical knowledge. His arguments reveal instead an alternative strand of thought within Enlightenment geography: Terrarossa was in favour of the advancement, the perfection of the science of geography, and he believed that this objective could be accomplished by the inclusion of ancient, old, and modern names together in a concordance – and, crucially, together on maps. The map, in other words, should act in the same manner as the encyclopedia, by holding old and new in parallel, and making both simultaneously visible. This at least would have the virtue of avoiding the danger that, in too readily discarding apparently fictitious regions, the praises due to discoverers may be omitted, important moments in the history of discoveries overlooked. However eccentric, however partisan, Terrarossa's outburst suggests that – as with the contemporary practice of distinguishing false from genuine charters – in 1686 the process of the exclusion of legendary places was not consensual, and that other models for the reformation of geographical knowledge existed.

Baudrand did not include *Terra Australis* in his list of places that were not, and had never been, in existence. He referred to it instead in the main section of his work as a most ample part of the world, 'but still unknown for the most part, and revealed only in some coastal places; thus even now little could be said about it' (pars amplissima orbis ... sed etiamnum pro majori parte incognita, & vix lustrata quibusdam in locis versus oram littoralem; ita ut de ea pauca admodum dici possint).[11] One hundred years later, the existence of the great southern land of sixteenth-century cartography was finally disproved by James Cook's relentless exploration of the south seas. Still, another geographer continued to search in the shadows of *Terra Australis*. As late as 1769, the hydrographer and historian of exploration Alexander Dalrymple had urged some

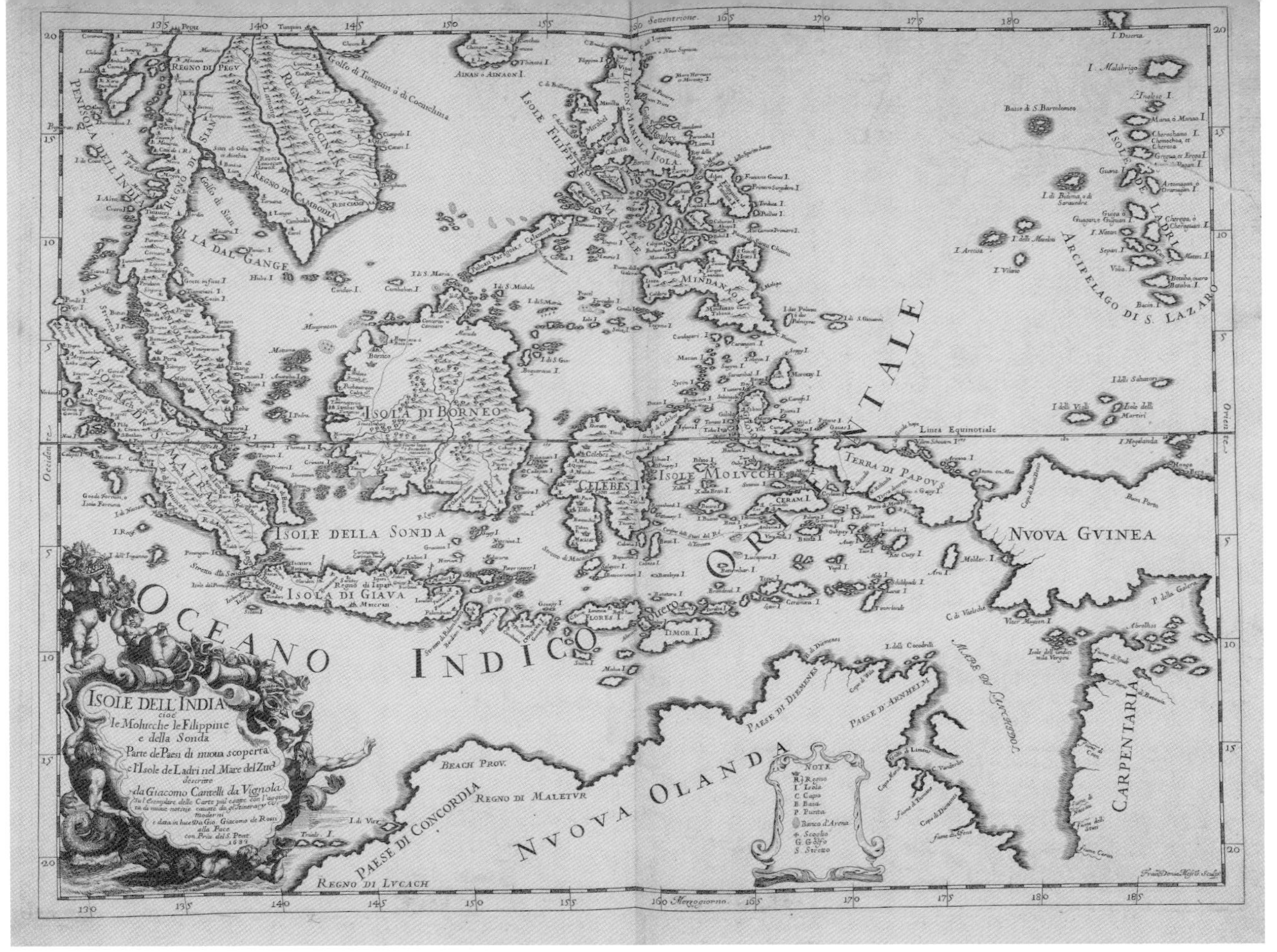

Fig. 47. Giacomo Cantelli da Vignola, 'Isole dell'Indie, e parte de'Paesi di nuoua scoperta', in G. Giacomo de Rossi, *Mercurio Geografico overo Guida Geografica in tutte le parti del mondo* (Rome, n.d.). London, BL, Maps C.39.f.7: map 63. In his *Riflessioni geografiche*, Vitale Terrarossa noted that Giacomo Cantelli da Vignola's map 'Isole dell'Indie, e parte de'Paesi di nuoua scoperta' shows the cape of *Terra Australis* seen by Marco Polo, with the toponyms Beach, Lucach, and Maletur alongside 'Paese di Concordia' and 'Nuoua Olanda'. The map does indeed locate these toponyms ('Regno di Lvcach', 'Beach Prov.', 'Regno di Maletvr'), in the north-west of the nascent continent of New Holland (i.e. present-day Australia).

emulator of Magellan to 'establish[] an intercourse' with the southern continent and open up its estimated 50 million inhabitants to the mutual benefits of British trade, urgings in part responsible for Cook's forays towards Antarctica in the following decade.[12] In 1786, with the thesis of a massive and abundantly populated Antarctic continent no longer tenable, Dalrymple – by now hydrographer to the East India Company – published a *Memoir concerning the Chagos and Adjacent Islands.* The purpose of this pamphlet was to clarify the position of the Chagos archipelago and certain other Indian Ocean islands, but in an aside Dalrymple noted that he had seen proof that Cook was not the first European to have charted the east coast of New Holland. Dalrymple had seen a map, then in the possession of Joseph Banks, now known as the 'Harleian' or 'Dauphin' world map (BL Additional MS 5413) (fig. 40). He realized that it dated from the sixteenth-century, and that it was earlier than André Thevet's chart of 1575:

> The very curious MS here mentioned is painted on Parchment, with the Dauphin's Arms, it contains *much lost knowledge*; Kerguelen's Land seems plainly denoted; The *East Coast* of New Holland, as we name it, is expressed with some curious circumstances of correspondence to Captain Cook's MS what he names

Bay of Inlets	is in the MS. called	*Baye Perdue*
Bay of *Isles*	* * *	R. de beaucoup disles.
Where the *Endeavour* struck		Coste dangereuse.

> So that we may say with Solomon 'There is nothing *new* under the Sun.' ... The correspondence above alluded to, shews, that where *Native Names* cannot be had, the fittest Names are *descriptive*, and not the nonsensical application of the *proper Names* of insignificant individuals to every Creek and Corner.[13]

Like Terrarossa, Dalrymple can be portrayed as an anachronism. Instead of giving full credit to Cook, that emblem of Enlightenment geography, he sought to undo his priority, to complicate the rhetoric of discovery. Instead of celebrating the erasure of the fictions of previous centuries, he attempted to preserve them. Instead of rationality, he championed proliferation; in the face of new knowledge, he adduced '*lost knowledge*'. But like Terrarossa, Dalrymple was concerned with acts of naming. The final comment of his note reveals that the primary purpose of his aside on the Harleian map was not to usurp Cook but to make a point about the assignation of nomenclature. The examples he cites – the parallels between Cook and the Harleian map – are designed to show that by descriptive naming of the kind practised by Cook lost knowledge could be retrieved. And coterminous with this question of naming practice was the issue – crucial for Dalrymple, as it was for Terrarossa – of priority. Later in his *Memoir* Dalrymple notes that a place named 'the Speaker's Bank, 1763, is the same passed over by the *Griffin*, 1749, and ought, therefore, properly to be named the Griffin's Bank, yet the *first name* having been adopted, I continue it, to prevent confusion, notwithstanding the kind of anticipation it produces in the *Griffin*'s Voyage'.[14] Both Dalrymple and Terrarossa wanted to maintain chronology in their naming practices, and for that reason they were concerned that old names may be replaced or, even more alarming, expunged by new ones. Avoidance of confusion was preferable to chronological accuracy if a choice had to be made, but even in this aside Dalrymple's concern to preserve a history of naming is notable. Geography in the seventeenth and eighteenth centuries was not without its paradoxes: a science of the earth, its practitioners sought perfection, greater accuracy, more information; some called for acts of collation and inclusion even as others expunged the false and the fictive from their maps. As their vision swept across the entire globe it also swept back in time, seeking to account for mutation, for the manifold ways in which places were known and marked, for the same place in different tongues and different voices. Back, but also forward: the persistent fear is of losing knowledge, of surrendering priority, of future generations inheriting only the most recent names.

The history of Marco Polo's Beach in itself reveals the historical transformations in antipodal representation. Polo's text originally appears to have described a region called 'Locac', but corruption by scribal error caused some sixteenth-century editions of the *Divisament*, including the one used by Mercator, to print 'Boeach'. Ortelius, drawing on an Italian edition of Polo, printed 'Locach' on his world map of 1564, causing Mercator to add 'Lucach' to 'Beach' and 'Maletur' on his 1569 map. In its original guise, then, the toponym was the product of the expansion of European contacts with the Asian world in the late thirteenth century. Transformed as the result of manuscript transmission, it was anthologized in the first half of the sixteenth century, and seized upon by a mapmaker eager for hard information about a land he had calculated must exist, but which, he thought, only a few Europeans had seen. The mapmaker's famous and influential maps were copied, and having lost a syllable, possibly for reasons of euphony, Beach was joined on the conjectured, partially uncovered, but not yet explored land by its doppelgänger, Locach. As the result of a long process of copying, a single toponym was doubled, then located on the world map, reproduced, authenticated – and eventually rejected.

There was no Beach on the southern temperate zone of medieval zonal maps, since there were no coastlines, no interior landscapes, and no peoples on these images. Its appearance in the sixteenth century clearly indicates that the possibilities of cartographic representation had shifted, and that earlier travel narratives could be read as part of the repository of European exploration from which maps were constructed. All the same, Mercator operated on principles not far removed from his medieval predecessors: a coastline and its toponyms were possible because they were, it seemed, supported by reliable authority. Mercator, unlike some of his contemporaries, stopped the toponyms where authority ran out, retaining the a-cartographic, interior function of unknown land in the expectation of further discoveries. Beach looked like the beginning of a process of attestation; in the end, it was the epitaph of a process of creation.

The story of the representation of the antipodes is the narrative of a tradition continually altered, reshaped, debated, and at no point maintained for its own sake. At some moments debate and reshaping of the tradition are particularly visible, and apparently epochal: Augustine's critique of antipodal habitation in *De civitate dei*; the fifteenth-century reformulation of *terra incognita* as contiguous with the known world, rather than separated by oceanic and climatic barriers; the invention of *Terra Australis Incognita*. Yet, at any given moment over the long history of antipodal representation, reinterpretation was occurring. The idea of antipodeans, inevitable if perspective-altering in classical thought, became heterodox to the Christian belief that the world was populated by the descendants of Adam. With the discovery of New Worlds, antipodeans became inevitable once more, if still puzzling and perspective-altering to theses of Christian monogenesis – until, gradually, specificity of ethnic, racial, and national description took the place of the generic antipodean, the unknown and unreachable person standing differently to us.[15] The forms taken by antipodal spaces on the world image changed over the centuries in which they were reproduced, but their political function continued to revolve around two poles, well established in classical literature. Unknown land could act as an emblem of imperial ambition, an incitement and promise of future conquest; it could serve to critique and rebuke political and intellectual folly. At a certain point, however, the antipodes gave rise to a less explicitly political meaning: they came to signify, in the thirteenth century, intellectuality itself; aligned to this meaning was the idea of the antipodes as a place of depth, of escape from the sins and corruption of the known world, a place where purgatory and the earthly paradise might be located. When conjoined with the discourse of New World exploration, that idea of the antipodes as *recessus* – as a non-place conjured by the mind, and subject to the laws of rational thought – was expressed anew in the form of More's Utopia, and thereafter in many reworkings. In the utopic literature that arose in the seventeenth and eighteenth centuries *Terra Australis* became the site for the discovery of lost Europeans;[16] or, following More's hints about the placement of Utopia, for a 'New Athens'.[17] Its remoteness lent itself to thoughts of the survival of pure strands of European (Athenian, Roman) identity, and to a purging of contemporary ills; but also to the imagination of a society closed in on itself, hermaphroditic, ultrarational.[18]

The increasingly sure status of the fictionality of *Terra Australis* enhanced its capacity to act as a mirror – the 'Speculum Orbis Terrarum' – well suited to satire in which Lucian rather than Cicero was the presiding spirit. Cicero gave to Scipio the vision of lands beyond the part of the earth known to Romans, in which people stood differently; but he did not permit the visionary to describe what he saw in those far-flung regions – it was enough simply to note their presence, their difference, and their unbridgeable distance. Lucian deployed two complementary modes of vision: the antipodes as reflection of the known world, as deception of the charlatan gazing into the well; and the antipodes as tall tale, a place conjured by verbal excess. Both these Menippean strategies lent themselves to the seventeenth- and eighteenth-century taste for the faux travel narrative, in which mirrors could be unexpectedly and unflatteringly turned on the Old World, and in which a variety of fantasies sexual, racial, religious, political and philosophical could be played out. Even as it slipped from the image of the world, the unknown space of the antipodes continued to signify a wealth of possibilities.

Representation of the antipodes is only part of the history of the world image, but it is critical in one regard: it shows that the very idea of a world image, whether a crude diagram or a mathematically-precise chart, made it possible to represent and therefore to see land that had never been seen, and that never would be seen. Once conjectured, space could be drawn, and filled with rationally supportable yet empirically unvalidated depictions of land and sea. Precisely because of its status as a particular category of land that could be represented on the world image as unknown, *terra incognita* signified the fundamental nature of land and its habitation. To show land was to raise the question of whether or not it was inhabited, and to think about its inhabitants was to consider human identities, and human history. The idea of the antipodes existed outside structures of faith and verifiable knowledge: it was an idea always on the peripheries of literate discussion, by its very nature uncertain, at times unwelcome, its many reincarnations testament to the immense fertility of space founded on the irony of vision beyond the visible – of land not yet, not ever known.

Notes

1 Michaelis Antonius Baudrand, *Geographia Ordine litterarum disposita*, 2 vols (Paris, 1681–2), vol. 1, sig. a.iii (verso).

2 Baudrand, *Geographia*, 'Praefatio ad Lectorem', n.p.

3 Baudrand, *Geographia*, vol. 2, pp. 442–3.

4 Baudrand, *Geographia*, vol. 2, p. 442.

5 Jean Mabillon, *De re diplomatica libri VI* (Paris, 1681).

6 Terrarossa, *Riflessioni geografiche*, fol. 3v: 'vna esatta, e perfetta Concordia degli antichi, e de i moderni racconti del Globo Terreno, anche in riuerenza della Nobilità, e Republica di Venezia'. Baudrand's *Geographia Ordine litterarum disposita* had already been severely criticized by the noted French geographer Guillaume Sanson. Sanson counted 500 errors under the letter A alone, and accused Baudrand of failing to acknowledge material taken from the work of his father, Nicolas Sanson: *In Geographiam Antiquam Michaelis Antonii Baudrand, Disquisitiones Geographicae* (Paris, 1683).

7 *Riflessioni geografiche*, pp. 51–2.

8 *Riflessioni geografiche*, pp. 118–59.

9 *Riflessioni geografiche*, pp. 53; 290.

10 *Riflessioni geografiche*, pp. 57, 95.

11 Baudrand, *Geographia*, vol. 2, p. 301.

12 Alexander Dalrymple, *An Historical Collection of the Several Voyages and Discoveries in the South Pacific Ocean*, 2 vols (London, 1770–1), pp. i–xxx. On the context of Dalrymple's interest in the southern continent see further Howard T. Fry, *Alexander Dalrymple (1737–1808) and the Expansion of British Trade* (London: Royal Commonwealth Society, 1970), pp. 94–135. Evidence of widespread contemporary interest in the southern continent, and an important source for Dalrymple, was the compilation of Charles De Brosses, *Histoire des Navigations aux Terres Australes*, 2 vols (Paris, 1756).

13 Alexander Dalrymple, *Memoir concerning the Chagos and Adjacent Islands* (London, 1786), p. 4 (continuation of note c).

14 Dalrymple, *Memoir concerning the Chagos and Adjacent Islands*, p. 14.

15 On the challenge of the New World discoveries to the theory of monogenesis see especially Giuliano Gliozzi, *Adamo e il Nuovo Mondo: la nascita dell'antropologia come ideologia coloniale: dalle genealogie bibliche alle teorie razziali, 1500–1700* (Florence: La nuova Italia, 1977).

16 Henry Neville, *The Isle of Pines, or, A late Discovery of a fourth Island near Terra Australis, Incognita* (London, 1668).

17 Anon, *A Description of New Athens in Terra Australis Incognita* (1720), in *Utopias of the British Enlightenment*, ed. Gregory Claeys (Cambridge: Cambridge University Press, 1994), pp. 27–53.

18 Gabriel de Foigny, *La Terre Australe connue* (Geneva, 1676); on Foigny see further Fausett's introduction to *The Southern Land, Known*, ed. and trans. David Fausett (Syracuse: Syracuse University Press, 1993), and Fausett, *Writing the New World*. Jonathan Swift's Gulliver arrives on the island of Lilliput in November, 1699, having been 'driven by a violent storm to the north-west of Van Diemen's Land': on the connections between *Gulliver's Travels* and Foigny's work see David Fausett, *Images of the Antipodes in the Eighteenth Century: A Study in Stereotyping* (Amsterdam: Rodopi, 1994), pp. 39–54.

Bibliography

Primary sources

Manuscripts

Basel, Universitätsbibliothek, MS F. IV. 24
Bern, Burgerbibliothek, MS 265
Brussels, Bibliothèque royale, MS 10146
Brussels, Bibliothèque royale, MS 18.210–15
Burgo de Osma, Archivo de la Catedral, MS 1
Cambridge, Trinity College, MS R.9.23
Chantilly, Bibliothèque Condé, MS 924
Cologny (Geneva), Fondation Martin Bodmer, MS 111
Florence, Biblioteca Laurenziana, MS Gaddi 91 inf. 7
Florence, Biblioteca Laurenziana, MS Plut. 51.14
Florence, Biblioteca Laurenziana, MS Plut. 77.9
Florence, Biblioteca Laurenziana, MS Plut. IX.28
Florence, Biblioteca Laurenziana, MS Santa Croce 22 sin. 9
Lisbon, Arquivo Nacional da Torre do Tombo, Lorvão
London, British Library, MS Additional 5413
London, British Library, MS Additional 24065
London, British Library, MS Additional 25691
London, British Library, MS Cotton Faustina C.I
London, British Library, MS Cotton Julius D.VII
London, British Library, MS Egerton 2976
London, British Library, MS Harley 2533
London, British Library, MS Harley 2772
London, British Library, MS Harley 3686
London, British Library, MS Royal 20.A.III
London, British Library, MS Royal 20.E.IX
Metz, Bibliothèque municipale, MS 271
Munich, Bayerische Staatsbibliothek, Clm 6362
Munich, Bayerische Staatsbibliothek, Clm 10268
Munich, Bayerische Staatsbibliothek, Clm 10663
Munich, Bayerische Staatsbibliothek, Clm 14436
Munich, Bayerische Staatsbibliothek, Clm 15738
Nancy, Bibliothèque municipale, MS 441
Naples, Biblioteca Nazionale, MS V.A.12
Oxford, Bodleian Library, MS Auct F.2.20
Oxford, Bodleian Library, MS Auct T.2.27
Oxford, Bodleian Library, MS D'Orville 77
Paris, Bibliothèque nationale de France, Cartes et Plans, Rés. Ge A 335

Paris, Bibliothèque nationale de France, MS lat. 6367
Paris, Bibliothèque nationale de France, MS lat. 6371
Paris, Bibliothèque nationale de France, MS lat. 6570
Paris, Bibliothèque nationale de France, MS lat. 6622
Paris, Bibliothèque nationale de France, MS lat. 6802
Paris, Bibliothèque nationale de France, MS lat. 7930
Paris, Bibliothèque nationale de France, MS lat. 10195
Paris, Bibliothèque nationale de France, MS lat. 12949
Paris, Bibliothèque nationale de France, MS lat. 15170
Paris, Bibliothèque nationale de France, MS lat. 16680
Paris, Bibliothèque nationale de France, MS lat. 18421
Paris, Bibliothèque nationale de France, MS nouv. acq. lat. 923
Prague, Státní knihovna, MS VIII.H.32 (1650)
Private collection, ex-Malibu, Getty Ludwig XII.4
Reims, Bibliothèque municipale, MS 1321
Rennes, Bibliothèque municipale, MS 256
St Gall, Stiftsbibliothek, MS 237
Stockholm, Kungliga Biblioteket, Cod. Holm. M.304
Vatican City, Biblioteca Apostolica Vaticana, Archivio di San Pietro, H 31
Vatican City, Biblioteca Apostolica Vaticana, Archivio di San Pietro, H 32
Vatican City, Biblioteca Apostolica Vaticana, MS Ottob. lat. 1939
Vatican City, Biblioteca Apostolica Vaticana, MS Palat. lat. 274
Vatican City, Biblioteca Apostolica Vaticana, MS Palat. lat. 1577
Vatican City, Biblioteca Apostolica Vaticana, MS Palat. lat. 1646
Vatican City, Biblioteca Apostolica Vaticana, MS Regin. lat. 123
Vatican City, Biblioteca Apostolica Vaticana, MS Urb. lat. 274
Vatican City, Biblioteca Apostolica Vaticana, MS Vat. lat. 1546
Vatican City, Biblioteca Apostolica Vaticana, MS Vat. lat. 1548
Vatican City, Biblioteca Apostolica Vaticana, MS Vat. lat. 1575
Vatican City, Biblioteca Apostolica Vaticana, MS Vat. lat. 5993
Vatican City, Biblioteca Apostolica Vaticana, MS Vat. lat. 6018
Wolfenbüttel, Herzog August Bibliothek, MS Gudeanus lat. 1
Worcester, Worcester Cathedral, MS F. 68
Zürich, Zentralbibliothek, MS Car.C.122

Printed Editions

Ælfric, *De temporibus anni*, ed. Heinrich Henel, EETS o.s. 213 (London: Oxford University Press, 1942)

Albertus Magnus, *De natura loci*, ed. Paul Hossfield, *Opera Omnia* v.ii (Aschendorf: Monasterii Westfalorum, 1980)

Alexander VI, *Inter cetera* (3 May 1493), *Inter cetera* (4 May 1493), *Eximiae devotionis* (3 May 1493), *Dudum siquidem* (25 September 1493), in *America Pontificia*, ed. Josef Metzler, 3 vols (Vatican City: Libreria editrice Vaticana, 1991–5), vol. 1, pp. 71–89

Alfonse, Jean: Jean Fonteneu dit Alfonse de Saintonge, *La Cosmographie avec l'espère et régime du soleil et du nord*, ed. Georges Musset (Paris: Leroux, 1904)

Almagià, Roberto, ed., *Monumenta Cartographica Vaticana*, 4 vols (Vatican City: Biblioteca Apostolica Vaticana, 1944–55)

Altolaguirre y Duvale, A., *et al.*, eds, *Mapas españoles de América: siglos XV–XVII* (Madrid: Real Academia de la Historia, 1951)

Ambrose, *Opera Omnia di Sant'Ambrogio, vol. 1: Hexameron (i sei giorni della creazione)*, ed. C. Schenkl, trans. Gabriele Banterle (Milan: Biblioteca Ambrosiana, 1979)

Anon, *A Description of New Athens in Terra Australis Incognita* (1720), in *Utopias of the British Enlightenment*, ed. Gregory Claeys (Cambridge: Cambridge University Press, 1994), pp. 27–53

Anon, *The Travels of Don Francisco de Quevedo Through Terra Australis Incognita. Discovering the Laws, Customs, Manners and Fashions Of The South Indians. A Novel. Originally in Spanish* (London, 1684)

Apian, Peter, *Cosmographicus Liber* (Landshut, 1524)

Aratus, *Phaenomena*, ed. and trans. Douglas Kidd (Cambridge: Cambridge University Press, 1997)

Aristotle, *De Caelo* (*On the Heavens*), trans. W.K.C. Guthrie (London: Heinemann, 1939)

Aristotle, *Meteorologica*, trans. H. D. P. Lee (London: Heinemann, 1952)

pseudo-Aristotle, *De mundo: Translationes Bartholomaei et Nicholai*, ed. William L. Lorimer, rev. L. Minio-Paluello (Bruges: De Brouwer, 1965)

Arnulf of Orléans, *Arnulfi Aurelianensis Glosule super Lucanum*, ed. Berthe M. Marti (Rome: American Academy in Rome, 1958)

Augustine of Hippo, *De civitate dei*, ed. B. Dombart and A. Kalb, Corpus Christianorum, Series Latina 47–8, 2 vols (Turnhout: Brepols, 1955)

Augustine, *Sancti Aureli Augustini De Genesi ad litteram libri duodecim*, ed. Joseph Zycha (Vienna: Tempsky, 1894)

Averroes (Ibn Rušd), *Averrois Cordubensis commentum magnum super libro De celo et mundo Aristotelis*, ed. Francis J. Carmody, Rüdiger Arnzen and Gerhard Endress, 2 vols (Leuven: Peeters, 2003)

Avicenna (Ibn Sīnā), *A Treatise on The Canon of Medicine of Avicenna Incorporating a Translation of the First Book*, trans. O. Cameron Gruner (London: Luzac, 1930)

Bacon, Roger, *The 'Opus Maius' of Roger Bacon*, ed. John Henry Bridges, 2 vols (Oxford: Clarendon Press, 1897)

Bartholomew of Parma, *I primi due libri del "Tractatus sphæræ" di Bartolomeo da Parma*, ed. Enrico Narducci, (Rome: Bullettino di bibliografia e di storia delle scienze matematiche e fisiche, 1885)

Baudrand, Michaelis Antonius, *Geographia Ordine litterarum disposita*, 2 vols (Paris, 1681–82)

Beazley, C. R., ed., *Directorium ad faciendum passagium transmarinum*, *American Historical Review* 12 (1907), 810–57; 13 (1907), 66–115

Bede, *De natura rerum*, in *Bedae Venerabilis Opera Didascalica*, ed. C. W. Jones, Corpus Christianorum, Series Latina 123A (Turnhout: Brepols, 1975)

Bede, *De temporum ratione*, in *Opera de temporibus*, ed. Charles W. Jones (Cambridge: Mediaeval Academy of America, 1943)

Bekynton, Thomas, *The Official Correspondence of Thomas Bekynton*, ed. George Williams, 2 vols, Rolls Series 56 (London: Longman, 1872)

Biondo, Flavio, *Scritti inediti e rari di Biondo Flavio*, ed. Bartolomeo Nogara (Rome: Tipografia poliglotta vaticana, 1927)

Blänsdorf, Jürgen, ed., *Fragmenta Poetarum Latinorum Epicorum et Lyricorum*, 3rd edn (Stuttgart: Teubner, 1995)

Bockwitz, H. H., ed., 'Die "Copia der Newen Zeytung auss Presillg Landt"', *Zeitschrift des Deutschen Vereins für Buchwesen und Schrifttum* 3 (1920), 27–35

Boethius, *De consolatione philosophiae*, ed. Claudio Moreschini (Munich: Saur, 2000)

Bracciolini, Poggio, *De varietate fortunae*, ed. Outi Merisalo (Helsinki: Suomalainen Tiedeakatemia, 1993)

Bracciolini, Poggio, *Lettere*, ed. Helene Harth, 3 vols (Florence: Olschki, 1984–7)

Broughton, Hugh, *A Concent of Scripture* (London, c. 1590)

Bünting, Heinrich, *Itinerarium Sacrae Scripturae. Das ist: Ein Reisebuch* (Helmstadt, 1581; Magdeburg, 1585)

Buridan, John, *Ioannis Buridani Expositio et Quæstiones in Aristotelis De cælo*, ed. Benoît Patar (Leuven: Peeters, 1996)

Cassiodorus, *Cassiodori Senatoris Institutiones*, ed. R. A. B. Mynors (Oxford: Clarendon Press, 1937)
Charles, R. H., ed., *The Apocrypha and Pseudepigrapha of the Old Testament in English*, 2 vols (Oxford: Clarendon Press, 1913)
Chrétien de Troyes, *Erec and Enide*, ed. and trans. Carleton W. Carroll (New York: Garland, 1987)
Cicero, Marcus Tullius, *Academicorum priorum*, ed. James S. Reid (London: Macmillan, 1885)
Cicero, Marcus Tullius, *De re publica*, ed. K. Ziegler (Leipzig: Teubner, 1964)
Cicero, Marcus Tullius, *Tusculan Disputations*, trans. J. E. King (London: Heinemann, 1927)
Claeys, Gregory, ed., *Utopias of the British Enlightenment* (Cambridge: Cambridge University Press, 1994)
Clement, *Épître aux Corinthiens*, ed. and trans. Annie Jaubert (Paris: Éditions du Cerf, 1971)
Cleomedes, *Caelestia (ΜΕΤΕΩΡΑ)*, ed. Robert Todd (Leipzig: Teubner, 1990)
Cleomedes, *Cleomedes' Lectures on Astronomy: A Translation of* The Heavens, trans. Alan C. Bowen and Robert B. Todd (Berkeley: University of California Press, 2004)
Columbus, Christopher, *Lettere e scritti (1495–1506)*, ed. Paolo Taviani and Consuelo Varela, 2 vols (Rome: Nuova Raccolta Colombiana, 1993)
Contarini, Giovanni Matteo, *A Map of the World Designed by Giovanni Matteo Contarini, Engraved by Francesco Roselli 1506*, 2nd edn (London: British Museum, 1926)
Cortesão, Armando, and Avelino Teixeira da Mota, eds, *Portugaliae Monumenta Cartographica*, 6 vols (Lisbon: Commissão Executiva das Comemorações do V Centenário da Morte do Infante D. Henrique, 1960)
Cosmas Indicopleustes, *Topographie chrétienne*, ed. and trans Wanda Wolska-Conus, 3 vols (Paris: Éditions du Cerf, 1968–73)
d'Ailly, Pierre, *Ymago Mundi*, ed. Edmond Buron, 3 vols (Paris: Maisonneuve Frères, 1930)
Dalrymple, Alexander, *An Historical Collection of the Several Voyages and Discoveries in the South Pacific Ocean*, 2 vols (London, 1770–1)
Dalrymple, Alexander, *Memoir concerning the Chagos and Adjacent Islands* (London, 1786)
Dante Alighieri, *Convivio*, ed. Franca Brambilla Ageno, 2 vols (Florence: Casa Editrice Le Lettere, 1995)
Dante Alighieri, *De situ et forma aque et terre*, ed. Giorgio Padoan (Florence: Le Monnier, 1968)
Dante Alighieri, *La Commedia secondo l'antica vulgata*, ed. Giorgio Petrocchi, 4 vols (Milan: Mondadori, 1966–7)
Dante Alighieri, *Questio de aqua et terra*, ed. Severino Ragazzini and Luigi Pescasio (Mantua: Editoriale Padus, 1978)
Dante Alighieri, *The Divine Comedy, vol. 1: Inferno*, trans. Charles S. Singleton, 2 vols (London: Routledge and Kegan Paul, 1970)
Dante Alighieri, *The Divine Comedy of Dante Alighieri*, ed. and trans. Robert M. Durling, 3 vols (Oxford: Oxford University Press, 2003)
Davenport, Frances Gardiner, ed., *European Treaties bearing on the History of the United States and its Dependencies to 1648* (Washington: Carnegie Institution, 1917)
De Brosses, Charles, *Histoire des Navigations aux Terres Australes*, 2 vols (Paris, 1756)
De Lollis, Cesare, ed., *Raccolta di documenti e studi pubblicati dalla R. Commissione Colombiana* (Roma: Ministero della pubblica istruzione, 1894)
Di Varthema, Lodovico, *Navigationes*, in Simon Grynaeus, *Novus Orbis Regionum ac Insularum Veteribus Incognitarum* (Paris and Basel, 1532)
Diogenes Laertius, *Lives of Eminent Philosophers*, trans. R. D. Hicks, 2 vols (London and New York: Heinemann, 1925)
Dümmler, E., ed., *Poetae latini aevi Carolini*, vol. 1 (Berlin: Weidmann, 1890)
Eriugena, Johannes Scottus, *Der Kommentar des Johannes Scottus zu den Opuscula Sacra des Boethius*, ed. E. K. Rand (Munich: Beck'sche Verlagsbuchhandlung, 1906)
Eriugena, Johannes Scottus, *Iohannis Scotti Annotationes in Marcianum*, ed. Cora E. Lutz (Cambridge: Medieval Academy of America, 1939)

Étienne de Rouen, *Le Dragon Normand et autres poèmes d'Étienne de Rouen*, ed. Henri Omont (Rouen: Société de l'histoire de Normandie, 1884)
Eugenius IV, *Romanus Pontifex* (September 15, 1436), in Charles-Martial de Witte, 'Les bulles pontificales et l'expansion portugaise au XVe siècle', *Revue d'histoire ecclésiastique* 48 (1953), 717–18
Finke, Heinrich, ed., *Acta Concilii Constanciensis*, 4 vols (Münster i. W.: Regensbergsche Buchhandlung, 1898–1928)
Foigny, Gabriel de, *La Terre Australe connue* (Geneva, 1676)
Foigny, Gabriel de, *The Southern Land, Known*, ed. and trans. David Fausett (Syracuse: Syracuse University Press, 1993)
Fra Mauro, *Fra Mauro's World Map*, ed. and trans. Piero Falchetta (Turnhout: Brepols, 2006)
Frisius, Gemma, *De principiis* (Leiden, 1544)
Gaude, F., and A. Tomassetti, eds, *Bullarium Romanum*, 27 vols (Turin: Augustae Taurinorum, 1857–85)
Geminus, *Introduction aux phénomènes*, ed. Germaine Aujac (Paris: Les Belles Lettres, 1975)
Gerald of Wales, *Itinerarium Kambriae*, ed. James F. Dimock, Rolls Series 21 (London: Longman, 1868)
Gervase of Tilbury, *Otia imperialia*, ed. and trans. S. E. Banks and J. W. Binns (Oxford: Clarendon Press, 2002)
Ghim, Walter, *Vita Mercatoris*, ed. Hans-Heinrich Geske, *Duisburger Forschungen* 6 (1962), 246–72
Girava, Hieronymo, *Dos Libros de Cosmographia* (Milan, 1556)
Godfrey of St Victor, *Microcosmus*, ed. Philippe Delhaye, 2 vols (Lille: Facultés Catholiques, 1951)
Gossouin of Metz, *L'Image du monde de Maitre Gossouin*, ed. O. H. Prior (Lausanne and Paris: Librairie Payot, 1913)
Grosseteste, Robert, *Die philosophischen Werke des Robert Grosseteste, Bischofs von Lincoln*, ed. Ludwig Baur (Münster i. W.: Aschendorffsche Verlagsbuchhandlung, 1912)
Grynaeus, Simon, *Novus Orbis Regionum ac Insularum Veteribus Incognitarum* (Paris and Basel, 1532)
'Gymnasium vosagense', *Globus mundi* (Strasbourg, 1509)
Hall, Joseph ('Mercurius Britannicus'), *Mundus Alter et Idem Siue Terra Australis ante hac semper incognita longis itineribus peregrini Academici nuperrime lustrata* (Frankfurt, [1605])
Hall, Joseph, *Another World and Yet the Same: Bishop Joseph Hall's* Mundus Alter et Idem, ed. and trans. John Millar Wands (New Haven: Yale University Press, 1981)
Hermann of Carinthia, *De essentiis*, ed. and trans. Charles Burnett (Leiden: Brill, 1982)
Hondius, Jodocus, *The Map of the World on Mercator's Projection by Jodocus Hondius*, ed. Edward Heawood (London: Royal Geographical Society, 1927)
Honorius Augustodunensis, *Imago Mundi*, ed. V. I. J. Flint, *Archives d'histoire doctrinale et littéraire du Moyen Âge* 49 (1983), 7–153
Irenaeus of Lyon, *Adversus haereses/Contre les hérésies*, ed. Adelin Rousseau and Louis Doutreleau, 10 vols (Paris: Éditions du Cerf, 1965–82)
Isidore of Seville, *De natura rerum (Traité de la nature)*, ed. Jacques Fontaine (Bordeaux: CNRS, 1960)
Isidore of Seville, *Etymologiarum sive originum libri xx*, ed. W. M. Lindsay, 2 vols (Oxford: Clarendon Press, 1911)
Isidore of Seville, *The* Etymologies *of Isidore of Seville*, trans. Stephen A. Barney, W. J. Lewis, J. A. Beach, Oliver Berghof (Cambridge: Cambridge University Press, 2006)
Jean de Hauville, *Architrenius*, ed. Paul Gerhard Schmidt (Munich: Wilhelm Fink, 1974)
Johannes de Hauvilla (Jean de Hauville), *Architrenius*, ed. and trans. Winthrop Wetherbee (Cambridge: Cambridge University Press, 1994)
John of Salisbury, *Metalogicon*, ed. J. B. Hall and K. S. B. Keats-Rohan, Corpus Christianorum, Continuatio Mediaevalis 98 (Turnhout: Brepols, 1991)
Julian the Emperor, *Œuvres complètes*, ed. and trans. Gabriel Rochefort *et al.*, 2 vols (Paris: Les Belles Lettres, 1924–63)

La Popelinière, Henri Lancelot-Voisin de, *Les Trois Mondes de La Popelinière*, ed. Anne-Marie Beaulieu (Geneva: Droz, 1997)

la Sale, Antoine de, *Œuvres complètes*, ed. Fernand Desonay, 3 vols (Paris: Les Belles Lettres, 1935)

la Sale, Antoine de, *Paradis de la Reine Sibylle*, ed. Fernand Desonay (Paris: Droz, 1930)

Lactantius Placidus, *Lactantii Placidi in Statii Thebaida Commentvm*, vol. 1, ed. Robert Dale Sweeney (Stuttgart: Teubner, 1997)

Lactantius, *Epitome divinarum institutionum*, ed. Eberhard Heck and Antonie Wlosok (Stuttgart: Teubner, 1994)

Lactantius, *Opera omnia*, ed. Samuel Brandt and Georg Laubmann, 2 vols (Vienna: Bibliopola Academiae Litterarum Caesareae Vindobonensis, 1890)

Latini, Brunetto, *Li Livres dou Tresor*, ed. Francis J. Carmody (Berkeley: University of California Press, 1949)

Lefèvre d'Etaples, Jacques, *The Prefatory Epistles of Jacques Lefèvre d'Etaples and related texts*, ed. Eugene F. Rice (New York: Columbia University Press, 1972)

Lilius, Zacharia, *Contra Antipodes* (Florence, 1496)

Lucan, *Bellum civile liber X*, ed. Emanuele Berti (Florence: Le Monnier, 2000)

Lucan, *De bello civili*, ed. D. R. Shackleton Bailey, 2nd edn (Stuttgart: Teubner, 1997)

Lucian, *Opera*, trans. A. M. Harmon *et al.*, 8 vols (London: Heinemann; New York: Macmillan, 1913–67)

Lucretius, *De rerum natura*, ed. Cyril Bailey, 3 vols (Oxford: Clarendon Press, 1947)

Lull, Ramon, *Arbor Scientiae*, ed. Pere Villalba Varneda, 3 vols, in *Raimundi Lulli Opera Latina* 65, Corpus Christianorum, Continuatio Mediaevalis 180 (Turnhout: Brepols, 2000)

Lull, Ramon, *De ascensu et descensu intellectus*, ed. Aloisius Madre, in *Raimundi Lulli Opera Latina 120–122*, Corpus Christianorum, Continuatio Mediaevalis 35 (Turnhout: Brepols, 1981)

Lull, Ramon, *Declaratio Raimundi*, ed. Theodor Pindl-Büchel, in *Raimundi Lulli Opera Latina 76–81*, Corpus Christianorum, Continuatio Mediaevalis 79 (Turnhout: Brepols, 1989)

Lull, Ramon, *Liber de centvm signis Dei*, ed. Aloisius Madre, in *Raimundi Lulli Opera Latina 131–133*, Corpus Christianorum, Continuatio Mediaevalis 114 (Turnhout: Brepols, 1998)

Lull, Ramon, *Liber de lumine*, ed. Jordi Gayà Estelrich, in *Raimundi Lulli Opera Latina 106–113*, Corpus Christianorum, Continuatio Mediaevalis 113 (Turnhout: Brepols, 1995)

Lull, Ramon, *Liber de regionibus sanitatis et informitatis*, ed. Aloisius Madre, in *Raimundi Lulli Opera Latina 120–122*, Corpus Christianorum, Continuatio Mediaevalis 35 (Turnhout: Brepols, 1981)

Lull, Ramon, *Tractatus novus de astronomia*, ed. Michela Pereira, in *Raimundi Lulli Opera Latina 76–81*, Corpus Christianorum, Continuatio Mediaevalis 79 (Turnhout: Brepols, 1989)

Maass, E., ed., *Commentariorum in Aratum reliquiae* (Berlin: Weidmann, 1958)

Mabillon, Jean, *De re diplomatica libri VI* (Paris, 1681)

Macrobius Ambrosius Theodosius, *Commentaire au Songe de Scipion*, ed. Mireille Armisen-Marchetti, 2 vols (Paris: Les Belles Lettres, 2001–3)

Macrobius Ambrosius Theodosius, *Commentarii in Somnium Scipionis*, ed. J. Willis (Leipzig: Teubner, 1963)

Macrobius Ambrosius Theodosius, *Commentary on the Dream of Scipio*, trans. William Harris Stahl (New York: Columbia University Press, 1952)

Macrobius Ambrosius Theodosius, *Commento al Somnium Scipionis*, ed. and trans. Mario Regali, 2 vols (Pisa: Giardini, 1990)

Macrobius Ambrosius Theodosius, *Saturnalia*, ed. J. Willis (Leipzig: Teubner, 1963)

Mandeville, John, *Mandeville's Travels*, ed. M. C. Seymour (Oxford: Clarendon Press, 1967)

Mandeville, John, *Mandeville's Travels: Texts and Translations*, ed. Malcolm Letts, 2 vols (London: Hakluyt Society, 1953)

Manegold of Lautenbach, *Liber contra Wolfelmum*, ed. Wilfried Hartmann (Weimar: Hermann Böhlaus Nachfolger, 1972)

Manilius, *Astronomica*, ed. G. P. Goold (Stuttgart: Teubner, 1998)

Manilius, *Astronomicon*, ed. A. E. Housman (London: Richards, 1903)
Map, Walter, *De nugis curialium*, ed. and trans. M. R. James, rev. C. N. L. Brooke and R. A. B. Mynors (Oxford: Clarendon Press, 1983)
Marino, Nancy F., ed. and trans., *El Libro del conoscimiento de todos los reinos (The Book of Knowledge of All Kingdoms)* (Tempe: Arizona Center for Medieval and Renaissance Studies, 1999)
Martianus Capella, *De nuptiis Mercurii et Philologiae*, ed. James Willis (Leipzig: Teubner, 1983)
Martianus Capella, *Le nozze di Filologia e Mercurio*, ed Ilaria Ramelli (Milan: Bompiani, 2001)
Mercator, Gerard, *Atlas, sive Cosmographicae meditationes de fabrica mundi et fabricati figura* (Duisburg, 1595)
Mercator, Gerard, *Correspondance Mercatorienne*, ed. M. Van Durme (Antwerp: Nederlandsche Boekhandel, 1959)
Mercator, Gerard, *Gerard Mercator's Map of the World (1569) in the Form of an Atlas in the Maritiem Museum 'Prins Hendrik' at Rotterdam*, reproduced and introduced by B. van't Hoff, *Imago Mundi supplement 6* (Rotterdam, 1961)
Mercator, Gerard, *Weltkarte ad usum navigantium, Duisburg 1569*, ed. Wilhelm Krücken and Joseph Milz (Duisburg: Mercator-Verlag, 1994)
Mercator, Gerard, *Les sphères terrestre et céleste de Gérard Mercator 1541 et 1551*, ed. A. De Smet (Brussels: Editions Culture et Civilisation, 1968)
Metzler, Josef, ed., *America Pontificia*, 3 vols (Vatican City: Libreria editrice vaticana, 1991–5)
Monachus, Franciscus, *De orbis situ ac descriptione* (Antwerp, c. 1527–30; Antwerp, 1565)
More, Thomas, *Utopia*, ed. and trans. George M. Logan, Robert M. Adams, and Clarence H. Miller (Cambridge: Cambridge University Press, 1995)
Münster, Sebastian, *Cosmographiae universalis* (Basel, 1550)
Neckham, Alexander, *De naturis rerum*, ed. Thomas Wright, Rolls Series 34 (London: Longman, 1863)
Neville, Henry, *The Isle of Pines, or, A late Discovery of a fourth Island near Terra Australis, Incognita* (London, 1668)
Nicholas V, *Romanus Pontifex* (January 8, 1455), in *Bullarium Romanum*, ed. Gaude and Tomassetti, vol. 5, pp. 110–15
Oresme, Nicole, *Le Livre du ciel et du monde*, ed. Albert D. Menut and Alexander J. Denomy, trans. Menut (Madison: University of Wisconsin Press, 1968)
Origen, *De principiis / Traité des principes*, ed. and trans. Henri Crouzel and Manlio Simonetti, 3 vols (Paris: Éditions du Cerf, 1978–84)
Ortelius, Abraham, *et al.*, *Ecclesiae Londino-Batavae Archivvm, vol. 1: Abrahami Ortelii et virorvm ervditorvm ad evndem et ad Jacobvm Colivm Ortelianvm Epistvlae*, ed. J. H. Hessels (Cambridge: University Press, 1887)
Ortelius, Abraham, *Theatrum Orbis Terrarum* (Antwerp, 1570; facsimile edition Amsterdam: Israel, 1964)
Ovid, *Metamorphoses*, ed. R. J. Tarrant (Oxford: Clarendon Press, 2004)
Pecham, John, *A Critical Edition and Translation, with Commentary, of John Pecham's* Tractatus de sphera, ed. and trans. Bruce MacLaren (University of Wisconsin-Madison, unpublished doctoral dissertation, 1978)
Peter Martyr, *Selections from Peter Martyr*, ed. and trans. Geoffrey Eatough, Repertorium Columbianum vol. 5 (Turnhout: Brepols, 1998)
Petrarch, Francis, *Canzoniere*, ed. Marco Santagata (Milan: Mondadori, 1996)
Petrarch, Francis, *Francisci Petrarchae Opera quae extant omnia*, 4 vols (Basel, 1554)
Petrarch, Francis, *L'Africa*, ed. Nicola Festa (Florence: Sansoni, 1926)
Petrarch, Francis, *Le Familiari (Familiarium Rerum Libri)*, ed. Vittorio Rossi, 4 vols (Florence: Sansoni, 1933–42)
Petrarch, Francis, *Lettere disperse*, ed. Alessandro Pancheri (Parma: Fondazione Pietro Bembo, 1994)
Petrarch, Francis, *Letters of Old Age: Rerum senilium libri I–XVIII*, trans. Aldo S. Bernardo, Saul Levin, and Reta A. Bernardo, 2 vols (Baltimore: Johns Hopkins Press, 1992)
Petrarch, Francis, *Secretum*, in *Opere Latine* vol. 1, ed. Antonietta Bufano (Turin: Unione Tipografico-Editrice

Torinese, 1975)
Petrus Alfonsi, *Dialogi, Patrologia Latina* 157
Pietro Alighieri, *Comentum super poema Comedie Dantis: A critical edition of the third and final draft of Pietro Alighieri's* Commentary *on Dante's* The Divine Comedy, ed. Massimiliano Chiamenti (Tempe, Arizona: Arizona Center for Medieval and Renaissance Studies, 2002)
Pietro Alighieri, *Il 'Commentarium' di Pietro Alighieri nelle redazioni ashburnhamiana e ottoboniana*, ed. R. Della Vedova and M. T. Silvotti (Florence: Olschki, 1978)
Plato, *Phaedo*, ed. C. J. Rowe (Cambridge: Cambridge University Press, 1993)
Plato, *Timaeus, Critias, Cleitophon, Menexenus, Epistles*, trans. R. G. Bury (Cambridge: Harvard University Press, 1929)
Pliny the Elder, *Naturalis historia*, ed. Jean Beaujeu (Paris: Les Belles Lettres, 1950)
Pliny the Elder, *Naturalis historiae libri xxxvii*, ed. C. Mayhoff, 6 vols (Leipzig: Teubner, 1892–1909)
Plutarch, *Moralia*, vol. 11, trans. Lionel Pearson and F. H. Sandbach (London: Heinemann, 1965)
Plutarch, *Moralia*, vol. 12, ed. and trans. Harold Cherniss and William C. Helmbold (London: Heinemann, 1984)
Plutarch, *Moralia*, vol. 13.1, trans. Harold Cherniss (London: Heinemann, 1976)
Plutarch, *Moralia*, vol. 13.2, trans. Harold Cherniss (London: Heinemann, 1976)
Plutarch, *Œuvres morales, vol. 12.2: Opinions des philosophes*, ed. and trans. Guy Lachenaud (Paris: Les Belles Lettres, 1993)
Polo, Marco, *Le divisament dou monde*, ed. Gabriella Ronchi (Milan: Mondadori, 1982)
Pomponius Mela, *De chorographia libri tres*, ed. Piergiorgio Parroni (Rome: Edizioni di storia e letteratura, 1984)
Posidonius: The Commentary, ed. I. G. Kidd, 2 vols (Cambridge: Cambridge University Press, 1988)
Posidonius: The Fragments, ed. L. Edelstein and I. G. Kidd (Cambridge: Cambridge University Press, 1972)
Postel, Guillaume, *Cosmographicae disciplinae compendium* (Basel, 1561)
Powell, J., ed., *Collectanea Alexandrina: Reliquiae minores Poetarum Graecorum* (Oxford: Clarendon Press, 1925)
Ptolemy, Claudius, *Cosmographia*, trans. Jacopo d'Angelo (Rome, 1478; reprinted Amsterdam: Theatrum Orbis Terrarum, 1966)
Ptolemy, Claudius, *Geographia* (Venice, 1511; facsimile ed. Amsterdam: Theatrum Orbis Terrarum, 1969)
Ptolemy, Claudius, *Geographię opus nouissima traductione e Gręcorum archetypis castigatissime pressum* (Strasbourg, 1513; facsimile edition Amsterdam: Theatrum Orbis Terrarum, 1966)
Ptolemy, Claudius, *Ptolemy's Almagest*, trans. G. J. Toomer (London: Duckworth, 1984)
Ptolemy, Claudius, *Ptolemy's Geography: An Annotated Translation of the Theoretical Chapters*, trans. J. Lennart Berggren and Alexander Jones (Princeton: Princeton University Press, 2000)
Pulci, Luigi, *Morgante*, ed. Franca Ageno (Milan-Naples: Ricciardi, 1955)
Pulci, Luigi, *Morgante: The Epic Adventures of Orlando and His Giant Friend Morgante*, trans. Joseph Tusiani (Bloomington: Indiana University Press, 1998)
Rabanus Maurus, *De universo, Patrologia Latina* 111.
Reisch, Gregor, *Margarita philosophica* (Freiburg, 1503)
Remigius of Auxerre, *Commentum in Martianum Capellam Libri III–IX*, ed. Cora E. Lutz (Leiden: Brill, 1965)
Ringmann, Matthias, *De ora antarctica* (Strasbourg, 1505)
Romer, F. E., *Pomponius Mela's Description of the World* (Ann Arbor: University of Michigan Press, 1998)
Rotz, Jean, *The Maps and Text of the Boke of Idrography presented by Jean Rotz to Henry VIII*, ed. Helen Wallis (Oxford: Roxburghe Club, 1981)
Sallust, *C. Sallusti Crispi Historiarum Reliquiae*, ed. Bertoldus Maurenbrecher (Leipzig: Teubner, 1891–3)
Sanson, Guillaume, *In Geographiam Antiquam Michaelis Antonii Baudrand, Disquisitiones Geographicae* (Paris, 1683)
Schöner, Johann, *Luculentissima descriptio* (Nuremburg, 1515)

Schöner, Johann, *Opusculum geographicum* (Nuremberg, 1533)
Seneca, *Epistulae morales*, ed. L. D. Reynolds, 2 vols (Oxford: Clarendon Press, 1965)
Seneca, *Naturalium quaestionum libri*, ed. Harry M. Hine (Leipzig: Teubner, 1996)
Servius, *Commento al libro VII dell'Eneide di Virgilio*, ed. Giuseppe Ramires (Bologna: Pàtron, 2003)
Servius, *In Vergilii carmina commentarii*, ed. Georgius Thilo and Hermannvs Hagen, 3 vols (Leipzig: Teubner, 1881–7)
Sextus Empiricus, *Adversus Mathematicos* 1, trans. D. L. Blank (Oxford: Clarendon Press, 1998)
Strabo, *Geography*, trans. H. L. Jones (Cambridge: Harvard University Press, 1917)
Suetonius, *De Grammaticis et Rhetoribus*, ed. and trans. Robert A. Kaster (Oxford: Clarendon Press, 1995)
Symcox, Geoffrey, and Luciano Formisano, eds, *Italian Reports on America 1493–1522: Accounts by Contemporary Observers*, trans. Theodore J. Cachey Jr and John C. McLucas, Repertorium Columbianum vol 12 (Turnhout: Brepols, 2002)
Tangl, Michael, ed., *Die Briefe des heiligen Bonifatius und Lullus* (Berlin: Weidmannsche Buchhandlung, 1916)
Terrarossa, Vitale, *Riflessioni geografiche circa le terre incognite* (Padua, 1686)
Tertullian, *Ad nationes*, ed. J. G. Ph. Borleffs, in *Tertulliani Opera*, Corpus Christianorum, Series Latina 1, 2 vols (Turnhout: Brepols, 1954)
Tertullian, *Le premier livre Ad Nationes de Tertullien*, trans. André Schneider (Neuchâtel: Institut Suisse de Rome, 1968)
Thevet, André, *Le Brésil d'André Thevet: Les Singularités de la France Antarctique (1557)*, ed. Frank Lestringant (Paris: Éditions Chandeigne, 1997)
Tränkle, Hermann, ed., *Appendix Tibulliana* (Berlin: de Gruyter, 1990)
van der Stock, Jan, and Marjolein Leesberg, eds, *The New Hollstein Dutch and Flemish Etchings, Engravings and Woodcuts 1450–1700: The Wierix Family Part VIII* (Rotterdam: Sound and Vision, 2004)
Vespucci, Amerigo, *Mundus Novus* (Rostock, ?1505)
Vespucci, Amerigo, *Der* Mundus Novus *des Amerigo Vespucci*, ed. Robert Wallisch (Vienna: Österreichischen Akademie der Wissenschaften, 2002)
Vespucci, Amerigo, *Letters from a New World: Amerigo Vespucci's Discovery of America*, ed. Luciano Formisano (New York: Masilio, 1992)
Virgil, *Aeneid*, in *Opera*, ed. R. A. B. Mynors (Oxford: Clarendon Press, 1969)
Virgil, *Georgics*, ed. R. A. B. Mynors (Oxford: Clarendon Press, 1990)
Virgil, *Georgics*, ed. Richard F. Thomas, 2 vols (Cambridge: Cambridge University Press, 1988)
Waldseemüller, Martin/'Gymnasium vosagense', *Cosmographiae introductio* (Saint-Dié, 1507)
Walter of Châtillon, *Alexandreis*, ed. Marvin L. Colker (Padua: Antenore, 1978)
Walter of Châtillon, *The* Alexandreis *of Walter of Châtillon: A Twelfth-Century Epic*, trans. David Townsend (Philadelphia: University of Pennsylvania Press, 1996)
Westrem, Scott, ed., *The Hereford Map* (Turnhout: Brepols, 2002)
William of Conches, *Dragmaticon Philosophiae*, ed. I. Ronca (Turnhout: Brepols, 1997)
William of Conches, *Glosae super Boetium*, ed. L. Nauta (Turnhout: Brepols, 1999)
William of Conches, *Philosophia*, ed. and trans. Gregor Maurach (Pretoria: University of South Africa, 1980)
Wright, Edward, *Certaine Errors in Navigation* (London, 1599; repr. Amsterdam: Theatrum Orbis Terrarum, 1974)
Yule, Henry, ed. and trans., *The Book of Ser Marco Polo the Venetian Concerning the Kingdoms and Marvels of the East*, rev. Henri Cordier, 2 vols (London: Murray, 1920)

Secondary Literature

Ackermann, Silke, 'Bartholomew of Parma, Michael Scot and the set of new constellations in Bartholomew's *Breviloquium de fructu tocius astronomie*', in *Seventh Centenary of the Teaching of Astronomy in*

Bologna 1297–1997, ed. Pierluigi Battistini, Fabrizio Bònoli, Alessandro Braccesi, Dino Buzzetti (Bologna: Clueb, 2001), pp. 77–98

Akçura, Yusuf, *Piri Reis haritasi* (Istanbul: Devlet Basimevi, 1935)

Almagià, Roberto, 'On the Cartographic Work of Francesco Rosselli', *Imago Mundi* 8 (1951), 27–34

Almagià, Roberto, 'Presentazione', in *Il Mappamondo di Fra Mauro*, ed. Tullia Gasparrini Leporace (Venice: Istituto Poligrafico dello Stato, 1956), pp. 5–10

Andrews, Michael C., 'The Study and Classification of Medieval Mappae Mundi', *Archaeologia* 75 (1924–5), 61–76

Anthiaume, A., 'Un pilote et cartographe havrais au XVIe siècle, Guillaume le Testu', *Bulletin de géographie historique et descriptive* 1–2 (1911), 135–202

Anthiaume, A., *Cartes marines*, 2 vols (Paris: Dumont, 1916)

Arentzen, Jörg-Geerd, *Imago Mundi Cartographica* (Munich: Fink, 1984)

Aujac, Germaine, 'Poseidonios et les zones terrestres, les raisons d'un échec', *Bulletin de l'Association Guillaume Budé* 1 (1976), 74–8

Aujac, Germaine, 'Greek Cartography in the Early Roman World', in *The History of Cartography*, ed. Harley and Woodward, vol. 1, pp. 161–76

Bagrow, Leo, *History of Cartography*, trans. D. L. Paisey, rev. R. A. Skelton (London: Watts, 1964)

Barber, Peter, 'The Christian Knight, the Most Christian King and the rulers of darkness', *The Map Collector* 52 (1990), 8–13

Barker-Benfield, Bruce, 'Macrobius', in *Texts and Transmission: A Survey of the Latin Classics*, ed. L. D. Reynolds (Oxford: Clarendon Press, 1983), pp. 222–35

Barker-Benfield, Bruce, 'The manuscripts of Macrobius' Commentary on the *Somnium Scipionis*', 2 vols (University of Oxford, unpublished D.Phil thesis, 1975–6)

Beagon, Mary, *Roman Nature: The Thought of Pliny the Elder* (Oxford: Clarendon Press, 1992)

Billanovich, G., 'Dall'antica Ravenna alle biblioteche umanistiche', in *Annuario della Università Cattolica del Sacro Cuore 1955–57* (Milan, 1957), pp. 73–107

Björnbo, Axel Anthon, and Carl S. Petersen, *Der Däne Claudius Claussøn Swart, der älteste Kartograph des Nordens, der erste Ptolemäusepigon der Renaissance* (Innsbruck: Wagner'schen Universitäts Buchhandlung, 1909)

Boffito, Giuseppe, 'La leggenda degli antipodi', in *Miscellanea di studi critici edita in onore di Arturo Graf* (Bergamo: Istituto Italiano d'Arti Grafici, 1903), 583–601

Bonner, Elizabeth, 'Did the French discover Australia? The first French scientific voyage of discovery, 1503–1505', in *Revolution, Politics, and Society: Elements in the Making of Modern France*, ed. David W. Lovell (Canberra: Australian Defence Force Academy, 1994), pp. 40–8

Bott, Gerhard, ed., *Focus Behaim Globus*, 2 vols (Nuremberg: Verlag des Germanischen Nationalmuseums, 1991)

Bouloux, Nathalie, *Culture et savoirs géographiques en Italie au XIVe siècle* (Turnhout: Brepols, 2002)

Bouwsma, William J., *Concordia Mundi: The Career and Thought of Guillaume Postel* (Cambridge: Harvard University Press, 1957)

Boyancé, Pierre, *Études sur le Songe de Scipion* (Bordeaux: Feret, 1936)

British Museum, *A Catalogue of the Manuscripts in the Cottonian Library deposited in the British Museum* (London, 1802)

Broc, Numa, *La géographie de la Renaissance 1420–1620* (Paris: Éditions du CTHS, 1980, repr. 1986)

Brotton, Jerry, *Trading Territories: Mapping the Early Modern World* (London: Reaktion, 1997)

Brubaker, Leslie, 'The Relationship of Text and Image in the Byzantine Manuscripts of Cosmas Indicopleustes', *Byzantinische Zeitschrift* 70 (1977), 42–57

Büchner, Karl, *De re publica: Kommentar* (Heidelberg: Winter, 1984)

Burnett, Charles, 'Arabic into Latin in Twelfth Century Spain: the Works of Hermann of Carinthia',

Mittellateinisches Jahrbuch 13 (1978), 100–34
Burnett, Charles, 'Hermann of Carinthia', in *A History of Twelfth-Century Western Philosophy*, ed. Peter Dronke (Cambridge: Cambridge University Press, 1988), pp. 386–404
Burnett, Charles, 'Michael Scot and the transmission of scientific culture from Toledo to Bologna via the court of Frederick II Hohenstaufen', *Micrologus* 2 (1994), 101–26
Burnett, Charles, '*Partim de suo et partim de alieno*: Bartholomew of Parma, the Astrological Texts in MS Bernkastel-Kues, Hospitalsbibliothek 209, and Michael Scot', in *Seventh Centenary of the Teaching of Astronomy in Bologna 1297–1997*, ed. Pierluigi Battistini, Fabrizio Bònoli, Alessandro Braccesi, Dino Buzzetti (Bologna: Clueb, 2001), pp. 37–76
Burnett, Charles, *The Introduction of Arabic Learning into England* (London: British Library, 1997)
Burton Russell, Jeffrey, *Inventing the Flat Earth: Columbus and Modern Historians* (New York: Praeger, 1991)
Cadden, Joan, 'Science and Rhetoric in the Middle Ages: The Natural Philosophy of William of Conches', *Journal of the History of Ideas* 56 (1995), 1–24
Caiazzo, Irene, *Lectures médiévales de Macrobe: Les* Glosæ Colonienses super Macrobium (Paris: Vrin, 2002)
Cameron, Alan, 'Paganism and literature in Late Fourth-Century Rome', in *Christianisme et formes littéraires de l'antiquité tardive en Occident* (Geneva: Foundation Hardt, 1977), pp. 1–30
Cameron, Alan, 'The Date and Identity of Macrobius', *Journal of Roman Studies* 56 (1966), 25–38
Campbell, Tony, 'Egerton MS 1513: A Remarkable Display of Cartographical Invention', *Imago Mundi* 48 (1996), 93–102
Campbell, Tony, 'Portolan Charts from the Late Thirteenth Century to 1500', in *The History of Cartography*, ed. Harley and Woodward, vol. 1 (1987), pp. 371–463
Caraci, Giuseppe, *Questioni e polemiche vespucciane*, 2 vols (Rome: Istituto di scienze geografiche e cartografiche, 1955–6)
Carey, John, 'Ireland and the Antipodes: The Heterodoxy of Virgil of Salzburg', *Speculum* 64 (1989), 1–10
Carey, Sorcha, *Pliny's Catalogue of Culture: Art And Empire in the* Natural History (Oxford: Oxford University Press, 2003)
Casella, Nicola, 'Pio II tra geografia e storia', *Archivio della Società romana di storia patria* 95 (1972), 35–112
Casey, Edward S., *The fate of place: a philosophical history* (Berkeley: University of California Press, 1997)
Cattaneo, Angelo, 'Fra Mauro Cosmographus incomparabilis and His Mappamundi: Documents, Sources, and Protocols for Mapping', in *La cartografia europea tra primo Rinascimento e fine dell'Illuminismo*, ed. Diogo Ramada Curto *et al.* (Florence: Olschki, 2003), pp. 19–48
Cattaneo, Angelo, 'Scritture di viaggio e scrittura cartografica. La *mappamundi* di Fra Mauro e i racconti di Marco Polo e Niccolò de' Conti', *Itineraria* 3–4 (2004–5), 157–202
Cerezo Martínez, Ricardo, 'La carta de Juan de la Cosa (I)', *Revista de Historia Naval* 10 (1992), 31–48
Chekin, Leonid S., *Northern Eurasia in Medieval Cartography: Inventory, Text, Translation, and Commentary* (Turnhout: Brepols, 2006)
Chiappelli, Fredi, *et al.*, eds, *First Images of America: The Impact of the New World on the Old*, 2 vols (Berkeley: University of California Press, 1976)
Clancy, Robert, *The Mapping of Terra Australis* (Sydney: Universal Press, 1995)
Cohen, Paul E., and Robert T. Augustyn, 'A Newly Discovered Hondius Map', *The Magazine* 153 (1999), 214–17
Conrad, Joseph, 'Geography and Some Explorers', *National Geographic Magazine* 45 (1924), 239–74
Cosgrove, Denis, *Apollo's Eye: A Cartographic Genealogy of the Earth in the Western Imagination* (Baltimore: Johns Hopkins University Press, 2001)
Courcelle, Pierre, 'La postérité chrétienne du *Songe de Scipion*', *Revue des Études Latines* 36 (1958), 205–34
Crinò, S., 'I mappamondi di Francesco Rosselli', *La Bibliofilia* 41 (1939), 381–405
Cumont, Franz, *Recherches sur le Symbolisme funéraire des Romains* (Paris: Geuthner, 1942)
d'Avezac, M., *Martin Hylacomylus Waltzemüller: ses ouvrages et ses collaborateurs* (Paris: Challamel aîné, 1867,

repr. Amsterdam: Meridian, 1963)
Danforth, Susan L., 'Notes on the Scientific Examination of the Wilczek-Brown Codex', *Imago Mundi* 40 (1988), 125
Dauge, Yves Albert, *Le Barbare: Recherches sur la conception romaine de la barbarie et de la civilisation* (Brussels: Latomus, 1981)
Davie, Mark, *Half-Serious Rhymes: The Narrative Poetry of Luigi Pulci* (Dublin: Irish Academic Press, 1998)
De Smet, Antoine, 'Jodocus Hondius, continuateur de Mercator', in *Album Antoine De Smet* (Brussels: Centre National d'Histoire des Sciences, 1974), pp. 305–27
De Smet, Antoine, 'L'orfevre et graveur Gaspar Vander Heyden et la construction des globes à Louvain dans le premier tiers du XVIe siècle', *Der Globusfreund* 13 (1964), 38–48
de Witte, Charles-Martial, 'Les bulles pontificales et l'expansion portugaise au XVe siècle', *Revue d'histoire ecclésiastique* 48 (1953), 683–718; 49 (1954), 438–61; 51 (1956), 413–53, 809–35; 53 (1958), 5–46, 443–71
Delisle, Léopold, *Notice sur les manuscrits du 'Liber Floridus' de Lambert, chanoine de Saint-Omer* (Paris: Imprimerie nationale, Klincksieck, 1906)
Deluz, Christiane, 'L'Europe selon Pierre d'Ailly ou selon Guillaume Fillastre? De l'*Ymago Mundi* aux légendes de la carte de Nancy', in *Humanisme et culture géographique à l'époque du concile de Constance*, ed. Marcotte, pp. 151–60
Deluz, Christiane, *Le livre de Jehan de Mandeville: Une 'géographie' au XIVe siècle* (Louvain: Institut d'Études Médiévales de l'Université Catholique de Louvain, 1988)
Derolez, Albert, *Lambertus qui librum fecit: een codicologische studie van de Liber Floridus-autograaf (Gent, Universiteitsbibliotheek, handschrift 92)* (Brussels: Verhandelingen van de Koninklijke Academie voor Wetenschappen, 1978)
Derolez, Albert, *The Autograph Manuscript of the* Liber Floridus*: A Key to the Encyclopedia of Lambert of Saint-Omer*, Corpus Christianorum, Autographa Medii Aevi 4 (Turnhout: Brepols, 1998)
Derrida, Jacques, *Of Grammatology* (De la grammatologie), trans. Gayatri Chakravorty Spivak (Baltimore: Johns Hopkins University Press, 1976)
Destombes, Marcel, 'Guillaume Postel cartographe', in *Guillaume Postel, 1581–1981* (Paris: Tredaniel, 1985), pp. 361–71
Destombes, M., *Mappemondes A.D. 1200–1500* (Amsterdam: Israel, 1964)
Dilke, O. A. W. and Margaret S., 'The Wilczek-Brown Codex of Ptolemy Maps', *Imago Mundi* 40 (1988), 118–24
Dilke, O. A. W., 'The Culmination of Greek Cartography in Ptolemy', in *The History of Cartography*, ed. Harley and Woodward, vol. 1 (1987), pp. 177–200
Diller, Aubrey, 'A Geographical Treatise by Georgius Gemistus Pletho', *Isis* 27 (1937), 441–51
Dörflinger, Johannes, 'Der Gemma Frisius-Erdglobus von 1536 in der Österreichischen Nationalbibliothek in Wien', *Der Globusfreund* 21–3 (1973), 81–99
Dronke, Peter, *Fabula: Explorations into the uses of myth in medieval Platonism* (Leiden: Brill, 1974)
Dubois, Claude-Gilbert, 'La fonction du roi de France dans la fantasmagorie politique de Guillaume Postel', in *L'Image du souverain dans les lettres françaises des guerres de religion à la révocation de l'Edit de Nantes*, ed. Noémi Hepp and Madeleine Bertaud (Paris: Klincksieck, 1985), pp. 141–51
Dubois, Claude-Gilbert, 'La mythologie nationaliste de Guillaume Postel', in *Guillaume Postel, 1581–1981* (Paris: Tredaniel, 1985), pp. 257–64
Duhem, Pierre, *Le système du monde: Histoire des doctrines cosmologiques de Platon à Copernic*, 10 vols (Paris: Libraire scientifique A. Hermann, 1915)
Durand, Dana Bennett, *The Vienna-Klosterneuburg Map Corpus of the Fifteenth Century* (Leiden: Brill, 1952)
Eastwood, Bruce, 'Manuscripts of Macrobius, *Commentarii in Somnium Scipionis*, before 1500', *Manuscripta* 38 (1994), 138–55
Edson, Evelyn, and Emilie Savage-Smith, 'An Astrologer's Map: A Relic of Late Antiquity', *Imago Mundi* 52

(2000), 7–29.
Edson, Evelyn, *Mapping Time and Space: How Medieval Mapmakers viewed their World* (London: British Library, 1997)
Edwards, Glenn M., 'The Two Redactions of Michael Scot's "Liber introductorius"', *Traditio* 41 (1985), 329–40
Edwards, Glenn Michael, 'The *Liber introductorius* of Michael Scot' (University of Southern California, unpublished doctoral dissertation, 1978)
Eisler, William, *The Furthest Shore: Images of Terra Australis from the Middle Ages to Captain Cook* (Cambridge: Cambridge University Press, 1995)
Elford, Dorothy, 'William of Conches', in *A History of Twelfth-Century Western Philosophy*, ed. Peter Dronke (Cambridge: Cambridge University Press, 1988), pp. 308–27
Elford, Dorothy, 'Developments in the natural philosophy of William of Conches: a study of his *Dragmaticon* and a consideration of its relationship to the *Philosophia*' (University of Cambridge, unpublished PhD dissertation, 1983)
Enterline, J., 'The Southern Continent and the False Strait of Magellan', *Imago Mundi* 26 (1972), 48–58
Fausett, David, *Writing the New World: Imaginary Voyages and Utopias of the Great Southern Land* (Syracuse: Syracuse University Press, 1993)
Fausett, David, *Images of the Antipodes in the Eighteenth Century: A Study in Stereotyping* (Amsterdam: Rodopi, 1994)
Fernández-Armesto, Felipe, *The Canary Islands after the Conquest: The Making of a Colonial Society in the Early Sixteenth Century* (Oxford: Clarendon Press, 1982)
Fischer, Joseph, *Claudii Ptolemaei Geographiae Codex Urbinas Graecus 82*, 2 vols in 4 (Leiden: Brill, 1932)
Flamant, Jacques, *Macrobe et le néo-Platonisme latin, à la fin du IVe siècle* (Leiden: Brill, 1977)
Flint, Valerie I.J., 'Monsters and the Antipodes in the Early Middle Ages and Enlightenment', *Viator* 15 (1984), 65–80
Flores, Enrico, 'Augusto nella visione astrologica di Manilio ed il problema della cronologia degli *Astronomicon libri*', *Annali della Facoltà di Lettere e Filosofia dell'Università di Napoli* 9 (1960–1), 5–66
Friedman, John Block, *The Monstrous Races in Medieval Art and Thought* (Cambridge: Harvard University Press, 1981; repr. Syracuse: Syracuse University Press, 2000)
Fry, Howard T., *Alexander Dalrymple (1737–1808) and the Expansion of British Trade* (London: Royal Commonwealth Society, 1970)
Gale, Monica R., *Virgil on the Nature of Things: The* Georgics, *Lucretius and the Didactic Tradition* (Cambridge: Cambridge University Press, 2000)
Gallois, L., *Les géographes allemands de la Renaissance* (Paris: Leroux, 1890, repr. Amsterdam: Meridian, 1963)
García y García, Antonio, 'Las donaciones pontificias de territorios y su repercusión en las relaciones entre Castilla y Portugal', in *Las relaciones entre Portugal y Castilla en la época de los descubrimientos y la expansión colonial*, ed. Ana María Carabias Torres (Salamanca: Ediciones Universidad de Salamanca, 1994), pp. 293–310
Gasti, Fabio, *L'antropologia di Isidoro: Le fonti del libro XI delle* Etimologie (Como: New Press, 1998)
Gautier Dalché, Patrick, 'A propos des antipodes. Note sur un critère d'authenticité de la *Vie de Constantin* slavonne', *Analecta Bollandiana* 106 (1988), 113–19
Gautier Dalché, Patrick, 'Entre le folklore et la science: la légende des antipodes chez Giraud de Cambrie et Gervais de Tilbury', in *La leyenda: antropología, historia, literatura* (Madrid: Universidad Complutense, 1989), pp. 103–14
Gautier Dalché, Patrick, 'L'œuvre géographique du cardinal Fillastre', *Archives d'histoire doctrinale et littéraire du Moyen Âge* 59 (1992), 319–83; reprinted in *Humanisme et culture géographique à l'époque du concile de Constance*, ed. Marcotte, pp. 293–355
Gautier Dalché, Patrick, 'L'influence de Jean Buridan: l'habitabilité de la terre selon Dominicus de

Clavasio', in *Comprendre et maîtriser la nature au Moyen Âge: Mélanges d'histoire des sciences offerts à Guy Beaujouan* (Geneva: Droz/Paris: Champion, 1994), pp. 101–13
Gautier Dalché, Patrick, 'De la glose à la contemplation. Place et fonction de la carte dans les manuscrits du haut Moyen Âge', *Settimane di studio del Centro italiano di studi sull'alto medioevo* 41 (1994), 693–771
Gautier Dalché, Patrick, 'Notes sur la "carte de Théodose II" et sur la "mappemonde de Théodulf d'Orléans"', *Geographia Antiqua* 3–4 (1994–5), pp. 91–106
Gautier Dalché, Patrick, 'Pour une histoire du regard géographique: conception et usage de la carte au XVe siècle', *Micrologus* 4 (1996), 77–103
Gautier Dalché, Patrick, 'Le renouvellement de la perception et de la representation de l'espace au XIIe siècle', in *Renovación intelectual del Occidente Europeo (siglo XII)* (Pamplona: Gobierno de Navarra, 1998), pp. 169–217
Gautier Dalché, Patrick, 'Mappae mundi anterieures au XIIIe siècle dans le manuscrits latins de la Bibliothèque nationale de France', *Scriptorium* 52 (1998), 102–62
Gautier Dalché, Patrick, 'Le paradis aux antipodes? Une *Distinctio divisionis terre et paradisi delitiarum* (XIVe siècle)', in *'Liber largitorius': Études d'histoire médiévale offertes à Pierre Toubert par ses élèves*, ed. Dominique Barthélemy and Jean-Marie Martin (Geneva: Libraire Droz, 2003), pp. 615–37
Gentile, Sebastiano, 'Emanuele Crisolora e la "Geographia" di Tolomeo', in *Dotti bizantini e libri greci nell'Italia del secolo XV*, ed. Mariarosa Cortesi and Enrico V. Maltese (Naples: D'Auria, 1992), pp. 291–308
Gentile, Sebastiano, 'Giorgio Gemisto Pletone e la sua influenza sull'umanesimo fiorentino', in *Firenze e il concilio del 1439*, ed. Paolo Viti, 2 vols (Florence: Olschki, 1994) vol. 2, pp. 813–32
Gentile, Sebastiano, 'L'ambiente umanistico fiorentino e lo studio della *Geografia* nel secolo XV', in *Amerigo Vespucci. La vita e i viaggi*, ed. L. Formisano *et al.* (Florence: Banca di Toscana, 1992), pp. 9–63
Gentile, Sebastiano, 'Toscanelli, Traversari, Niccoli e la geografia', *Rivista Geografica Italiana* 100 (1993), 113–31
Gillies, John, *Shakespeare and the Geography of Difference* (Cambridge: Cambridge University Press, 1994)
Gingerich, Owen, 'Sacrobosco as a Textbook', *Journal for the History of Astronomy* 19 (1988), 269–73
Gliozzi, Giuliano, *Adamo e il Nuovo Mondo: la nascita dell'antropologia come ideologia coloniale: dalle genealogie bibliche alle teorie razziali, 1500–1700* (Florence: La nuova Italia, 1977)
Goldstein, Thomas, *Merchants and Scholars: Essays in the History of Exploration and Trade* (Minneapolis: University of Minnesota Press, 1965)
Gormley, Catherine M., Mary A. Rouse and Richard H. Rouse, 'The Medieval Circulation of the *De chorographia* of Pomponius Mela', *Mediaeval Studies* 46 (1984), 266–320
Grant, Robert M., *Irenaeus of Lyons* (London: Routledge, 1997)
Gregory, Tullio, *Anima Mundi: La filosofia di Guglielmo di Conches e la scuola di Chartres* (Florence: Sansoni, 1956)
Gruber, Joachim, *Kommentar zu Boethius de consolatione Philosophiae* (Berlin: de Gruyter, 1978)
Haase, W., and M. Reinhold, eds, *The Classical Tradition and the Americas*, 6 vols, vol. 1: *European Images of the Americas and the Classical Tradition* (Berlin: de Gruyter, 1994)
Hall, Edith, *Inventing the Barbarian: Greek Self-Definition through Tragedy* (Oxford: Clarendon Press, 1989)
Hamilton, Alastair, *The Family of Love* (Cambridge: Clarke, 1981)
Harley, J. B., and David Woodward, eds, *The History of Cartography*, vol. 1: *Cartography in Prehistoric, Ancient, and Medieval Europe and the Mediterranean* (Chicago: University of Chicago Press, 1987)
Harley, J. B., and David Woodward, eds, *The History of Cartography*, vol. 2.1: Cartography in the Traditional Islamic and South Asian Societies (Chicago: University of Chicago Press, 1992)
Harley, J. B., 'Silences and Secrecy: the Hidden Agenda of Cartography in Early Modern Europe', *Imago Mundi* 40 (1988), 57–76
Hartog, François, *The Mirror of Herodotus: The Representation of the Other in the Writing of History*, trans. Janet

Lloyd (Berkeley: University of California Press, 1988)
Haskins, Charles Homer, *Studies in the History of Mediaeval Science* (Cambridge: Harvard University Press, 1924)
Hennig, Richard, *Terrae incognitae*, 2nd edn, 4 vols (Leiden: Brill, 1944–56)
Herren, Michael, 'The Commentary on Martianus Attributed to John Scottus: its Hiberno-Latin Background', in *Jean Scot écrivain*, ed. G.-H. Allard (Montreal: Bellarmin, 1986), pp. 265–86
Higgins, Iain, *Writing East: The 'Travels' of Sir John Mandeville* (Philadelphia: University of Pennsylvania Press, 1997)
Hirsch, Bertrand, 'L'espace nubien et éthiopien sur les cartes portulans du XIVe siècle', *Médiévales* 18 (1990), 69–92
Holtz, Louis, 'L'école d'Auxerre', in *L'école carolingienne d'Auxerre de Murethach à Remi 830–908* (Paris: Beauchesne, 1991), pp. 131–46.
Humboldt, Alexander von, *Kritische Untersuchungen über die historische Entwickelung der geographischen Kenntnisse von der neuen Welt*, 3 vols (Berlin: Nicolai'schen Buchhandlung, 1852)
Hüttig, Albrecht, *Macrobius im Mittelalter: Ein Beitrag zur Rezeptionsgeschichte der Commentarii in Somnium Scipionis* (Frankfurt am Main: Lang, 1990)
Huygens, R. B. C., 'Mittelalterliche Kommentare zum *O qui perpetua* ...', *Sacris Erudiri* 6 (1954), 373–427
Iwańczak, Wojciech, 'Entre l'espace ptolémaïque et l'empire: les cartes de Fra Mauro', *Médiévales* 18 (1990), 53–68
Jacob, Christian, 'Il faut qu'une carte soit ouverte ou fermée: le tracé conjectural', *Revue de la Bibliothèque Nationale* 45 (1992), 34–41
Jantz, Harold, 'Images of America in the German Renaissance', in *First Images of America*, ed. Chiappelli, vol. 1, pp. 91–106
Jeauneau, Edouard, *'Lectio Philosophorum': Recherches sur l'Ecole de Chartres* (Amsterdam: Hakkert, 1973)
Jordan, Constance, *Pulci's* Morgante*: Poetry and History in Fifteenth-Century Florence* (Washington: Folger Books, 1986)
Kamal, Youssouf, *Monumenta Cartographica Africae et Aegypti*, 5 vols (Cairo, 1926–51)
Karamustafa, Ahmet T., 'Cosmographical Diagrams', in *The History of Cartography*, ed. Harley and Woodward, vol. 2.1 (1992), pp. 71–89
Karrow, Robert W., *Mapmakers of the Sixteenth Century and their Maps: Bio-bibliographies of the Cartographers of Abraham Ortelius, 1570* (Chicago: University of Chicago Press, 1993)
Kaster, Robert A., *Guardians of Language: The Grammarian and Society in Late Antiquity* (Berkeley: University of California Press, 1988)
Kaster, Robert, 'Macrobius and Servius', *Harvard Studies in Classical Philology* 84 (1980), 219–62
Kenney, James F., *The sources for the Early History of Ireland: Ecclesiastical* (New York: Columbia University Press, 1929)
Kremers, Dieter, *Rinaldo und Odysseus: Zur Frage der Diesseitserkenntnis bei Luigi Pulci und Dante Alighieri* (Heidelberg: Winter, 1966)
Laboulais-Lesage, Isabelle, ed., *Combler les blancs de la carte: Modalités et enjeux de la construction des savoirs géographiques (XVIe–XXe siècle)* (Strasbourg: Presses Universitaires de Strasbourg, 2004)
Lafferty, Maura K., 'Nature and an unnatural man: Lucan's influence on Walter of Châtillon's concept of nature', *Classica et mediaevalia* 46 (1995), 285–300
Lafferty, Maura K., *Walter of Châtillon's* Alexandreis*: Epic and the Problem of Historical Understanding* (Turnhout: Brepols, 1998)
Lakowsky, Romuald, 'Geography and the More Circle: John Rastell, Thomas More, and the "New World"', *Renaissance Forum: An Electronic Journal of Early-Modern Literary and Historical Studies* 4 (1999)
Larner, John, *Marco Polo and the Discovery of the World* (New Haven: Yale University Press, 1999)
Laubenberger, Franz, 'The Naming of America', *Sixteenth Century Journal* 13 (1982), 91–113

Lavezzo, Kathy, *Angels on the Edge of the World: Geography, Literature, and English Community, 1000–1534* (Ithaca: Cornell University Press, 2006)

Lecoq, Danielle, 'La Mappemonde du Liber Floridus ou La Vision du Monde de Lambert de Saint-Omer', *Imago Mundi* 39 (1987), 9–49

Lecoq, Danielle, 'Au delà des limites de la terre habitée. Des îles extraordinaires aux terres antipodes (XIe–XIIIe siècles)', in *Terre à découvrir, terres à parcourir: Exploration et connaissance du monde XIIe–XIXe siècles*, ed. Danielle Lecoq and Antoine Chambard (Paris: L'Harmattan, 1998), pp. 14–41

Lecoq, Danielle, 'Des antipodes au Nouveau Monde ou de la difficulté de l'Autre', in *La France-Amérique*, ed. Lestringant, pp. 65–102

Lefèvre, Yves, 'Le Liber Floridus et la littérature encyclopédique au Moyen Âge', in *Liber Floridus Colloquium: Papers read at the international meeting held in the University Library Ghent on 3–5 September 1967*, ed. Albert Derolez (Ghent: E. Story-Scientia, 1973), pp. 1–9

Leonardi, Claudio, 'I codici di Marziano Capella', *Aevum* 33 (1959), 443–89, and *Aevum* 34 (1960), 1–99, 411–524

Leonardi, Claudio, 'Illustrazioni e glosse in un codice di Marziano Capella', *Bulletino dell' Archivio Palaeografico Italiano* n.s. 2–3 (1956–7), 39–60

Leonardi, Claudio, 'Martianus Capella et Jean Scot: nouvelle présentation d'un vieux problème', in *Jean Scot écrivain*, ed. G.-H. Allard (Montreal: Bellarmin, 1986), pp. 187–207

Leonardi, Claudio, 'Nota introduttiva per un'indagine sulla fortuna di Marziano Capella nel Medioevo', *Bullettino dell'Istituto Storico Italiano per il Medio Evo e Archivio Muratoriano* 67 (1955), 265–88

Lestringant, Frank, 'Fictions cosmographiques à la Renaissance', in *Philosophical Fictions and the French Renaissance*, ed. Neil Kenny (London: Warburg Institute, 1991), pp. 101–25

Lestringant, Frank, ed., *La France-Amérique (XVIe–XVIIIe siècles)* (Paris: Champion, 1988)

Lestringant, Frank, *Le Huguenot et le Sauvage: L'Amérique et la controverse coloniale, en France, au temps des Guerres de Religion (1555–1589)* (Paris: Amateurs de livres, 1990)

Lestringant, Frank, *Mapping the Renaissance World: The Geographical Imagination in the Age of Discovery*, trans. David Fausett (Berkeley: University of California Press, 1994)

Ligota, C. R., 'L'influence de Macrobe pendant la Renaissance', in *Le soleil e la Renaissance ... Colloque international* (Brussels: Presses universitaires de Bruxelles, 1965), pp. 465–82

Loomis, R. S., 'King Arthur and the Antipodes', *Modern Philology* 38 (1941), 289–304

Lozovsky, Natalia, *'The Earth Is Our Book': Geographical Knowledge in the Latin West ca. 400–1000* (Ann Arbor: University of Michigan Press, 2000)

Lutz, Cora E., 'Martianus Capella', in *Catalogus translationum et commentariorum: Medieval and Renaissance Latin Translators and Commentaries*, ed. P. O. Kristeller and F. Edward Cranz, vol. 2 (Washington: Catholic University of America Press, 1971), pp. 367–81

Luzzana Caraci, Ilaria, *Amerigo Vespucci*, 2 vols (Rome: Nuova Raccolta Colombiana, 1996–9)

Luzzana Caraci, Ilaria, *Colombo vero e falso* (Genoa: Sagep Editrice, 1989)

Luzzana Caraci, Ilaria, *The Puzzling Hero: Studies on Christopher Columbus and the Culture of his Age*, trans. Mayta Munson (Rome: Carocci, 2002)

McCabe, Richard A., *Joseph Hall: A Study in Satire and Meditation* (Oxford: Clarendon Press, 1982)

McCready, W. D., 'Bede and the Isidorian Legacy', *Mediaeval Studies* 57 (1995) 41–73

McCready, William D., 'Isidore, the Antipodeans, and the Shape of the Earth', *Isis* 87 (1996), 108–27

McIntosh, Gregory C., *The Piri Reis Map of 1513* (Athens: University of Georgia Press, 2000)

Magnaghi, Alberto, *Amerigo Vespucci*, 2 vols (Rome: Istituto Cristoforo Colombo, 1924)

Mangani, Giorgio, 'Abraham Ortelius and the Hermetic Meaning of the Cordiform Projection', *Imago Mundi* 50 (1998), 59–83

Mangani, Giorgio, *Il 'mondo' di Abramo Ortelio: Misticismo, geografia e collezionismo nel Rinascimento dei Paesi Bassi* (Modena: Panini, 1998)

Marcon, Susy, 'Leonardo Bellini and Fra Mauro's World Map: the *Earthly Paradise*', in *Fra Mauro's World Map*, ed. and trans. Piero Falchetta (Turnhout: Brepols, 2006), pp. 135–61

Marcotte, Didier, ed., *Humanisme et culture géographique à l'époque du concile de Constance autour de Guillaume Fillastre* (Turnhout: Brepols, 2002)

Marenbon, John, *Aristotelian Logic, Platonism, and the Context of Early Medieval Philosophy in the West* (Aldershot: Ashgate, 2000)

Marenbon, John, *From the circle of Alcuin to the school of Auxerre* (Cambridge: Cambridge University Press, 1981)

Mattiacci, Silvia, *I carmi e frammenti di Tiberiano* (Florence: Olschki, 1990)

Mazzoli, Giancarlo, 'Riflessioni sulla semantica ciceroniana della gloria', in *Cicerone tra antichi e moderni*, ed. Emanuele Narducci (Florence: Le Monnier, 2004)

Mazzotta, Giuseppe *The Worlds of Petrarch* (Durham: Duke University Press, 1993)

Merisalo, Outi, 'Le prime edizioni stampate del *De varietate fortunae* di Poggio Bracciolini', *Arctos* 19 (1985), 81–102; 20 (1986), 101–29

Mette, Hans, *Sphairopoiia: Untersuchungen zur Kosmologie des Krates von Pergamon* (Munich: Beck'sche, 1936)

Milanesi, Marica, 'A Forgotten Ptolemy: Harley Codex 3686 in the British Library', *Imago Mundi* 48 (1996), 43–64

Milanesi, Marica, 'Il commento al "Dittamondo" di Guglielmo Capello (1435–37)', in *Alla corte degli Estensi: Filosofia, arte e cultura a Ferrara nei secoli XV e XVI*, ed. Marco Bertozzi (Ferrara: Università degli Studi, 1994), pp. 365–88

Miller, Konrad, *Mappae Arabicae*, 6 vols (Stuttgart: Miller, 1927)

Miller, Konrad, *Mappaemundi: Die ältesten Weltkarten*, 6 vols (Stuttgart: Roth'sche Verlagshandlung, 1895–8)

Minnis, Alistair, *Chaucer and Pagan Antiquity* (Cambridge: Cambridge University Press, 1982)

Miquel, André, *La géographie humaine du monde musulman jusqu'au milieu du 11e siècle*, 3 vols (Paris: Mouton, 1967–80)

Mollat du Jourdin, Michelle, and Monique de la Roncière, *Les portulans: cartes marines du XIIIe au XVIIe siècle* (Fribourg: Office du Livre, 1984)

Moretti, Gabriella, *Agli antipodi del mondo* (Trent: Dipartimento di Scienze Filologiche e Storiche, 1990)

Moretti, Gabriella, *Gli antipodi: avventure letterarie di un mito scientifico* (Parma: Pratiche Editrice, 1994)

Moretti, Gabriella, 'The Other World and the "Antipodes". The Myth of the Unknown Countries between Antiquity and the Renaissance', in *The Classical Tradition and the Americas*, ed. Haase and Reinhold, vol. 1 (1994), pp. 241–84.

Morpurgo, Piero, 'Fonti di Michele Scoto', *Accademia nazionale dei Lincei* ser. 8, 38 (1983), 59–71

Mudimbe, V. Y., *The Idea of Africa* (Bloomington: Indiana University Press, 1994)

Muldoon, James, 'Papal Responsibility for the Infidel: Another Look at Alexander VI's *Inter Cetera*', *The Catholic Historical Review* 64 (1978), 168–84

Muldoon, James, 'The struggle for justice in the conquest of the New World', in *Proceedings of the Eighth International Congress of Medieval Canon Law*, ed. Stanley Chodorow (Vatican City: Biblioteca Apostolica Vaticana, 1992), pp. 707–20

Muldoon, James, *Popes, Lawyers, and Infidels* (Philadelphia: University of Pennsylvania Press, 1979)

Muris, Oswald, and Gert Saarmann, *Der Globus im Wandel der Zeiten* (Berlin: Columbus Verlag Paul Oestergaard, 1961)

Murphy, Trevor, *Pliny the Elder's* Natural History*: the Empire in the Encyclopedia* (Oxford: Oxford University Press, 2004)

Nachrodt, Hans Werner, 'Martin Behaim und sein "Erdapfel": Bemerkungen zur Lebens- und Wirkungsgeschichte', *Jahrbuch für Frankische Landesforschung* 55 (1995), 45–64

Nardi, Bruno, *La caduta di Lucifero e l'autenticità della 'Quaestio de aqua et terra'* (Turin: Editrice Internazionale, 1959)

Nebenzahl, Kenneth, *Maps of the Holy Land: Images of* Terra Sancta *through Two Millennia* (New York: Abbeville Press, 1986)
Nicolet, Claude, *L'inventaire du monde: Géographie et politique aux origines de l'Empire romain* (Paris: Fayard, 1988)
Nolhac, Pierre de, *Pétrarque et l'humanisme*, 2 vols (Paris: Champion, 1907)
Nordenskïold, A. E., *Facsimile-Atlas till Kartografiens äldsta historia* (Stockholm, 1889)
Nuti, Lucia, 'The World Map as an Emblem: Abraham Ortelius and the Stoic Contemplation', *Imago Mundi* 55 (2003), 38–55
Obrist, Barbara, 'Wind Diagrams and Medieval Cosmology', *Speculum* 72 (1997), 33–84
Obrist, Barbara, *La cosmologie médiévale. Textes et images*, vol. 1: *Les fondements antiques* (Florence: Edizioni del Galluzzo, 2004)
Osley, A. S., *Mercator* (London: Faber, 1969)
Paolis, Paolo De, 'Macrobio 1934–1984', *Lustrum* 28–9 (1986–7), 107–249
Pastoureau, Mireille, 'The 1569 World Map', in *The Mercator Atlas of Europe*, ed. Marcel Watelet (Pleasant Hill: Walking Tree Press, 1998), pp. 79–88
Peden, Alison M., 'Echternach as a Cultural Entrepôt. The case of Macrobius', in *Willibrord: Apostel der Niederlande, Gründer der Abtei Echternach* (Luxembourg: Editions Saint-Paul, 1989), pp. 166–70.
Pedersen, Olaf, 'In Quest of Sacrobosco', *Journal for the History of Astronomy* 16 (1985), 175–221
Pelletier, Monique, 'Le globe vert et l'œuvre cosmographique du gymnase vosgien', *Bulletin du Comité français de cartographie* 163 (2000), 17–31
Pellizzari, Andrea, *Servio: Storia, cultura e istituzioni nell'opera di un grammatico tardoantico* (Florence: Olschki, 2003)
Petry, Yvonee, *Gender, Kabbalah and the Reformation: The Mystical Theology of Guillaume Postel* (Leiden: Brill, 2004)
Philipp, Hans, *Die historisch-geographischen Quellen in den etymologiae des Isidorus von Sevilla*, 2 vols (Berlin: Weidmannsche Buchhandlung, 1912–13)
Phillips, Carla Rahn, 'The Portraits of Columbus: Heavy Traffic at the Intersection of Art and Life', *Terrae Incognitae* 24 (1992), 1–18
Rainaud, Armand, *Le Continent Austral: Hypothèses et Découvertes* (Paris: Colin, 1893)
Ramaswamy, Sumathi, *The Lost Land of Lemuria: Fabulous Geographies, Catastrophic Histories* (Berkeley: University of California Press, 2004)
Randles, W. G. L., 'Le Nouveau Monde, l'Autre Monde et la Pluralité des Mondes', in *Actas do Congresso Internacional de História dos Descobrimentos* (Lisbon: Comissão Executiva das Comemorações do V Centenário da Morte do Infante D. Henrique, 1961), vol. 4, pp. 347–82
Ravenstein, E. G., *Martin Behaim: His life and his globe* (London: Philip, 1908)
Reichert, Folker, 'Geographie und Weltbild am Hofe Friedrichs II', *Deutsches Archiv für Erforschung des Mittelalters* 51 (1995), 433–91
Relaño, Francesc, 'Cartography and Discoveries: the Re-Definition of the Ptolemaic Model in the First Quarter of the Sixteenth Century', in *La cartografia europea tra primo Rinascimento e fine dell'Illuminismo*, ed. Diogo Ramada Curto *et al.* (Florence: Olschki, 2003), pp. 49–61
Rennie, Neil, *Far-Fetched Facts: The Literature of Travel and the Idea of the South Seas* (Oxford: Clarendon Press, 1995)
Ribémont, Bernard, *Les origines des encyclopédies médiévales: D'Isidore de Séville aux Carolingiens* (Paris: Honoré Champion, 2001)
Richardson, W. A. R., 'Jave-la-Grande: a case study of place-name corruption', *The Globe* 22 (1984), 9–32
Richardson, W. A. R., 'Enigmatic Indian Ocean Coastlines on Early Maps and Charts', *The Globe* 46 (1998), 21–41
Richardson, W.A.R., 'Mercator's Southern Continent: Its Origins, Influence and Gradual Demise', *Terrae*

Incognitae 25 (1993), 67–98
Romm, James S., *The Edges of the Earth in Ancient Thought* (Princeton: Princeton University Press, 1992)
Ronca, Italo, 'Reason and Faith in the *Dragmaticon*: The Problematic Relation between *philosophica ratio* and *diuina pagina*', in *Knowledge and the sciences in medieval philosophy*, ed. Simo Knuuttila *et al.*, 3 vols (Helsinki: Luther-Agricola Society, 1990), vol. 2, pp. 331–41
Ronsin, A., 'L'Amérique du Gymnase vosgien de Saint-Dié-des-Vosges: invention et postérité', in *La France-Amérique*, ed. Lestringant, pp. 37–64
Santarem, Manuel de, *Atlas, composé de mappemondes, de portulans, et de cartes hydrographiques et historiques, depuis le VIe jusqu'au XVIIe siècle* (Paris, 1842–53)
Santarem, Manuel de, *Essai sur l'histoire de la cosmographie et de la cartographie*, 3 vols (Paris: Maulde et Renou, 1849–52)
Scafi, Alessandro, *Mapping Paradise: A History of Heaven on Earth* (London: British Library, 2006)
Schilder, Günter, 'Willem Jansz. Blaeu's Wall Map of the World, on Mercator's Projection, 1606–07 and its Influence', *Imago Mundi* 31 (1979), 36–54
Schilder, Günter, *Australia Unveiled: the share of the Dutch navigators in the discovery of Australia*, trans. Olaf Richter (Amsterdam: Theatrum Orbis Terrarum, 1976)
Schilder, Günter, *Monumenta Cartographica Neerlandica*, 7 vols (Alphen aan den Rijn: Canaletto, 1987–2003)
Schilder, Günter, *Three World Maps by Francois van den Hoeye of 1661, Willem Janszoon (Blaeu) of 1607, Claes Janszoon Visscher of 1650* (Amsterdam: Israel, 1981)
Schmidt, Charles, *Histoire littéraire de l'Alsace*, 2 vols (Paris: Sandoz and Fischbacher, 1879)
Schmidt, Thomas Christian, 'Die Entdeckung des Ostens und der Humanismus: Niccolò de' Conti und Poggio Bracciolinis *Historia de Varietate Fortunae*', *Mitteilungen des Instituts für österreichische Geschichtsforschung* 103 (1995), 392–418
Shirley, Rodney W., *The Mapping of the World: Early Printed World Maps 1472–1700*, 2nd edn (Riverside: Early World Press, 2001)
Simek, Rudolf, *Altnordische Kosmographie: Studien und Quellen zu Weltbild und Weltbeschreibung in Norwegen und Island vom 12. bis zum 14. Jahrhundert* (Berlin: de Gruyter, 1990)
Skelton, R. A., 'Bibliographical Note' in Claudius Ptolemy, *Geographię opus nouissima traductione e Gręcorum archetypis castigatissime pressum* (Strasbourg, 1513; facsimile edition, Amsterdam: Theatrum Orbis Terrarum, 1966), pp. v–xxii
Skelton, R. A., 'Bibliographical Note', in Claudius Ptolemy, *Geographia* (Venice, 1511; facsimile ed. Amsterdam: Theatrum Orbis Terrarum, 1969), pp. v–xi
Smyth, Marina, *Understanding the Universe in Seventh-Century Ireland* (Woodbridge: Boydell Press, 1996)
Soucek, Svat, *Piri Reis and Turkish Mapmaking after Columbus: The Khalili Portolan Atlas* (Oxford: Oxford University Press, 1996)
Southern, R. W., *Medieval Humanism and Other Studies* (Oxford: Blackwell, 1970)
Stahl, William H., 'To a better understanding of Martianus Capella', *Speculum* 40 (1965), 102–15
Stevenson, Edward Luther, *Terrestrial and Celestial Globes: Their history and construction including a consideration of their value as aids in the study of geography and astronomy*, 2 vols (New Haven: Yale University Press, 1921)
Sturtevant, William C., 'First Visual Images of Native America', in *First Images of America*, ed. Chiappelli, vol. 1, pp. 417–54
Sumien, N., *La correspondance du savant florentin Paolo dal Pozzo Toscanelli avec Christophe Colomb* (Paris: Société d'Éditions, 1927)
Tatlock, J. S. P., 'Geoffrey and King Arthur in *Normannicus Draco*', *Modern Philology* 31 (1933), 1–18, 113–25
Tattersall, Jill, '"Terra incognita": allusions aux extrêmes limites du monde dans les anciens textes français jusqu'en 1300', *Cahiers de civilisation médiévale* 24 (1981), 247–55
Tedeschi, Salvatore, 'Etiopi e Copti al Concilio di Firenze', *Annuarium historiae conciliorum* 21 (1989), 380–407

Thomas, Richard F., *Lands and Peoples in Roman Poetry: The Ethnographical Tradition* (Cambridge: Cambridge Philological Society, 1982)

Thomas, Richard F., 'Virgil's *Georgics* and the Art of Reference', *Harvard Studies in Classical Philology* 90 (1986), 171–98

Thorndike, Lynn, *Michael Scot* (London: Nelson, 1965)

Thorndike, Lynn, *The Sphere of Sacrobosco and Its Commentators* (Chicago: Chicago University Press, 1949)

Tibbetts, Gerald R., 'Later Cartographic Developments', in *The History of Cartography*, ed. Harley and Woodward, vol. 2.1 (1992), pp. 137–55

Tibbetts, Gerald R., 'The Beginnings of a Cartographic Tradition', in *The History of Cartography*, ed. Harley and Woodward, vol. 2.1 (1992), pp. 90–107

Uhden, Richard, 'Die Weltkarte des Isidorus von Sevilla', *Mnemosyne*, 3rd ser. 3.1 (1935–6), 1–28

Uhden, Richard, 'Die Weltkarte des Martianus Capella', *Mnemosyne*, 3rd ser. 3.1 (1935–6), 97–124

Uzielli, G., *Paolo dal Pozzo Toscanelli, iniziatore della scoperta d'America* (Florence, 1892)

van den Broecke, Marcel P.R., 'Introduction to the Life and Works of Abraham Ortelius (1527–1598)', in *Abraham Ortelius and the First Atlas: Essays commemorating the Quadricentennial of his Death 1598–1998*, ed. Marcel van den Broecke, Peter van der Krogt, and Peter Meurer (Utrecht: HES, 1998), pp. 29–54

van der Heijden, Henk A. M., 'Heinrich Büntings Itinerarium Sacrae Scripturae, 1581: Ein Kapitel der biblischen Geographie', *Cartographica Helvetica* 23 (2001), 5–14

van der Krogt, Peter, *Globi Neerlandici: The production of globes in the Low Countries* (Utrecht: HES, 1993)

Van Duzer, Chet, 'Cartographic Invention: The Southern Continent on Vatican MS Urb. Lat. 274, fols 73v–74r (c. 1530)', *Imago Mundi* 59 (2007), 193–222

Van Duzer, Chet, 'The Cartography, Geography, and Hydrography of the Southern Ring Continent, 1515–1760', *Orbis Terrarum* 8 (2002), 115–58

Vaughan, Richard, *Matthew Paris* (Cambridge: Cambridge University Press, 1958)

Vidier, A., 'La mappemonde de Théodulfe et la mappemonde de Ripoll (IXe–XIe siècle)', *Bulletin de géographie historique et descriptive* 3 (1911), 285–313

Vignaud, Henry, *The Letter and Chart of Toscanelli* (London: Sands, 1902)

von den Brincken, Anna-Dorothee, 'Die Klimatenkarte in der Chronik des Johann von Wallingford – ein Werk des Matthaeus Parisiensis?', *Westfalen* 51 (1973), 47–56

von den Brincken, Anna-Dorothee, *Fines Terrae: Die Enden der Erde und der vierte Kontinent auf Mittelalterlichen Weltkarten*, MGH Schriften 36 (Hanover: Hahnsche Buchhandlung, 1992)

von den Brincken, Anna-Dorothee, 'Mappa mundi und chronographia: Studien zur imago mundi des abendländischen Mittelalters', *Deutsches Archiv für Erforschung des Mittelalters* 24 (1968), 118–86

von den Brincken, Anna-Dorothee, '*Occeani Angustior Latitudo*: Die Ökumene auf der Klimatenkarte des Pierre d'Ailly', in *Studien zum 15. Jahrhundert: Festschrift für Erich Meuthen*, ed. Heribert Müller, 2 vols (Munich: Oldenbourg, 1994), vol. 1, pp. 565–82

von den Brincken, Anna-Dorothee, '*Terrae Incognitae.* Zur Umschreibung empirisch noch unerschlossener Räume in lateinischen Quellen des Mittelalters bis in die Entdeckungszeit', in *Raum und Raumvorstellungen im Mittelalter*, ed. Jan A. Aertsen and Andreas Speer (Berlin: de Gruyter, 1998), pp. 557–72

Wallis, Helen, 'Java-la-Grande: the first sight of Australia?', in *The Maps and Text of the Boke of Idrography*, ed. Wallis, pp. 58–67

Wallis, Helen, 'The Boke in the context of the Dieppe school of hydrography', in *The Maps and Text of the Boke of Idrography*, ed. Wallis, pp. 38–9

Wallis, Helen, 'Visions of Terra Australis in the Middle Ages and Renaissance', in *Terra Australis: The Furthest Shore* (Sydney: International Cultural Corporation of Australia, 1988), pp. 35–8

Westrem, Scott D., 'Against Gog and Magog', in *Text and Territory: Geographical Imagination in the European Middle Ages*, ed. Sylvia Tomasch and Sealy Gilles (Philadelphia: University of Pennsylvania Press, 1998), pp. 54–75

Whitfield, Peter, *New Found Lands: Maps in the History of Exploration* (London: British Library, 1998)

Wieder, F.C., *Monumenta Cartographica*, 5 vols (The Hague: Nijhoff, 1925–33)

Wieser, Franz von, *Magalhães-Strasse und Austral-Continent auf den Globen des Johannes Schöner* (Innsbruck: Verlag der Wagner'schen Universitäts-Buchhandlung, 1881)

Wilke, Jürgen, *Die Ebstorfer Weltkarte*, 2 vols (Bielefeld: Verlag für Regionalgeschichte, 2001)

Willers, Johannes, 'Leben und Werk des Martin Behaim' and 'Die Geschichte des Behaim-Globus', in *Focus Behaim Globus*, ed. Bott, pp. 173–88; 209–16

Williams, Glyndwr, *The Great South Sea: English Voyages and Encounters 1570–1750* (New Haven: Yale University Press, 1997)

Williams, John, 'Isidore, Orosius and the Beatus Map', *Imago Mundi* 49 (1997), 7–32

Williams, John, *The Illustrated Beatus: A Corpus of the Illustrations of the Commentary on the Apocalypse*, 5 vols (London: Harvey Miller, 1994–2003).

Wolska-Conus, Wanda, 'La "Topographie Chrétienne" de Cosmas Indicopleustès: hypothèses sur quelques thèmes de son illustration', *Revue des Études Byzantines* 48 (1990), 155–91

Woodward, David, 'Medieval *Mappaemundi*', in *The History of Cartography*, ed. Harley and Woodward, vol. 1 (1987), pp. 286–370

Woodward, David, and Herbert M. Howe, 'Roger Bacon on Geography and Cartography', in *Roger Bacon and the Sciences: Commemorative Essays*, ed. Jeremiah Hackett (Leiden: Brill, 1997), pp. 199–222

Woodward, David, *Maps as Prints in the Italian Renaissance: Makers, Distributors and Consumers* (London: British Library, 1995)

Woodward, David, ed., *The History of Cartography*, vol. 3: *Cartography in the European Renaissance* (Chicago: University of Chicago Press, 2007)

Wright, John Kirtland, '*Terrae Incognitae*: The Place of the Imagination in Geography', in *Human Nature in Geography: Fourteen Papers, 1925–1965* (Cambridge: Harvard University Press, 1966), pp. 68–88

Wright, John Kirtland, *The Geographical Lore at the Time of the Crusades: A Study in the History of Medieval Science and Tradition in Western Europe* (New York: American Geographical Society, 1925)

Zumthor, Paul, *La Mesure du monde: Représentation de l'espace au Moyen Âge* (Paris: Éditions du Seuil, 1993)

Index of Manuscripts

Page numbers in italic refer to illustrations; numbers in bold refer to colour plates.

General Index

Page numbers in italic refer to illustrations; numbers in bold refer to colour plates.